T0139813

Methods in Molecular Biology

Series Editor
John M. Walker
School of Life and Medical Sciences
University of Hertfordshire
Hatfield, Hertfordshire, AL10 9AB, UK

For further volumes:
http://www.springer.com/series/7651

The Bacterial Nucleoid

Methods and Protocols

Edited by

Olivier Espéli

CIRB, Collège de France
UMR CNRS 7241, INSERM U1050
Paris, France

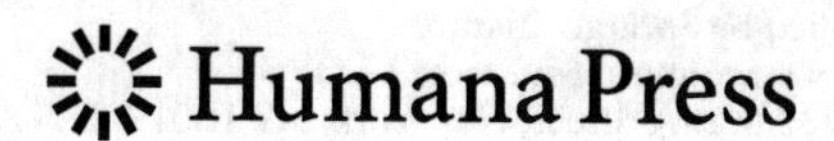

Editor
Olivier Espéli
CIRB, Collège de France
UMR CNRS 7241, INSERM U1050
Paris, France

ISSN 1064-3745 ISSN 1940-6029 (electronic)
Methods in Molecular Biology
ISBN 978-1-4939-8384-1 ISBN 978-1-4939-7098-8 (eBook)
DOI 10.1007/978-1-4939-7098-8

Printed on acid-free paper

This Humana Press imprint is published by Springer Nature
The registered company is Springer Science+Business Media LLC
The registered company address is: 233 Spring Street, New York, NY 10013, U.S.A.

Preface

Despite long-standing assumptions, the work of many investigators over the past 10 years has demonstrated that the bacterial chromosome is highly organized. The precise dynamic organization of genes and domains has far reaching consequences for a wide range of biological processes from those required for the successful transmission of heredity, i.e., DNA replication and chromosome segregation in coordination with cell division, to processes ranging from gene regulation to pathogenicity. Bacterial chromosomes are condensed by several factors including molecular crowding of the cytoplasm, DNA supercoiling, nucleoid associated proteins, and condensins. These factors define layers of organization of different sizes such as plectonemic DNA loops (few kb), microdomains or chromosome interaction domains (CIDs) (few tens of kb), and macrodomains (hundreds of kb). Genetics and molecular biology methods have paved the way for the understanding of chromosome structuring and remain extremely powerful to reveal the molecular mechanisms involved in this structuring. During the last 15 years new approaches, technologies, and insights have revolutionized the field of chromosome folding. Imaging technologies have changed our perception of the nucleoid; they revealed the spatial organization of the chromosome, the dynamics of DNA and bacterial chromatin, and the coupling of nucleoid dynamics with cell architecture and cell cycle. Moreover these observations opened a door for quantitative analysis by biophysicists. More recently, super-resolution microscopy has been used to visualize previously unresolved structure essential to understanding the bacterial chromosome. From the sequencing perspective, ChIP-seq, capture of chromosome conformation (3C), and Hi-C, its deep-sequencing derivative, have enabled the capture of new types of structural information on a genomic scale. These techniques, combined with state-of-the-art genetic, genomic, molecular, and cell biology approaches, have provided a wealth of new information about the chromosome. Meanwhile polymer physics models and physical nanomanipulation of cells, protein, and DNA allow researchers to tackle questions that go far beyond the traditional biological description of chromosome structure. This issue of Methods in Molecular Biology will propose state-of-the-art protocols for these key experiments that have, over the last decade, revolutionized our understanding of the bacterial nucleoid.

Paris, France | ***Olivier Espéli***

Contents

Contributors

FRANÇOIS-XAVIER BARRE • *Institute for Integrative Biology of the Cell (I2BC), Université Paris-Saclay, CEA, CNRS, Université Paris Sud, Gif sur Yvette, France*
DAVID BATES • *Integrative Molecular and Biomedical Sciences, Baylor College of Medicine, Houston, TX, USA; Department of Molecular and Human Genetics, Baylor College of Medicine, Houston, TX, USA*
FABRIZIO BENEDETTI • *Center for Integrative Genomics, University of Lausanne, Lausanne, Switzerland; Vital-IT, SIB Swiss Institute for Bioinformatics, Lausanne, Switzerland*
VÉRONIQUE ANTON LE BERRE • *Laboratoire d'Ingénierie des Systèmes Biologiques et des Procédés, Université de Toulouse, UPS, INSA, INP, CNRS, Toulouse, France*
JEAN-YVES BOUET • *Laboratoire de Microbiologie et Génétique Moléculaires, Centre de Biologie Intégrative (CBI), Centre National de la Recherche Scientifique (CNRS), Université de Toulouse, UPS, Toulouse, France*
PETER BRAZDA • *Department of Bionanoscience, Kavli Institute of Nanoscience, Faculty of Applied Sciences, Delft University of Technology, Delft, The Netherlands*
JACK A. BRYANT • *Institute of Microbiology and Infection, University of Birmingham, Birmingham, UK*
YANNIS BURNIER • *Center for Integrative Genomics, University of Lausanne, Lausanne, Switzerland; Institute of Theoretical Physics, Ecole Polytechnique Federal de Lausanne, Lausanne, Switzerland*
DIEGO I. CATTONI • *Centre de Biochimie Structurale, CNRS UMR5048, INSERM U1054, Université de Montpellier, Montpellier, France*
MARCO COSENTINO LAGOMARSINO • *Sorbonne Universit´es, UPMC Univ Paris 06, UMR 7238, Computational and Quantitative Biology, Paris, France; FIRC Institute of Molecular Oncology (IFOM), Milan, Italy; CNRS, UMR 7238, Paris, France*
REMUS T. DAME • *Gorlaeus Laboratories, Leiden Institute of Chemistry, Leiden University, Leiden, The Netherlands*
NYNKE H. DEKKER • *Department of Bionanoscience, Kavli Institute of Nanoscience, Faculty of Applied Sciences, Delft University of Technology, Delft, The Netherlands*
GAËLLE DEMARRE • *CIRB, Collège de France, UMR CNRS 7241, INSERM U1050, Paris, France*
ROXANNE E. DIAZ • *Laboratoire de Microbiologie et Génétique Moléculaires, Centre de Biologie Intégrative (CBI), Centre National de la Recherche Scientifique (CNRS), Université de Toulouse, UPS, Toulouse, France*
JULIEN DORIER • *Center for integrative Genomics, University of Lausanne, Lausanne, Switzerland; Vital-IT, SIB Swiss Institute for Bioinformatics, Lausanne, Switzerland*
TOBIAS DÖRR • *Department of Microbiology, Weill Institute for Cell and Molecular Biology, Cornell University, Ithaca, NY, USA*
OLIVIER ESPÉLI • *CIRB, Collège de France, UMR CNRS 7241, INSERM U1050, Paris, France*
YONG HWEE FOO • *Mechanobiology Institute, T-Lab, National University of Singapore, Singapore, Singapore*
ERIC FOURMENTIN • *Fourmentin-Guilbert Scientific Foundation, Noisy-le-Grand, France*

Rodrigo Galindo-Murillo • *Department of Medicinal Chemistry, College of Pharmacy, University of Utah, Salt Lake City, UT, USA*
Antoine Le Gall • *Centre de Biochimie Structurale, CNRS UMR5048, INSERM U1054, Université de Montpellier, Montpellier, France*
Elisa Galli • *Institute for Integrative Biology of the Cell (I2BC), Université Paris-Saclay, CEA, CNRS, Université Paris Sud, Gif sur Yvette, France*
Yunfeng Gao • *Mechanobiology Institute, T-Lab, National University of Singapore, Singapore, Singapore*
Marco Gherardi • *Sorbonne Universités, UPMC Univ Paris 06, UMR 7238, Computational and Quantitative Biology, Paris, France; FIRC Institute of Molecular Oncology (IFOM), Milan, Italy*
Mathilda Glaesmann • *Institute of Physical and Theoretical Chemistry, Goethe-University Frankfurt, Frankfurt, Germany*
Stephan Gruber • *Research Group 'Chromosome Organization and Dynamics', Max Planck Institute of Biochemistry, Martinsried, Germany; Department of Fundamental Microbiology, University of Lausanne, Lausanne, Switzerland*
Mike Heilemann • *Institute of Physical and Theoretical Chemistry, Goethe-University Frankfurt, Frankfurt, Germany*
Samuel Hornus • *Equipe Alice, Inria Nancy – Grand Est, Villers-lès-Nancy, France*
Mohan C. Joshi • *Department of Molecular and Human Genetics, Baylor College of Medicine, Houston, TX, USA*
Suckjoon Jun • *UCSD Physics and Molecular Biology, La Jolla, CA, USA*
Ivan Junier • *CNRS, TIMC-IMAG, Grenoble, France; University of Grenoble Alpes, TIMC-IMAG, Grenoble, France*
Linda J. Kenney • *Mechanobiology Institute, T-Lab, National University of Singapore, Singapore, Singapore; Jesse Brown VA Medical Center, University of Illinois - Chicago, Chicago, IL, USA*
J.W.J. Kerssemakers • *Department of Bionanoscience, Kavli Institute of Nanoscience, Faculty of Applied Sciences, Delft University of Technology, Delft, The Netherlands*
Romain Koszul • *Département Génomes et Génétique, Groupe Régulation Spatiale des Génomes, Institut Pasteur, Paris, France; CNRS, UMR 3525, Paris, France; Centre de Bioinformatique, Biostatistique et Biologie Intégrative, Institut Pasteur, Paris, France*
Valérie Lamour • *IGBMC, Integrated Structural Biology Department, UMR7104 CNRS, U964 INSERM, Université de Strasbourg, Strasbourg, France; Hôpitaux Universitaires de Strasbourg, Strasbourg, France*
Marko Lampe • *Advanced Light Microscopy Facility, European Molecular Biology Laboratory, Heidelberg, Germany*
Damien Larivière • *Fourmentin-Guilbert Scientific Foundation, Noisy-le-Grand, France*
Niels Laurens • *Leiden Institute of Physics, Leiden University, Leiden, The Netherlands*
David J. Lee • *Department of Life Sciences, School of Health Sciences, Birmingham City University, Birmingham, UK*
Roy de Leeuw • *Department of Bionanoscience, Kavli Institute of Nanoscience, Faculty of Applied Sciences, Delft University of Technology, Delft, The Netherlands*
Thibaut Lepage • *CNRS, TIMC-IMAG, Grenoble, France; University of Grenoble Alpes, TIMC-IMAG, Grenoble, France*
Bruno Lévy • *Equipe Alice, Inria Nancy – Grand Est, Villers-lès-Nancy, France*

MARTIAL MARBOUTY • *Département Génomes et Génétique, Groupe Régulation Spatiale des Génomes, Institut Pasteur, Paris, France; CNRS, UMR 3525, Paris, France; Centre de Bioinformatique, Biostatistique et Biologie Intégrative, Institut Pasteur, Paris, France*
SANJA MEHANDZISKA • *School of Engineering and Science, Jacobs University Bremen, Bremen, Germany*
JEAN-FRANÇOIS MÉNÉTRET • *IGBMC, Integrated Structural Biology Department, UMR7104 CNRS, U964 Inserm, Université de Strasbourg, Strasbourg, France*
M. CHARL MOOLMAN • *Department of Bionanoscience, Kavli Institute of Nanoscience, Faculty of Applied Sciences, Delft University of Technology, Delft, The Netherlands*
GEORGI MUSKHELISHVILI • *School of Engineering and Science, Jacobs University Bremen, Bremen, Germany*
MARCELO NOLLMANN • *Centre de Biochimie Structurale, CNRS UMR5048, INSERM U1054, Université de Montpellier, Montpellier, France*
JULIE PAPILLON • *IGBMC, Integrated Structural Biology Department, UMR7104 CNRS, U964 INSERM, Université de Strasbourg, Strasbourg, France*
N. PATRICK HIGGINS • *Department of Biochemistry and Molecular Genetics, University of Alabama at Birmingham, Birmingham, AL, USA*
JAMES PELLETIER • *Center for Bits and Atoms, Massachusetts Institute of Technology, Cambridge, MA, USA; Department of Physics, Massachusetts Institute of Technology, Cambridge, MA, USA*
ALEXANDER M. PETRESCU • *School of Engineering and Science, Jacobs University Bremen, Bremen, Germany*
ZOYA M. PETRUSHENKO • *Department of Chemistry and Biochemistry, University of Oklahoma, Norman, OK, USA*
CHARLÈNE PLANCHENAULT • *CIRB, Collège de France, UMR CNRS 7241, Paris, France*
MICKAËL POIDEVIN • *Institute for Integrative Biology of the Cell (I2BC), Université Paris-Saclay, CEA, CNRS, Université Paris Sud, Gif sur Yvette, France*
VICTORIA PRUDENT • *CIRB, Collège de France, UMR CNRS 7241, INSERM U1050, Paris, France*
DUSAN RACKO • *Center for Integrative Genomics, University of Lausanne, Lausanne, Switzerland; SIB Swiss Institute for Bioinformatics, Lausanne, Switzerland; Polymer Institute of the Slovak Academy of Sciences, Bratislava, Slovak Republic*
VALENTIN V. RYBENKOV • *Department of Chemistry and Biochemistry, University of Oklahoma, Norman, OK, USA*
AURORE SANCHEZ • *Laboratoire de Microbiologie et Génétique Moléculaires, Centre de Biologie Intégrative (CBI), Centre National de la Recherche Scientifique (CNRS), Université de Toulouse, UPS, Toulouse, France*
RUPA SARKAR • *Department of Chemistry and Biochemistry, University of Oklahoma, Norman, OK, USA*
ASWIN SAI NARAIN SESHASAYEE • *National Centre for Biological Sciences, Tata Institute of Fundamental Research, Bangalore, Karnataka, India*
PARUL SINGH • *National Centre for Biological Sciences, Tata Institute of Fundamental Research, Bangalore, Karnataka, India; SASTRA University, Thanjavur, Tamil Nadu, India*
BELEN SOLANO • *Department of Bionanoscience, Kavli Institute of Nanoscience, Faculty of Applied Sciences, Delft University of Technology, Delft, The Netherlands*
CHRISTOPH SPAHN • *Institute of Physical and Theoretical Chemistry, Goethe-University Frankfurt, Frankfurt, Germany*

ANDRZEJ STASIAK • *Center for integrative Genomics, University of Lausanne, Lausanne, Switzerland; SIB Swiss Institute for Bioinformatics, Lausanne, Switzerland*

RAMON A. VAN DER VALK • *Gorlaeus Laboratories, Leiden Institute of Chemistry, Leiden University, Leiden, The Netherlands*

ELISE VICKRIDGE • *CIRB, Collège de France, UMR CNRS 7241, Paris, France*

BRYAN J. VISSER • *Integrative Molecular and Biomedical Sciences, Baylor College of Medicine, Houston, TX, USA*

LARISSA WILHELM • *Research Group 'Chromosome Organization and Dynamics', Max Planck Institute of Biochemistry, Martinsried, Germany*

RICKSEN S. WINARDHI • *Department of Physics, National University of Singapore, Singapore, Singapore; Mechanobiology Institute, National University of Singapore, Singapore, Singapore; Centre for Bioimaging Sciences, National University of Singapore, Singapore, Singapore*

YOSHIHARU YAMAICHI • *Institute for Integrative Biology of the Cell (I2BC), Université Paris-Saclay, CEA, CNRS, Université Paris Sud, Gif sur Yvette, France*

JIE YAN • *Department of Physics, National University of Singapore, Singapore, Singapore; Mechanobiology Institute, National University of Singapore, Singapore, Singapore; Centre for Bioimaging Sciences, National University of Singapore, Singapore, Singapore*

Part I

Molecular Genetic Methods to Study Bacterial Nucleoid

Chapter 1

Homologous Recombineering to Generate Chromosomal Deletions in *Escherichia coli*

Jack A. Bryant and David J. Lee

Abstract

Homologous recombination methods enable modifications to be made to the bacterial chromosome. Commonly, the λ phage RED proteins are employed as a site-specific recombinase system, to facilitate recombination of linear DNA fragments with targeted regions of the chromosome. Here we describe methods for the efficient delivery of linear DNA segments containing homology to the chromosome into the cell as substrates for the λRED proteins. Combined with antibiotic selection and counterselection, we demonstrate that using this method facilitates accurate, rapid editing of the chromosome.

Key words Recombineering, Lambda RED, *Escherichia coli*, Chromosome, Deletion

1 Introduction

The λRED recombinase system consists of three gene products [1]. The *gam* gene, encodes a protein that protects linear, double-stranded DNA from host cell-mediated degradation. The *exo* gene product generates single-stranded DNA overhangs at each end of the linear fragment, and the *bet* gene codes for a recombinase for which the linear DNA fragment is a substrate for homologous recombination with targeted regions of the chromosome [2]. Several methods have been developed that exploit the λRED recombinase system, which essentially all have the same requirements [3–10]: The first is the requirement to have controlled, inducible expression of the λRED genes. The second requirement is the efficient delivery of a linear DNA fragment that contains the desired homology to the chromosome target into the cell. The third is the inclusion of a selectable marker, typically an antibiotic resistance cassette, for the selection of recombinants. Finally, a PCR screen to identify correctly modified chromosome as desired.

Here we describe a method for recombineering, termed Gene Doctoring that we have previously employed to edit the chromosome of laboratory and pathogenic strains of *Escherichia coli* [6].

Olivier Espéli (ed.), *The Bacterial Nucleoid: Methods and Protocols*, Methods in Molecular Biology, vol. 1624,
DOI 10.1007/978-1-4939-7098-8_1,

This method utilizes the λRED genes and, due to the incorporation of counterselection markers in the system, results in the rapid identification of recombinant clones. This technique has been used previously to delete targeted regions of the chromosome [11], to label proteins with epitope tags [12] and to insert promoter–reporter gene fusions into different regions of the chromosome while simultaneously deleting large sections of DNA [13]. The gene doctoring method has also been used to modify pathogenic bacteria, including *Pseudomonas putida*, *Pseudomonas syringae*, an outbreak strain of *Klebsiella pneumoniae* and *Escherichia coli* strains UPEC CFT073, ETEC H10407, EHEC O157:H7 [6, 14–19]. Here, we provide a worked example of how to generate a gene deletion in the *E. coli* chromosome.

1.1 Two-Plasmid System for Gene Deletion

A λRED based recombination technique developed by Datsenko and Wanner [3], relies upon electroporation of double stranded DNA donor molecule, carrying an antibiotic resistance cassette flanked by regions of homology to the chromosomal target, into cells in which the expression of the λRED genes has been induced [3]. This method yields a low recombination efficiency of 1 in every 3.5×10^7 cells, which correlates with the proportion of cells that survive electroporation [3, 6]. The efficiency of λRED recombination was improved, by Herring and coworkers [5], who supplied the target donor DNA for recombination by excising it from a donor plasmid in vivo. This increases the efficiency of recombination to around 1–15%. However, since the target donor DNA is supplied on a plasmid, any uncut plasmid retains the antibiotic selection marker. Thus, the identification of recombinants that have retained the antibiotic resistance marker, whilst having lost the donor plasmid, requires the screening of a large number (hundreds, sometimes thousands) of candidates [6].

The Gene Doctoring system was developed to increase the efficiency of λRED-mediated recombination method of Herring et al. [5]. The system is reliant on two plasmids: the first is a pDONOR plasmid (Fig. 1a), which contains DNA that is homologous to the region of the chromosome where modifications are to be made (Fig. 1a, b). These regions of DNA homology are cloned either side of a kanamycin resistance cassette. Immediately adjacent to the homology sequences are restriction sites for meganuclease I-SceI. This pDONOR plasmid also carries an ampicillin resistance cassette and the *sacB* gene from *Bacillus subtilis.* The gene doctoring method has the same recombination efficiency as the method of Herring et al. [5], but has improved selection efficiency due to the incorporation of the sucrose sensitivity counterselective marker on the donor plasmid [6]. The second plasmid is the gene recombineering plasmid pACBSce, which carries resistance to chloramphenicol, the λRED genes as well as the gene coding for the meganuclease I-SceI and an I-SceI restriction site (Fig. 1c).

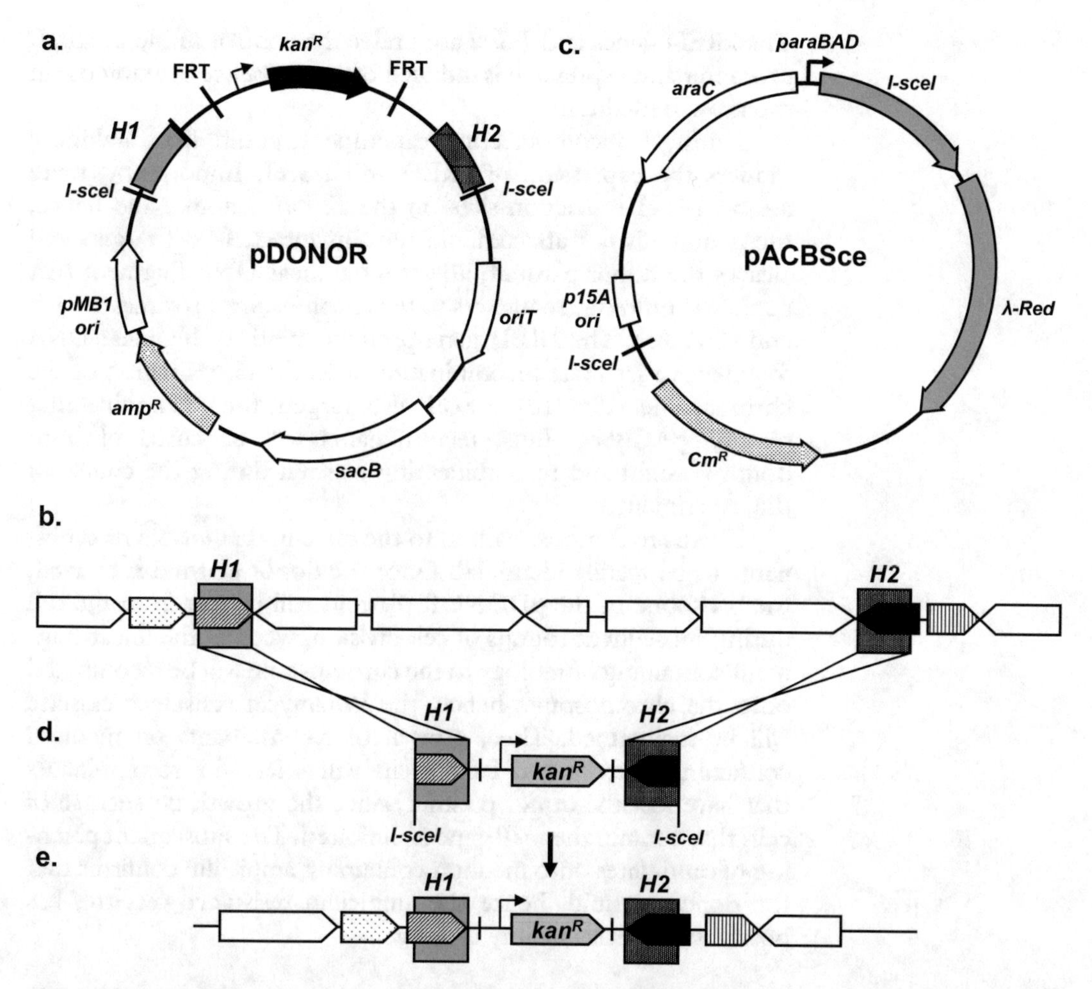

Fig. 1 Gene Doctoring two-plasmid system for genomic deletions. Schematic representations of the two-plasmid system used for homologous recombineering of chromosomal gene deletions. (**a**) Map of the pDONOR plasmid used for creating chromosomal deletions. The vector encodes a kanamycin resistance marker, kan^R, flanked by *flp* recombinase recognition target sequences, FRT, and two regions of homology to the chromosomal target, *H1* and *H2*. These features are all flanked by two target sequences for the yeast meganuclease I-SceI, which do not occur within the *E. coli* genome. Also represented are the origin of replication, *pMB1 Ori*, ampicillin resistance marker, amp^R, and the sucrose sensitivity determinant for counterselection, *sacB*. (**b**) Representation of a typical chromosomal target sequence for creating genomic deletions. Genes are represented by *boxed arrows*. *H1* and *H2* homology regions are represented by *shaded boxes*. These regions can be separated by 0 kb or anywhere up to 40 Kbp apart (*see* Subheading 3.7). (**c**) Schematic representation of the mutagenesis plasmid, pACBSce, which encodes the I-SceI meganuclease and the λRED recombination system under control of the arabinose inducible *araBAD* promoter. Further to this, the vector encodes a chloramphenicol resistance marker, Cm^R, and an I-SceI recognition target sequence to allow curing of the vector during recombineering. (**d**) The homology regions to the target sequence, *H1* and *H2* are cloned in to the pDONOR vector flanking the kan^R resistance marker and the linear donor molecule is liberated from the plasmid in vivo by the I-SceI meganuclease. (**e**) The λRED recombination system facilitates homologous recombination between the donor molecule and the chromosomal target sequence leading to deletion of the region between *H1* and *H2*

The λRED genes and *I-SceI* are under the control of the *araBAD* promoter and expression is induced by the presence of arabinose in the growth medium.

During a recombineering experiment, addition of arabinose induces the expression of λRED and I-SceI. Importantly, there are no I-SceI restriction sites on the *E. coli* genome, and hence, the genome is not affected and remains intact. I-SceI targets and cleaves the donor plasmid, liberating a linear DNA fragment that contains homology sequences to the chromosome proximal to each end (Fig. 1d). The λRED gene products bind to the linear DNA fragment to facilitate recombination with the target region of the chromosome (Fig. 1e). I-SceI also targets the recombineering plasmid pACBSce, thus, recombinants will be cured of both donor plasmid and recombineering plasmid during the course of the experiment.

There are features, built in to the system, that enable recombinants to be readily identified. Once the donor plasmid is cleaved, the backbone of the pDONOR plasmid will be lost from the cell during subsequent rounds of cell division, whereas the linear fragment containing homology to the chromosome will be recombined onto the chromosome; hence, the kanamycin resistance cassette will be maintained. Thus, growth of recombinants on medium containing sucrose and kanamycin will select for recombinants that have lost the donor plasmid, since the growth on sucrose of cells that contain the *sacB* gene is inhibited. The subsequent patching of candidates onto medium containing ampicillin confirms that the donor plasmid, hence the ampicillin resistance cassette, has been lost by their inability to grow.

1.2 pDONOR Plasmid Preparation

For homologous recombination to generate a gene deletion, regions of homology to the target region of chromosome are cloned into the donor plasmid (Fig. 2). Two pDOC plasmids are required to generate the final donor plasmid [6]. The first is pDOC-K (Fig. 2a). This plasmid contains a kanamycin cassette, flanked by target sites for Flp recombinase. This plasmid is used as a template in a PCR reaction with oligonucleotides that anneal to pDOC-K and contain 40-bp regions of homology to the chromosomal target (sequences H1 and H2). Once the PCR product has been generated (Fig. 2b), it is cloned into plasmid pDOC-C (Fig. 2c). In the example shown, restriction sites *Eco*RI and *Spe*I have been used for cloning. This positions the homology cassette between two I-SceI target sites generating the pDONOR plasmid. Clones can be selected on LB agar supplemented with kanamycin, as the parent pDOC-C plasmid does not encode a kanamycin resistance cassette. The example given below demonstrates how a gene is deleted, by being replaced by the kanamycin resistance cassette.

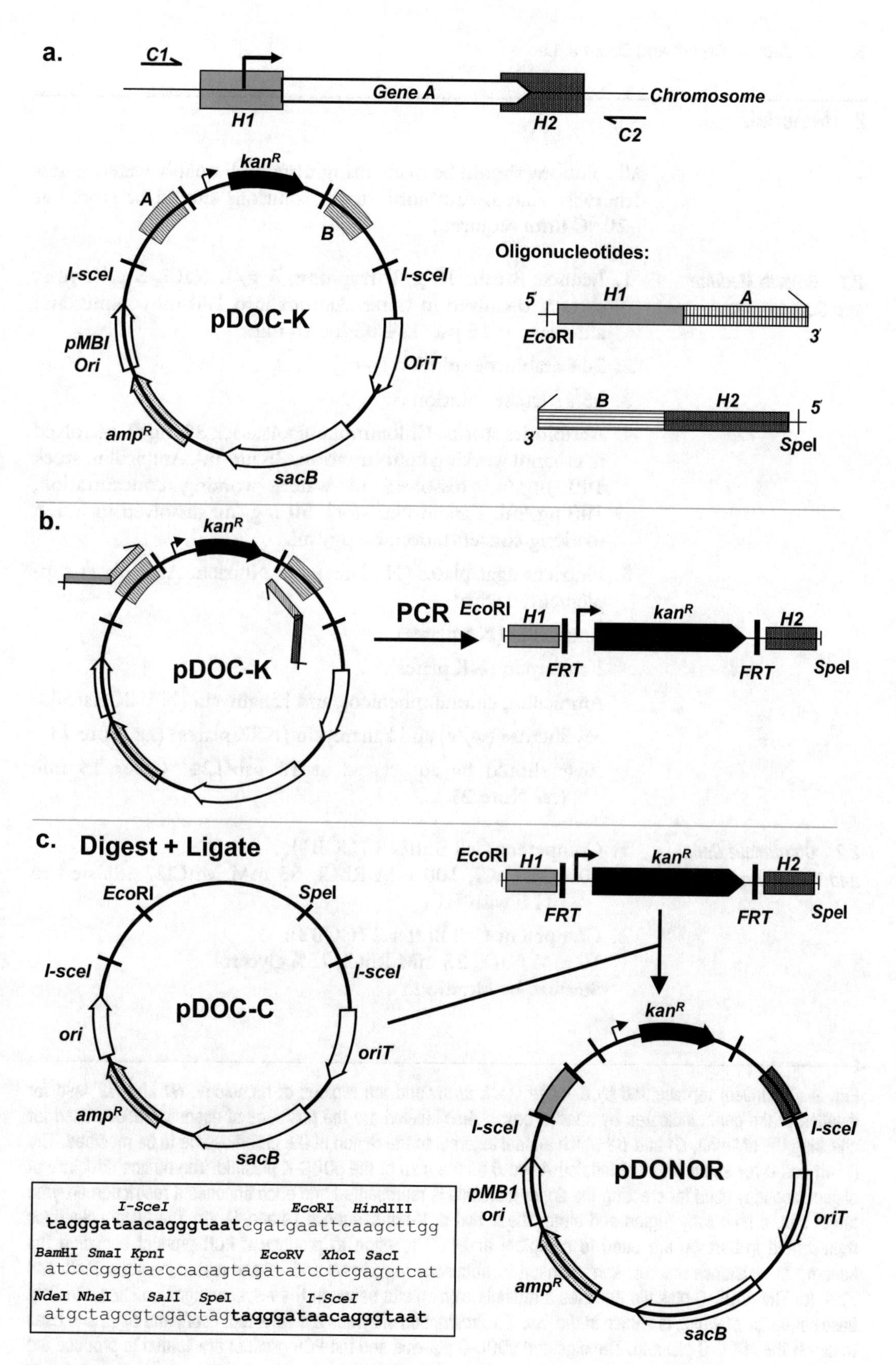

Fig. 2 Construction method for the pDONOR plasmid. Schema for construction of the pDONOR plasmid for deletion of chromosomal genes. (**a**) Representation of a typical chromosomal target region with the gene

2 Materials

All solutions should be made using ultra high quality water, unless otherwise stated. Antibiotic stocks solutions should be stored at −20 °C until required.

2.1 Growth Medium and Supplements

1. Lennox Broth: 10 g/L tryptone, 5 g/L NaCl, 5 g/L yeast extract, dissolved in water. Aliquot into 100 ml volumes and autoclave at 15 psi/124 °C for 15 min.
2. 20% arabinose solution (w/v).
3. 55% sucrose solution (w/v).
4. Antibiotics stocks. Chloramphenicol: stock 35 mg/L dissolved in ethanol working concentration, 35 μg/ml. Ampicillin: stock 100 mg/ml dissolved in water, working concentration, 100 μg/ml. Kanamycin: stock 50 mg/ml dissolved in water, working concentration, 50 μg/ml.
5. Nutrient agar plates (N plates) and Nutrient Agar plates supplemented with:

   Ampicillin (NA plates).

   Kanamycin (NK plates).

   Ampicillin, chloramphenicol, and kanamycin (NACK plates).

   5% Sucrose (w/v) and kanamycin (NSK plates) (*see* **Note 1**).

   Agar should be autoclaved at 15 psi/124 °C for 15 min (*see* **Note 2**).

2.2 Competent Cells and Transformation

1. Competent Cell Buffer 1 (CCB1):
   10 mM CaCl, 100 mM RbCl, 55 mM MnCl2. Adjusted to pH 6 with HCl.
2. Competent Cell Buffer 2 (CCB2):
   10 mM CaCl, 25 mM RbCl, 15% glycerol.
   Sterilize by filtration.

Fig. 2 (continued) represented by a *white block arrow* and the regions of homology, *H1* and *H2* used for deletion of the gene, indicated by *shaded boxes*. Also labeled are the positions of oligonucleotides used for checking the deletion, *C1* and *C2* which anneal external to the region of the chromosome to be modified. The k-fwd and k-rev sequences are labeled *A* and *B* on the map of the pDOC-K plasmid. The outline structure of oligonucleotides used for creating the donor molecule is represented and each encodes a restriction enzyme target site, a homology region and either the k-fwd or k-rev sequence (*A* and *B*). (**b**) The oligonucleotides represented in part (**a**) are used to complete and PCR reaction to produce a PCR product encoding the kanamycin resistance marker, *kan*^{R}, flanked by homology regions *H1* and *H2* and restriction sites *Eco*RI and *Spe*I. (**c**) The pDOC-C plasmid encodes a multiple cloning site between the *I-sceI* recognition sites, the DNA base sequence of which is shown in the *box*. Two restriction enzymes are selected, *Eco*RI and *Spe*I, and used to digest the pDOC-C plasmid. The digested pDOC-C plasmid and the PCR product are ligated to produce the pDONOR plasmid

2.3 PCR and Cloning

1. For PCR cloning, use a high-fidelity DNA Polymerase. For recombinant checking, use a standard Taq Polymerase.
2. dNTPs: concentration as per DNA Polymerase requirements.
3. Restriction enzymes and DNA ligase as per manufacturers' instruction.

3 Methods

3.1 pDONOR Plasmid Preparation

1. Design oligonucleotide primers that contain the pDOC-K priming sequences (sequences K-fwd and K-rev: *see* **Note 3**) at the 3′, homology DNA sequences for the chromosomal target (sequences H1 and H2) and an appropriate restriction enzyme site at the 5′. This example uses restriction enzymes EcoRI and SpeI (Fig. 2a). Other restriction sites, unique to plasmid pDOC-K are available for cloning (Fig. 2c). Typically, for a gene deletion, 40 bp of DNA homology is sufficient for efficient recombination. Homology region H1 is the DNA sequence immediately upstream of the start codon. Homology region H2 is the DNA sequence including the last 10–15 codons of the gene (*see* **Note 4**).
2. In addition, design 2 oligonucleotide primers that will be used to check that recombination has succeeded as expected (check primers C1 and C2: Fig. 1a). These primers should be designed to anneal to the region flanking the gene targeted for deletion, such that they can be used in a PCR reaction to amplify the target region. These primers will be used after recombination, to amplify the target region to confirm recombination.
3. Amplify the kanamycin cassette using the guidelines provided by the manufacture for the high-fidelity DNA Polymerase that is used.
4. Purify the PCR product, Digest both pDOC-C and the PCR product with EcoRI and SpeI restriction enzymes and ligate, to create the final pDONOR plasmid (*see* **Note 5**).
5. Competent cells of the desired *E. coli* strain into which the chromosomal modification is to be made are then cotransformed with the final pDONOR plasmid and the recombineering plasmid pACBSce, as described below.

3.2 Preparation of Competent Cells

1. Inoculate 5 ml of LB with one colony of bacteria of the *E. coli* strain to be modified and incubate overnight at 37 °C with aeration.
2. The following day, subculture 100 μl into 10 ml of fresh LB and incubate at 37 °C with aeration until the OD_{650} reaches 0.5 (*see* **Note 6**).

3. At this stage, rapidly chill the culture by swirling in iced water bath for 2 min, then continue to incubate on ice for 5 min.
4. Transfer the culture to a precooled centrifuge tube and harvest the cells at by centrifugation at 4000 × *g* for 10 min.
5. Dispose of all of the supernatant: gently tap the inverted tube on a paper towel to remove any remaining LB.
6. Gently resuspend the cells in 5 ml of ice-cold CCB1. Do this by pipetting up and down, whilst keeping the suspension ice-cold.
7. Incubate on ice for 30 min.
8. Harvest the cells at by centrifugation at 4000 × *g* for 10 min.
9. Dispose of all of the supernatant: gently tap the inverted tube on a paper towel to remove any remaining CCB1
10. Gently resuspend the cells in CCB2, to a final volume of 350 μl. Do this by using 250 μl of CCB2 ml to resuspend the cell pellet. Once resuspended, determine the volume of the cell suspension using a pipette, then add the appropriate amount of CCB2 to make the volume up to 350 μl.
11. Aliquot the cells into 50 μl samples and store at −70 °C until required. Each 50 μl aliquot is sufficient for two transformations, and is stable at −70 °C for several months.

3.3 Transformation of Competent Cells

1. Aliquot the plasmid DNA to be transformed into a 1.5 ml tube and incubate on ice. Typically, 1–100 ng of DNA should be transformed, but the amount of pACBSce and pDONOR plasmid should be roughly equivalent (*see* **Note 7**).
2. Add 25 μl of defrosted competent cells to the DNA and incubate on ice for 15 min.
3. Heat-shock by placing the tube in a water bath set at 42 °C for 30 s; then return the tube to the ice for 5 min.
4. Add 750 μl of SOC solution, prewarmed to 37 °C, and incubate at 37 °C for 1 h. During incubation, place the tubes, in a shaking incubator set at 37 °C. Set the shaker so that the cell suspension is gently swirling: this prevents the cells from settling to the bottom of the tube during this outgrowth step.
5. After 1 h, harvest the cells by centrifugation and remove all but 100 μl of the supernatant.
6. Resuspend the cells in the remaining supernatant.
7. Pipette the cell suspension onto the surface of an NACK plate. Spread the cells evenly across the surface of the agar plate.
8. Finally, invert the agar plate and incubate at 37 °C overnight, or until colonies become visible (*see* **Note 8**).

3.4 Prerecombination Quality Control

Before commencing with the recombination protocol, it is recommended that the transformants are checked. Once the following criteria have been confirmed, proceed with the recombineering protocol.

1. Patch individual colonies onto nutrient agar plates containing the following antibiotics:

   Ampicillin, chloramphenicol, and kanamycin (NACK plates).

   Kanamycin plus 5% sucrose (NSK plates).

   Use a sterile loop to pick a colony and patch onto the NACK plates first, then without flaming the loop, patch onto the NSK plate.
2. The expected outcome is that colonies should be able to grow on Chloramphenicol, due to the resistance cassette carried on plasmid pACBSce, and also on ampicillin and kanamycin, due to the resistance cassettes carried on the pDONOR plasmids.
3. Colonies should not be able to grow on the kanamycin + sucrose plate because, despite having resistance to kanamycin, carried by the pDONOR plasmid, the presence of *sacB* in the pDONOR plasmid prevents growth on sucrose.

3.5 Recombineering

1. Inoculate 500 μl LB, supplemented with ampicillin and chloramphenicol, with a single ampicillin, chloramphenicol, and kanamycin-resistant, sucrose-sensitive colony.
2. Incubate at 37 °C with shaking for 2–4 h until the culture is turbid (within the OD_{650} range of 0.2–0.5) (*see* **Note 9**).
3. Remove 1 μl of culture and dilute with 1 ml of LB (10^{-3} dilution). Make a tenfold serial dilution to 10^{-6}, then plate 100 μl of the 10^{-4}, 10^{-5}, and 10^{-6} dilutions onto both an N plate, and an NA plate. This step ensures that the pDONOR plasmid has been maintained in the culture. Thus, the same number of colonies, for each of the serial dilutions, should grow on the N and NA plates (*see* **Note 10**).
4. Harvest cells by centrifugation and wash by resuspension with 500 μl 0.1 × LB. Repeat three times to remove any residual antibiotics (*see* **Note 11**).
5. Resuspend cells in 500 μl 0.1 × LB supplemented with 0.3% arabinose to induce the expression of the λRED genes and SceI meganuclease from plasmid pACBSce. 0.1 × LB is used during this step as the cells are no longer required to divide rapidly.
6. Incubate the culture at 37 °C with shaking for a further 2–3 h.
7. Remove 1 μl of culture and dilute with 1 ml of LB (10^{-3} dilution). As before, make a tenfold serial dilution to 10^{-6}, then plate 100 μl of the 10^{-4}, 10^{-5} and 10^{-6} dilutions onto both an N plate, and an NA plate. This step is to ensure that the

pDONOR plasmid has been digested by the I-SceI meganuclease. There should be significantly less colonies on the NA plate than on the N plate. Typically, >95% of the pDONOR plasmid is digested, and hence the ampicillin resistance cassette has been lost.

8. Spread 125 μl of the remaining culture onto each of four NSK plates.
9. Incubate plates at 30 °C or room temperature (preferable) until colonies are visible (*see* **Note 12**).

3.6 Checking Candidate Colonies

1. Identify candidate colonies on the NSK plate to be checked. Patch the individual colonies first onto an NA plate, then onto an NSK plate. To do this, use a sterile loop to pick a colony and streak onto the NA plate. Then without flaming the loop, streak immediately onto the NSK plate.
2. True recombinants will have lost the pDONOR plasmid backbone, hence will have lost the amp resistance cassette and the *sacB* gene, and will therefore be able to grow on sucrose, but not on ampicillin. True recombinants will have retained the kanamycin cassette since it will have been inserted into the chromosome through homologous recombination.
3. True recombinants should be purified to single colonies by streaking onto NK plates. To do this, a sample from the patched colonies which grew on NSK, but not on NA, is picked with a sterile loop and streaked onto a fresh NK plate to single colonies.

3.7 Confirmation of Recombinants by PCR

1. Select candidate colonies from the NK plate and resuspend each separately in 100 μl of sterile water. Spot 2 μl of the resuspension onto an NK plate for storage. Boil the remaining solution for 5 min and briefly centrifuge the suspension to pellet cell debris.
2. Repeat the same procedure for a colony of the parent *E. coli* strain. Comparison of the PCR product generated from the parent strain, and the recombineered candidates, will identify successful clones (Fig. 3).
3. Use 2–10 μl of cell solution from the candidate colonies and the parent strain in a PCR reaction, using oligonucleotide primers C1 and C2 to amplify the target region of the chromosome. The annealing temperature used for the PCR will depend on the sequences of primers C1 and C2 and the extension time for the PCR will depend on the bp distance between the annealing sites for primers C1 and C2 on the chromosome.
4. The expected outcome is dependent upon the size of the gene that was deleted. The length of the kanamycin insertion

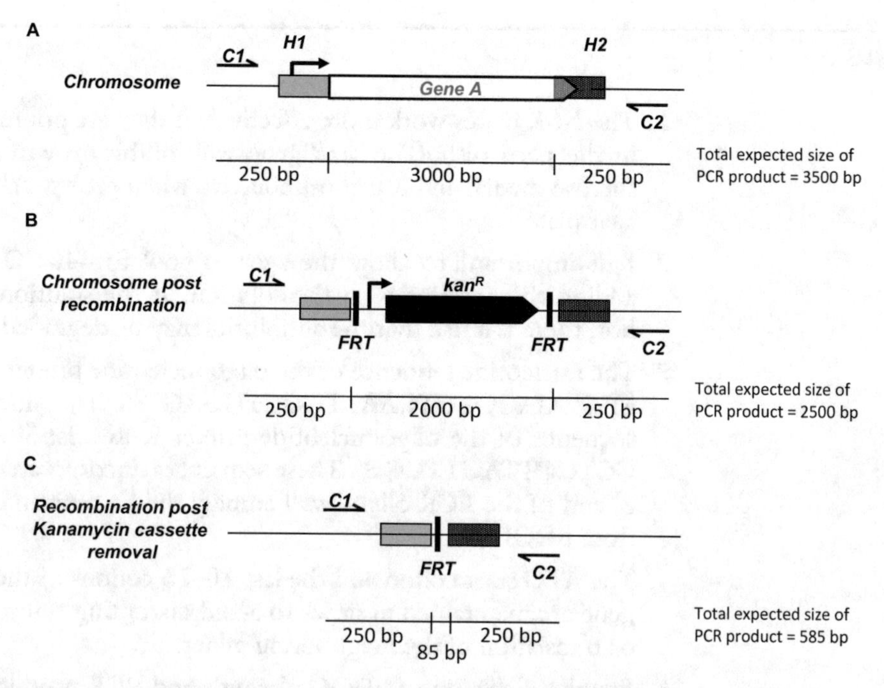

Fig. 3 Example of the expected fragment size after checking recombinant candidates by PCR. Schematic representation of the expected outcome of check PCR reactions using oligonucleotide primers C1 and C2. (**a**) The expected product size from the parent strain is 3500 bp. (**b**) The expected product size from a recombineered candidate strain is 2500 bp. (**c**) The expected product size from a recombineered candidate strain post kanamycin cassette removal is 585 bp

cassette between primer annealing sequences A and B on the pDOC-K plasmid is 1327 bp. An example of the expected outcomes from the check PCR is shown in Fig. 3.

5. Post recombination, the kanamycin cassette can be excised from the chromosome (Fig. 3c). This is facilitated by the FLP sites that flank the cassette. A plasmid, encoding the FLP recombinase, is used to transform cells. Expression of FLP recombinase results in excision of the kanamycin cassette, leaving a "scar" of 85 bp. The procedure for removal is described by Datsenko and Wanner [3].

3.8 Additional Uses of the Gene Doctoring Technique

1. The gene doctoring method can be exploited to edit the genome to introduce DNA gene tags and to delete large portions of the chromosome. Lengths of DNA, up to 11.5 Kbp were deleted by Bryant et al. [13], and a 40-Kbp region of the *E. coli* chromosome has been removed by the Grainger laboratory (unpublished personal communication). Introduction of DNA gene tags is facilitated by a set of pDOC plasmids which contain coding sequences for a 6xHistidine tag, a 3xFLAG tag, a 4xProteinA tag, and a GFP tag. Using these plasmids, any gene product can be tagged at the C-terminus [6].

4 Notes

1. The NSK plates work more effectively if they are poured thick in the petri dish. The *sacB* gene will inhibit growth on the sucrose media, and it is most effective when grown on a thick agar plate.
2. It is important to allow the agar to cool to ~40 °C before adding the antibiotics to the solution. If the solution is too hot, there is a risk that the antibiotics may be degraded.
3. The nucleotide sequence of the oligonucleotide primer K-Fwd is: 5′ GACCGGTCAATTGGCTGGAG 3′. The nucleotide sequence of the oligonucleotide primer K-Rev. is: 5′ AATAT CCTCCTTAGTTCC 3′. These sequences, incorporated at the 3′ end of the PCR oligos, will amplify the kanamycin cassette from pDOC-K.
4. The ATG start codon and the last 10–15 codons of the target gene are maintained in order to avoid disrupting transcription or translation of the neighboring genes.
5. For the digestion of pDOC plasmids and PCR products, and the subsequent ligation, it is recommended that the manufacturers' optimal enzyme working conditions are followed.
6. At this stage it is recommended that buffers CCB1 and CCB2 are incubated on ice, so that when required, they will be ice cold. Maintaining the cells at ice cold conditions during preparation increases competency.
7. It is sufficient to quantify the DNA, using gel electrophoresis, against commercial DNA ladders containing known quantities of DNA.
8. If cotransformation of both pACBSce and the pDONOR plasmid is unsuccessful, it is recommended that cells are transformed with pACBSce alone. Once transformants are confirmed, competent cells can be prepared that contain pACBSce. These cells can then be stored at −70 °C and used as competent cells for transformation with pDONOR plasmids.
9. This recombineering technique is most efficient when the cells are harvested in the early to mid-logarithmic phase of growth.
10. It is recommended that, when following the procedure for the first time, that all of the dilutions are plated. This is to ensure that coverage of growth to single countable colonies is achieved.
11. This step is included to ensure that traces of antibiotics are removed from the growth medium. After induction with arabinose, which will result in plasmid loss and consequently loss of

antibiotic resistance, the presence of antibiotics may be detrimental to bacterial growth.

12. Some *E. coli* strains may begin to overcome the counterselection and grow on sucrose containing media after long periods of incubation, despite containing the *sacB* gene. This is typical of wild-type strains, such as O157:H7 strains, whereas laboratory strains, such as MG1655, do not grow through. For this reason, the agar plates are poured thick and slower colony growth is favorable by incubation at a lower temperature.

References

1. Court DL, Sawitzke JA, Thomason LC (2002) Genetic engineering using homologous recombination. Annu Rev Genet 36:361–388
2. Mosberg JA, Lajoie MJ, Church GM (2010) Lambda red recombineering in *Escherichia coli* occurs through a fully single-stranded intermediate. Genetics 186:791–799
3. Datsenko KA, Wanner BL (2000) One step inactivation of chromosomal genes in *Escherichia coli* K-12 using PCR products. Proc Natl Acad Sci U S A 97:6640–6645
4. Ellis HM, Yu D, DiTizio T, Court DL (2001) High efficiency mutagenesis, repair and engineering of chromosomal DNA using using single-stranded oligonucleotides. Proc Natl Acad Sci U S A 98:6742–6746
5. Herring CD, Glasner JD, Blattner FR (2003) Gene replacement without selection: regulated suppression of amber mutations in *Escherichia coli*. Gene 311:153–163
6. Lee DJ, Bingle LEH, Heurlier K, Pallen MJ, Penn CW, Busby SJW, Hobman JL (2009) Gene doctoring: a method for recombineering in laboratory and pathogenic *Escherichia coli* strains. BMC Microbiol 9:252
7. Murphy KC (1998) Use of bacteriophage lambda recombination functions to promote gene replacement in *Escherichia coli*. J Bacteriol 180:2063–2071
8. Stringer AM, Singh N, Yermakova A, Petrone BL, Amarainghe JJ, Reyes-Diaz L, Mantis NJ, Wade JT (2012) Fruit, a scar-free system for targeted chromosomal mutagenesis, epitope tagging and promoter replacement in *Escherichia coli* and *Salmonella enterica*. PLoS One 7: e44841
9. Warming S, Costantino N, Court DL, Jenkins NA, Copeland NG (2005) Simple and highly efficient BAC recombineering using galK selection. Nucleic Acids Res 33:e36
10. Yu D, Ellis HM, Lee EC, Jenkins NA, Copeland NG, Court DL (2000) An efficient recombination system for chromosome engineering in *Escherichia coli*. Proc Natl Acad Sci U S A 97:5978–5983
11. Bingle LEH, Constantinidou C, Shaw RK, Islam MS, Patel M, Snyder LAS, Lee DJ, Penn CW, Busby SJW, Pallen MJ (2014) Microarray analysis of the Ler regulon in enteropathogenic and enterohaemorrhagic *Escherichia coli* strains. PLoS One 9:e80160
12. Li G, Young KD (2012) Isolation and identification of new inner membrane-associated proteins that localize to the cell poles in *Escherichia coli*. Mol Microbiol 84:276–295
13. Bryant JA, Sellars LE, Busby SJ, Lee DJ (2014) Chromosome position effects on gene expression in *Escherichia coli* K-12. Nucleic Acids Res 42:11383–11392
14. Moor H, Teppo A, Lahesaare A, Kivisaar M, Teras R (2014) Fis overexpression enhances *Pseudomonas putida* biofilm formation by regulating the ratio of LapA and LapF. Microbiology 160:2681–2693
15. Rufian JS, Sanchez-Romero MA, Lopez-Marquez D, Macho AP, Mansfield JW, Arnold DL, Ruiz-Albert J, Casadesus J, Beuzon CR (2016) *Pseudomonas syringae* differentiates into phenotypically distinct subpopulations during colonization of a plant host. Environ Microbiol 18(10):3593–3605
16. Mosberg JA, Yep A, Meredith TC, Smith S, Wang PF, Holler TP, Mobley HLT, Woodard RW (2011) A unique arabinose 5-phosphate isomerase found within a genomic island associated with the uropathogenicity of *Escherichia coli* CFT073. J Bacteriol 193:2981–2988
17. Haines S, Arnaud-Barbe N, Poncet D, Reverchon S, Wawrzyniak J, Nasser W, Renauld-Mongenie G (2015) IscR regulates synthesis of colonization factor antigen 1 fimbriae in response to iron starvation in enterotoxigenic *Escherichia coli*. J Bacteriol 197:2896–2907

18. Alsharif G, Ahmad S, Islam MS, Shah R, Busby SJ, Krachler AM (2015) Host attachment and fluid shear are integrated into a mechanical signal regulating virulence in *Escherichia coli* O157:H7. PNAS 112:5502–5508

19. Jorgensen SB, Bojer MS, Boll EJ, Martin Y, Helmersen K, Skogstad M, Struve C (2016) Heat-resistant, extended-spectrum β-lactamase-producing *Kelbsiella pneumonia* in endoscope-mediated outbreak. J Hosp Infect 93:57–62

Chapter 2

Measuring In Vivo Supercoil Dynamics and Transcription Elongation Rates in Bacterial Chromosomes

N. Patrick Higgins

Abstract

DNA gyrase is the only topoisomerase that can catalytically introduce negative supercoils into covalently closed DNA. The enzyme plays a critical role in many phases of DNA biochemistry. There are only a few methods that allow one to measure supercoiling in chromosomal DNA and analyze the role of gyrase in transcription and its interaction with the other three bacterial topoisomerases. Here, we provide molecular tools for measuring supercoil density in the chromosome and for connecting the dots between transcription and DNA topology.

Key words Supercoiling, Transcription, Resolvase, Gyrase, Salmonella, Chromosome

1 Introduction

Site-specific recombination assays based on the γδ resolution system have provided information about nucleoid structure and quantitative estimates of supercoil density in living bacterial cells [1]. The γδ resolvase system requires supercoiled substrates, as do many transposons and other site-specific recombination systems. There are three requirements for the γδ system. First, two 114 bp γδ Res sites must be present in the DNA substrate, arranged as direct repeats. Second, sufficient resolvase must be present to saturate all chromosomal Res sites. A dimer of resolvase must bind to Res sub-sites I, II, and III. Third, the DNA substrate must be negatively supercoiled. Supercoiling is needed to drive formation of a 3-node synapse (Fig. 1a), which precisely aligns two directly repeated Res I sites for DNA breakage and reunion cycles [2, 3]. Only resolvase dimers bound to Res sub-site I can catalyze strand exchanges. Once the supercoiled resolvase synapse is formed, the complex deletes the intervening circular DNA segment and rejoins the substrate using no external energy and leaving a single Res site scar. Formation of a three-node

Olivier Espéli (ed.), *The Bacterial Nucleoid: Methods and Protocols*, Methods in Molecular Biology, vol. 1624,
DOI 10.1007/978-1-4939-7098-8_2, © Springer Science+Business Media LLC 2017

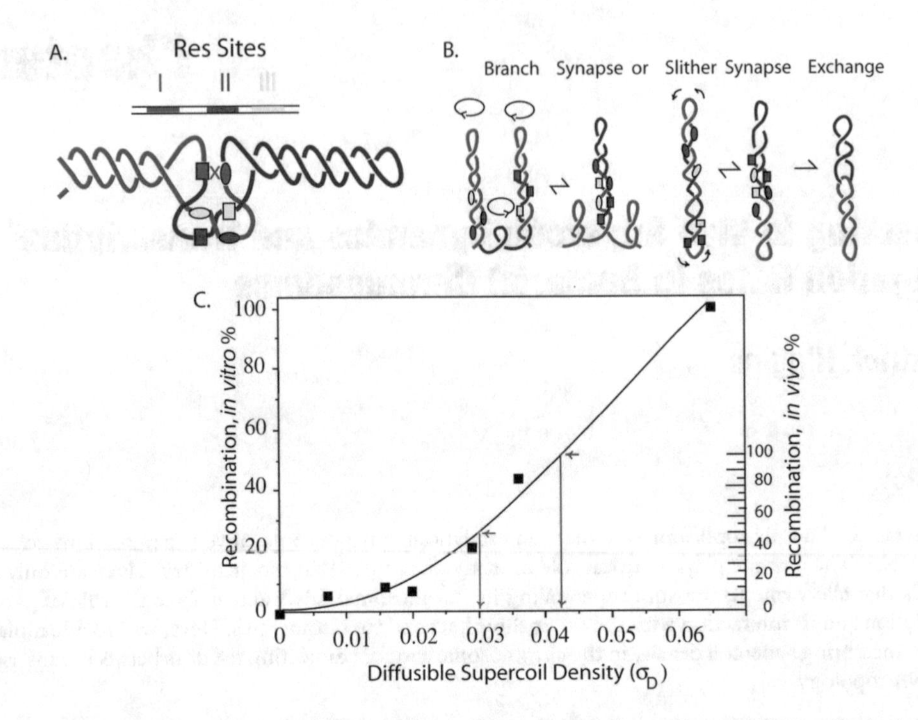

Fig. 1 Supercoil dynamics of the autocatalytic reaction of the γδ resolution system. (**a**) The topological structure required for resolution juxtaposes two blue I sub-sites that become cleaved and rejoined by resolvase within the three-supercoil synapse. (**b**) Dynamic DNA movements of branching and slithering promote the three-node tangle (**a**) and release two singly catenated supercoiled circles as end products at *right*. (**c**) The negative supercoil dependence is shown for in vitro (*left*) and in vivo (*right*) resolution reactions

synapse involves movements called slithering and branching (Fig. 1b). Branching rearranges the interwound DNA structure by starting new loops that grow and recede laterally. Slithering is a reptilian mode of writhe that displaces two opposing strands along an axis of interwound loops. When branching and slithering are unobstructed, resolution efficiency increases with the level of diffusible negative supercoiling in vitro [4] and in vivo [1, 5] for distances of 10 kb in log phase cells and up to 100 kb in stationary phase cells [6]. This in vitro/in vivo relationship provides the basis for estimating the in vivo supercoil density from resolution frequencies [7].

Initially, the γδ resolvase system for in vivo chromosome analysis had three of four characteristics that were essential for measuring accurate supercoil differences over a 100-fold range of sensitivity: 1—Tight repression of resolvase synthesis that generates a low background in resolution assays; 2—A temperature-sensitive repressor that dissociates from DNA at 42° and promotes rapid high-level resolvase expression; 3—A cI^{857} repressor that refolds rapidly to reestablish repression when cultures are shifted back to 30°.

The fourth property required for an optimal system became clear when cells induced for resolvase expression were tested for the ability to make deletions in a plasmid substrate that was introduced into cells by electroporation. WT resolvase catalyzed resolution reactions for several hours after induction followed by growth at 30°. The enzyme was immortal and only became depleted by cell division dilution. The significance of this result was that we could not detect barriers that form and disappear over the time frame of a cell generation, much less measure the time frame of gene expression [6]. Colony assays carried out with a WT resolvase were blind to many of the most interesting temporal supercoil changes that occur in a specific genome region.

Robert Stein solved this problem by making 5 min, 15 min, and 30 min time-restricted derivatives of γδ resolvase. He first added a C-terminal extension to WT resolvase, which is the 11-amino acid sequence that the SsrA system adds to translation fragments that result from a stalled ribosome [8]. This sequence normally releases the stalled ribosome by making ribosomes hop to a special RNA that has a translation stop codon at the end of an 11 amino acid sequence. Proteins with the 11 amino acid tag are efficiently targeted for degradation by the ClpXP protease. Stein's mutant resolvase with C-terminal extension of Ala-Ala-Asn-Asp-Glu-Asn-Tyr-Ala-Leu-Ala-Ala had a 5 min half-life in both *E. coli* and *Salmonella*. He made a second mutant resolvase with the change in the SsrA tag from A8-D, which produced a 15 min resolvase. A third change of L9-D yielded a 30 min resolvase. The 30-min resolvase provided a method to analyze the consequences of transcription throughout a chromosome by transiently halting most transcription for 30 min with rifampicin.

γδ resolvase experiments have provided critical data for five discoveries in bacterial chromosome physiology. (1) Resolution studies using the 5-, 15-, and 30-min derivatives provided the compelling evidence that bacterial chromosomes are organized into about 400 domains (@ 10 kb), which are stochastic in both length and boundary positions during normal cell growth [1, 5, 9]. (2) Experiments using the 5-min resolvase showed that gene transcription creates a supercoil diffusion barrier and that temporal appearance and disappearance of boundaries coincide precisely with transcription [10, 11]. (3) Resolution studies at the rrnG operon proved that the model of transcription proposed by Liu and Wang [12] generates a gradient with excess (−) supercoils upstream and depleted levels of (−) supercoils in a 10 kb domain downstream of ribosomal RNA operons in WT cells [7, 13]. Experiments with the 30 min resolvase demonstrated that the loss of (−) supercoiling generated by RNA polymerase downstream of a transcription terminator must match the rate of (−) supercoiling by gyrase to maintain supercoil density throughout a chromosome [7]. In this paper, we describe the strategy, basic modules, and methods that can be extended to other bacteria for systematic analyses of chromosome mechanics [7, 14].

2 Materials

2.1 Res Modules

Four selectable drug modules containing Res sites are illustrated in Fig. 2. Modules in A, B, and C include IS10 terminal ends [15]. These modules can be introduced in to bacterial chromosomes using *Tn10* transposition reactions (1) Alternatively, they can be introduced using homology-dependent λ red recombineering [16]. Recombineering of modules allows one to perform chromosome walks along the bacterial genomes to search for supercoil barriers [10, 11]. A chromosome walk can also identify essential genes that block recovery of recombinants because the deletion is lethal. With long inexpensive DNA synthetic blocks and Gibson assembly [17], one can easily create new Res modules for many purposes.

The fourth element in Fig. 2 (D) is an 8998 bp supercoil sensor that can be used to measure supercoil density of WT or mutant strains at a specific point and to determine the speed of RNA elongation at that same chromosomal location [18]. The sensor has an entire lac operon (*lacIZYA*) plus a selectable Gentamycin resistance gene (Gn) flanked by directly repeated res sites. Directly repeated FRT sites make the ends of the element. the supercoil sensor can be inserted into any chromosomal locus that has a single FRT site using the yeast 2 μ FLP recombinase [18, 19]. The resolution efficiency of the sensor increases in proportion to local negative supercoil density, and the in vivo range of the sensor spans

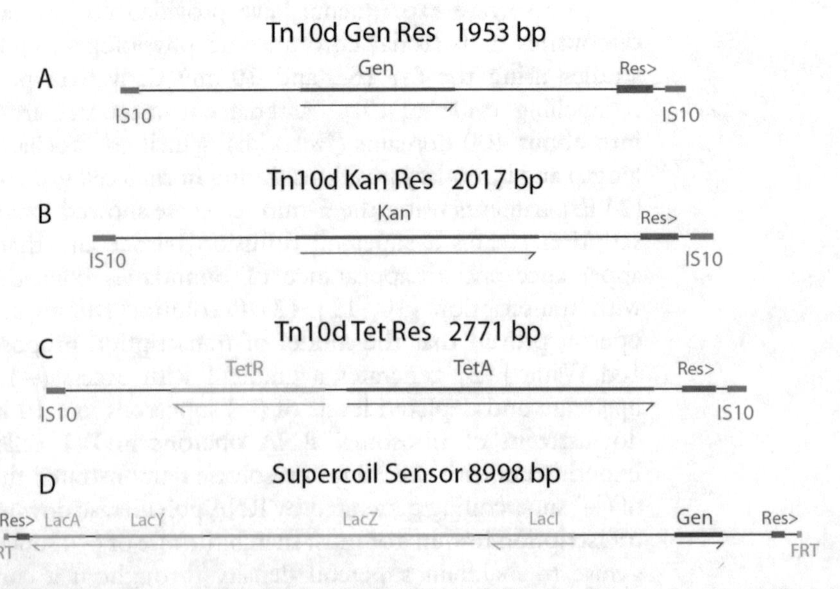

Fig. 2 Modules for chromosome tagging with a single 114 bp γδ Res site (**a**–**c**) or the supercoil sensor which has two Res sites (**d**)

2 orders of magnitude from <1% to 100% resolution (Fig. 1). This sensor is superior to supercoil-dependent promoter modules and supercoil-dependent psoralen crosslinking methods that have a 2 or 3-fold range of detection [20–23].

2.2 Resolvase Expression Plasmids

Four plasmids that encode time-restricted or WT γδ Resolvase under λ cI^{857} are: pJB Res cI^{857}-SsrA (5″) Cm; pJB Res $c^{857}I$-SsrA-A8D (15″) Cm; pJB Res cI^{857}-SsrA-L9D.
(30 “) Cm; and pJB RES cI^{857} (> 2 h).

2.3 Growth Chambers, Culture Media, and Biochemical Solutions

Most of our experiments have been done with LB medium, which results in very rapid growth. Supercoiling activity at certain chromosomal positions changes dramatically when cells are grown in minimal medium (*see* **Note 1**).

1. LB Medium: Dissolve 5 g NaCl, 10 g Tryptone, 5 g yeast extract in 1 L of deionized H_2O. Add 15 G Difco Agar to 1 L of LB in a 2 L flask for plates. Autoclave liquid and solid LB for 20–30 min and pour plates when the agar is cool enough to be handled without a glove (about 45 °C).
2. AB minimal medium: **5 ×** ***A*** = 2 g $(NH_4)_2SO_4$; 6 g Na_2HPO_4; 3 g KH_2PO_4; 3 g NaCl dissolved in 200 ml H_2O and autoclaved for 30 min. ***B*** = 797 ml H_2O autoclaved for 30 min. Add 1 ml 0.1 M $CaCl_2$ (sterile); 1 ml 1.0 M $MgCl_2$ (sterile); and 1 ml 0.003 M $FeCl_3$ (sterile). Mix the 200 ml of 5 × *A* with 800 ml of *B*. Add a sterile carbon source of 0.05% Glucose, 0.2% Alanine, or 0.2% Succinate plus any strain-required amino acids and vitamins to support growth rate doubling times of 90 min, 125 min, and 300 min, respectively.
3. Antibiotics Add antibiotics to LB liquid and solid medium at a concentration of 20 μg/ml Tetracycline HCl; 50 μg/ml Kanamycin SO_4; 20 μg/ml Chloramphenicol; and10 μg/ml Gentamycin. In liquid and solid minimal media, add drugs at concentrations of 10 μg/ml Tetracycline HCl; 125 μg/ml Kanamycin SO_4; 5 μg/ml Chloramphenicol; and 10 μg/ml Gentamycin.
4. Culture and spectrometer equipment and supplies. For culture growth and easy measuring of cell concentration, 125 ml side arm flasks and a Klett spectrophotometer permit rapid monitoring of culture growth rates without removing flask lids or caps. However, one can monitor culture density using any type of culture flask or tube and read the A_{650} in a standard spectrophotometer with glass or plastic cuvettes or a nanodrop instrument. An OD of 0.4 is a reasonable target for mid-log stage experiments. For experiments that demand short time sampling (transcription elongation), 50 ml media bottles with reusable polypropylene screw tops removed during fast manipulation are ideal.

5. Fisher blue cap snap on culture tubes (17 × 100 mm) are ideal for growing overnight cultures and β-Gal assays.
6. Two shaking incubators are required, preferably near each other. Set one at 30° and set a shaking water bath incubator at 42°.

3 Methods

γδ Resolution Assays.

3.1 Drug Module Deletion Protocol

1. Streak out strains to be tested on LB plates containing the drugs that select for Res module(s) plus chloramphenicol, which selects for stable maintenance of the appropriate Resolvase expression plasmid.
2. For each strain to be tested, inoculate three blue cap tubes containing 2 ml of the same medium in step one. Incubate overnight at 30°.
3. For each test culture, preload the appropriate number of flasks with 5–10 ml of LB + Cm medium. Inoculate each flask by adding a portion of the fresh overnight culture at a ratio of 1:100 (50 μl overnight culture into 5 ml of LB + Cm).
4. Incubate cultures in a 30° shaking incubator and monitor the cell growth by measuring culture OD_{600}. For strains like WT *E. coli* or *Salmonella*, it takes 4–5 h for cultures to reach mid log phase, which equates to an OD_{600} of 0.4–0.5.
5. When cells reach the desired OD, take a sample of cells for an un-induced control and place each flask in the 42° incubator for 10 min.
6. Return each temperature-induced flask to the 30° incubator and incubate for at least 2 h to allow recombinant chromosome to segregate into daughter cells. Alternatively, let the flasks incubate overnight, since the ratio of recombinants is stable after the Resolvase enzyme is degraded.
7. Spread cells on plates after making appropriate dilutions to get plates with 100–300 colonies/plate and incubate at 30°.
8. Use one sterile toothpick to patch a single colony on two plates that contain the antibiotic for each drug module. Patch 100 colonies in this way and calculate the deletion efficiency.

3.2 Lac Deletion Protocol

Steps 1–6 are the same.

7. Plate diluted cells on Lac indicator plates. Resolution efficiency is determined by color. On XGal, whites are recombinant and blues are non-recombinants. On MacConkey the pattern is white and

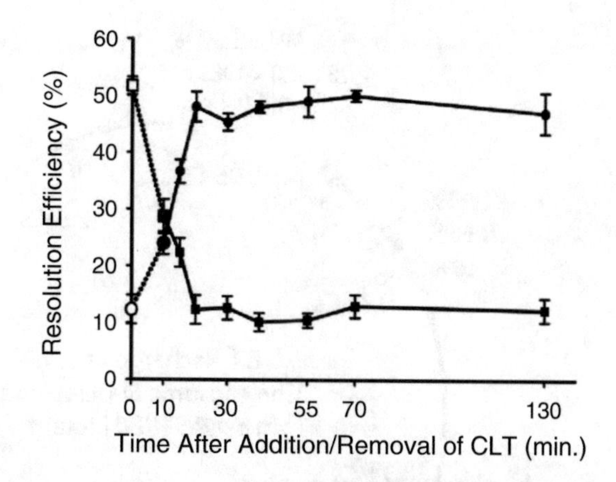

Fig. 3 Find supercoil barriers caused by gene expression [10]. An example of the lac method for measuring deletion efficiencies is shown above. The *Salmonella* strain NH3492 has a 12 kD chromosomal segment flanked by a Gn Res module on the left and a Lac Res module on the right in addition to the 5″ resolvase plasmid. *Tn10* repressor, TetR, regulates expression of β-Gal. The *lacZ* gene becomes deleted by γδ resolution). In the experiment shown in Fig. 3, three cultures (*open squares*) were initiated by diluting a fresh overnight culture 100-fold into LB + Cm medium and grown to 50 Klett in the absence of the inducer chlorotetracycline (CLT). Each culture had a γδ resolution efficiency of 50%. CLT was added to the cultures (filled squares) and thermo-induced at 5 min intervals. Resolution rates dropped fivefold to 10%. The reciprocal experiment was carried out by growing three cultures to 50 Klett in LB + CLT. Assays of resolution in these cells occurred at a rate near 10% (*open circles*). CLT was washed out of the cells and incubation continued at 30° in LB for 130 min (*filled circles*). The resolution efficiency rose quickly to 50% and remained at this level for 2 h. This experiment was the first demonstration that transcription creates its own barrier to supercoil diffusion in a bacterial chromosome

red, but colonies can be counted in 20 h whereas it takes 2–3 days to develop good color at 30° on XGal (*see* **Note 2**).

An example of the lac method for measuring deletion efficiencies is shown above (Fig. 3) (*see* **Notes 3** and **4**).

3.3 Measure Supercoil Dynamics with the Supercoil Sensor (Fig. 2d)

The supercoil sensor provides a way to monitor topological dynamics at multiple positions in a chromosome. The sensor enabled us to demonstrate that the Liu and Wang model of twin domains of supercoiling [12] causes a gradient of supercoiling with an excess upstream and a deficit downstream of highly expressed genes even in WT cells [13]. The 30″ resolvase (pJB Res cI^{857}-SsrA-L9D Cm) is ideal for this experiment. One variation in the standard experiment showed that the enzymatic rates of RNA polymerase elongation and gyrase supercoiling are linked [7].

To test the hypothesis that transcription can produce a flat chromosome without diffusible supercoils, a simple modification

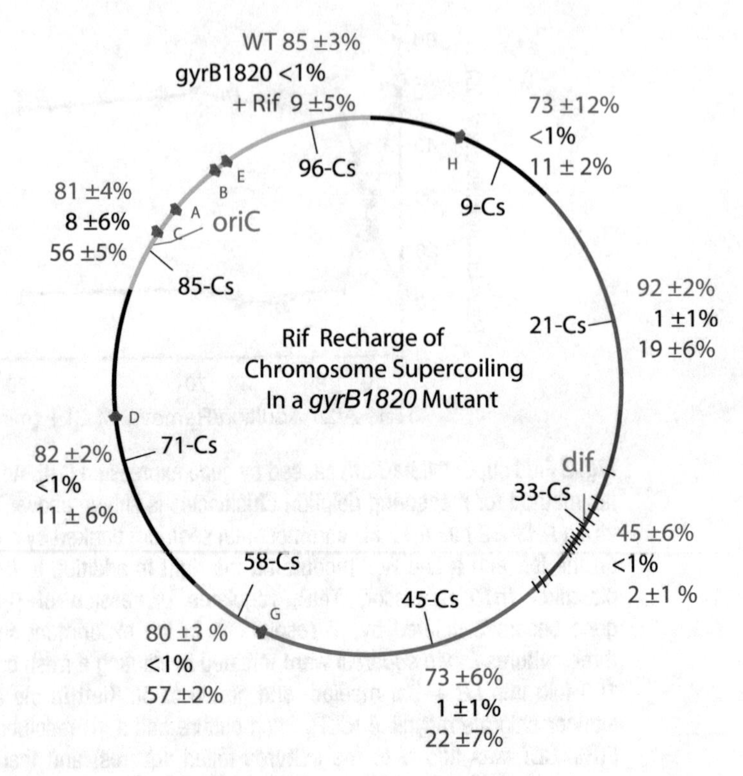

Fig. 4 Chromosome supercoiling linked to RNA transcription and gyrase

after the 10 min incubation at 42° provides the experimental answer (Fig. 4). The test is to split the culture and add rifampicin to one half during the post-induction 30 min incubation at 30°. Rifampicin blocks initiation of RNA synthesis but not elongation and termination. If transcription is responsible for running the chromosome flat, supercoiling might restore resolution during the 30 min half-life of resolvase. At 30 min, rifampicin was removed from cells by centrifugation and resuspension in LB + Cm. Then both cultures were handled with the normal protocol. For the gyrB1820 strain in Fig. 5, 60% of the resolution efficiency was restored by the Rif treatment. Although not all parts of the genome recovered equally, this result proves that gyrase catalytic rates and RNA polymerase rates must match to maintain WT supercoil levels in a *Salmonella* chromosome.

3.4 Measuring Transcription Elongation

The observation that transcription can run the chromosome flat in cells with a slow gyrase leads to a question. Does a mutant gyrase affect the rate of transcription?

To measure transcription elongation rates, the Lac operons at each location provides a way to test transcription elongation rates at the same sites that were used to measure supercoiling.

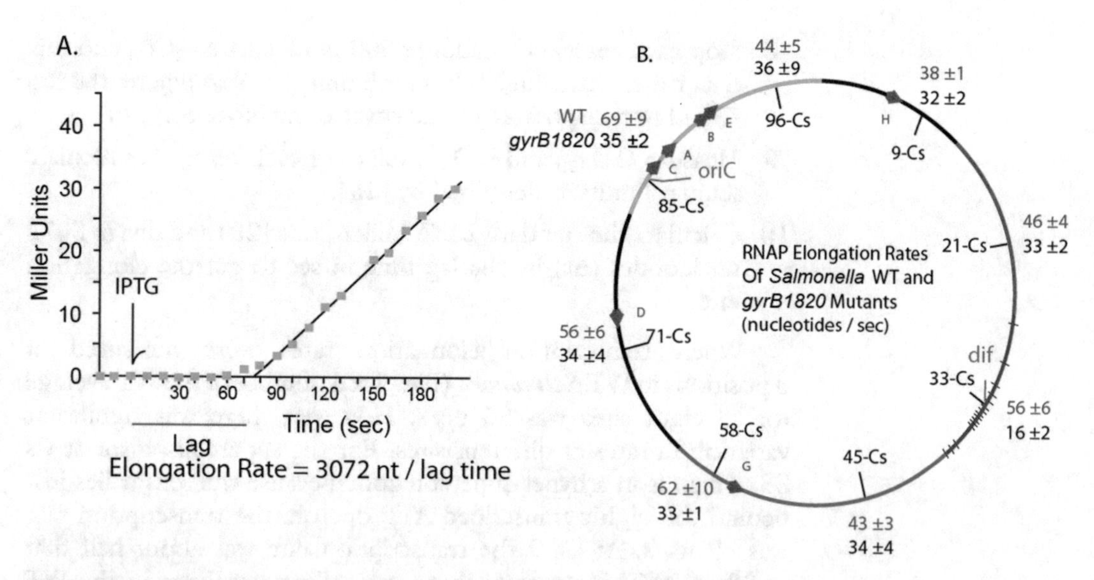

Fig. 5 Transcription elongation rates measured in WT and gyrB1820 strains of *Salmonella*

3.5 Protocol to Test RNA Elongation Rates, Adapted from Vogel [24]

1. Inoculate a flask with 20 ml of minimal AB medium supplemented with selective drugs for the modules + Cm and 0.02% glucose with 1 ml of an overnight culture.
2. Measure cell density and carry out the assay when the OD_{600} is between 0.2 and 0.4. Record the OD_{600}. It is important to use an open culture vessel that allows rapid pipetting of 500 μl samples from a culture every 10 s for 4 min.
3. Prepare enough blue cap tubes for the number of samples needed plus 10% extras for mistakes. Add 500 μl of ice-cold ZS-Cm buffer =60 mM Na_2HPO_4; 40 mM NaH_2PO_4 10 mM KCl; 1 mM $MgSO_4$; 50 mM β-Mercaptoethanol; 0.01% SDS; and 200 μg/ml chloramphenicol to block further protein synthesis at 4°. Keep these tubes in an ice-cold bath.
4. Start by recording the culture OD600 and take three 500 μl samples at 10 s intervals to establish the background subtraction. Place these tubes in a separate ice bath.
5. Add IPTG to the remaining culture in a shaking water bath at a final concentration of 1.5 mM.
6. Remove 500 μl aliquots at 10 s intervals, placing each tube in the second ice bath for the next 4 min. Once all samples are in the ice bath, add 100 μl of chloroform, vortex, and incubate the tubes for 2 min at 30°.
7. Start timed B-Gal assays on each set by adding 200 μl of ONPG 4 mg/ml. Place the tubes in a stationary 30° incubator for 1.5–4 h to allow development of appropriate levels of color. Note the start time.

8. Stop each reaction by adding 500 μl of 1 M Na_2Co_3 and mix each tube. Centrifuge tubes for 5 min and then pipette the top 750 μl to a disposable plastic cuvette (*see* **Note 5**).
9. Measure OD_{420} and OD_{550} values of each reaction. Calculate standard units as described by [25].
10. Calculate the standard LacZ Miller units [25] and divide 3072 nucleotides (nt) by the lag time in sec to get the elongation rate.

When transcription elongation rates were measured at 8 positions in WT *Salmonella* (Fig. 5 red numbers), the WT average for all eight sites was 55 nr/s. However, there was significant variation in rates at different sites. For the supercoil sensor at Cs 85, which is in a hyper-supercoil zone because the sensor lies just behind the highly transcribed ATP operon, the transcription rate was 69 nt/s. At Cs 9 the transcription rate was about half that at 38 nt/s. This suggests that supercoiling gradients in the WT chromosome influence RNA polymerase. Data from the flat chromosome gyrB1820 proves the case (Fig. 5 black numbers). Transcription rates at 7 positions fell to 32 nt/s. The sensor at Cs 33 is very near the Dif site, and at this position transcription fells to 16 nt/s. This position may become positively supercoiled due to convergent replication forks.

4 Notes

1. Use fresh medium made within 1 month of an experiment. Complex media like LB stored on lab a shelf develops toxic-free radicals that result from exposure to fluorescent light. Diluting cells into such medium changes cellular metabolism, and cells can become induced for the RecA SOS response that can cleave λ repressor.
2. Most strains show ratios of recombinant to non-recombinant cells that remain stable after the resolvase is degraded. Therefore, resolution rates measured 2 h after induction agree with rates measured the following day in stationary culture.
3. Most strains give similar efficiencies from 1 h at 30° until they reach mid log phase. As culture densities become high, growth rates slow as cells enter stationary phase. Resolution efficiency increases as transcription rates decrease going into stationary phase.
4. When experiments are carried out with *E. coli*, the high supercoil density of this organism (15% higher than *Salmonella* Typhimurium) results in average dilution efficiencies of the 9 kD sensor of >80% throughout most of the genome. To lower this rate it is advisable to use the 15″ or 5″ resolvase and construct supercoil sensors longer (15–20 kD) than the one used in *Salmonella*.

5. For the LacZ-transcription elongation assay, disposable 1.5 ml plastic cuvettes for measuring a large number of color changes are useful. Fisher supplies these cuvettes in convenient boxes that are easy to handle (Cat no. 14–955-127 includes 500 cuvettes).

References

1. Higgins NP, Yang X, Fu Q, Roth JR (1996) Surveying a supercoil domain by using the gd resolution system in *Salmonella typhimurium*. J Bacteriol 178:2825–2835
2. Stark MW, Boocock MR (1995) Topological selectivity in site-specific recombination. In: Sherratt D (ed) Mobile genetic elements. IRL Press, Oxford, p 179
3. Grindley ND, Whiteson KL, Rice PA (2006) Mechanisms of site-specific recombination. Annu Rev Biochem 75:567–605
4. Benjamin KR, Abola AP, Kanaar R, Cozzarelli NR (1996) Contributions of supercoiling to Tn*3* resolvase and phage Mu Gin site-specific recombination. J Mol Biol 256:50–65
5. Stein R, Deng S, Higgins NP (2005) Measuring chromosome dynamics on different timescales using resolvases with varying half-lives. Mol Microbiol 56:1049–1061
6. Staczek P, Higgins NP (1998) DNA gyrase and Topoisomerase IV modulate chromosome domain size in vivo. Mol Microbiol 29:1435–1448
7. Rovinskiy N, Agbleke AA, Chesnokova O, Pang Z, Higgins NP (2012) Rates of gyrase supercoiling and transcription elongation control supercoil density in a bacterial chromosome. PLoS Genet 8:e1002845
8. Keiler KC, Waller PRH, Sauer RT (1996) Role of a peptide tagging system in degradation of proteins synthesized from damaged messenger RNA. Science 271:990–993
9. Postow L, Hardy CD, Arsuaga J, Cozzarelli NR (2004) Topological domain structure of the *Escherichia coli* chromosome. Genes Dev 18:1766–1779
10. Deng S, Stein RA, Higgins NP (2004) Transcription-induced barriers to supercoil diffusion in the *Salmonella typhimurium* chromosome. Proc Natl Acad Sci U S A 101:3398–3403
11. Deng S, Stein RA, Higgins NP (2005) Organization of supercoil domains and their reorganization by transcription. Mol Microbiol 57:1511–1521
12. Liu LF, Wang JC (1987) Supercoiling of the DNA template during transcription. Proc Natl Acad Sci U S A 84:7024–7027
13. Booker BM, Deng S, Higgins NP (2010) DNA topology of highly transcribed operons in Salmonella enterica serovar Typhimurium. Mol Microbiol 78:1348–1364
14. Higgins NP (2016) Species-specific supercoil dynamics of the bacterial nucleoid. Biophysical Rev 8:113–121
15. Raleigh EA, Kleckner N (1986) Quantitation of insertion sequence IS10 transposase gene expression by a method generally applicable to any rarely expressed gene. Proc Natl Acad Sci U S A 83:1787–1791
16. Datta S, Costantino N, Court DL (2006) A set of recombineering plasmids for gram-negative bacteria. Gene 379:109–115
17. Gibson DG, Young L, Chuang RY, Venter JC, Hutchison CA 3rd, Smith HO (2009) Enzymatic assembly of DNA molecules up to several hundred kilobases. Nat Methods 6:343–345
18. Pang Z, Chen R, Manna D, Higgins NP (2005) A gyrase mutant with low activity disrupts supercoiling at the replication terminus. J Bacteriol 187:7773–7783
19. Senecoff JF, Cox MM (1986) Directionality in FLP protein-promoted site-specific recombination is mediated by DNA-DNA pairing. J Biol Chem 261:7380–7386
20. Miller WG, Simons RW (1993) Chromosomal supercoiling in *Escherichia coli*. Mol Microbiol 10:675–684
21. Pavitt GD, Higgins CF (1993) Chromosomal domains of supercoiling in Salmonella typhimurium. Mol Microbiol 10:685–696
22. Mojica FJM, Higgins CF (1997) In vivo supercoiling of plasmid and chromosomal DNA in an *Escherichia coli hns* mutant. J Bacteriol 179:3528–3533
23. Moulin L, Rahmouni AR, Boccard F (2005) Topological insulators inhibit diffusion of transcription-induced positive supercoils in the chromosome of *Escherichia coli*. Mol Microbiol 55:601–610
24. Vogel U, Sorensen M, Pedersen S, Jensen KF, Kilstrup M (1992) Decreasing transcription elongation rate in *Escherichia coli* exposed to amino acid starvation. Mol Microbiol 6:2191–2200
25. Miller JH (1972) Experiments in molecular genetics. Cold Spring Harbor Laboratory, Cold Spring Harbor, NY

Chapter 3

Revealing Sister Chromatid Interactions with the *loxP*/Cre Recombination Assay

Elise Vickridge, Charlène Planchenault, and Olivier Espéli

Abstract

Sister chromatid cohesion is a transient state during replication in bacteria. It has been recently demonstrated that the extent of contact between cohesive sisters during the cell cycle is dependent on topoisomerase IV activity, suggesting that topological links hold sister chromatids together. In the present protocol, we describe a simple method to quantify the frequency of the contacts between two cohesive sister chromatids. This method relies on a site specific recombination assay between loxP sites upon Cre induction.

Key words *loxP*/Cre, Site-specific recombination, Sister chromatids, Cohesion, Segregation

1 Introduction

Site specific recombination methods have been used in several occasions to probe chromosome structure. Site specific recombinases only recognize DNA in an adequate structure. These properties have been used to reveal different DNA structuring features. The supercoiling density of E. coli plasmids has been measured in vivo and in vitro according to the recombination frequency of λ Int/Xis recombinase on attL/R sites [1]. Biases in λ Int/Xis mediated recombination also revealed macrodomain structures in the E. coli chromosome [2]. Tn3/ gd recombination has been used to monitor supercoiling microdomains [3, 4] in S. typhimurium. In yeast, the loxP/Cre recombination system has allowed to measure interactions between homologous and ectopic loci [5]. Here we describe a simple assay based on the loxP/Cre recombination system that reveals interactions between sister chromatids and allows to follow segregation of loci in a quantitative manner [6]. This method can be used to study sister chromatid interactions in a

Olivier Espéli (ed.), *The Bacterial Nucleoid: Methods and Protocols*, Methods in Molecular Biology, vol. 1624, DOI 10.1007/978-1-4939-7098-8_3, © Springer Science+Business Media LLC 2017

normal cell cycle, in response to genotoxic stress or in various mutants. Technically, two consecutive *loxP* sites, spaced by only 20 bp, are cloned at a given site on the chromosome. In our original setup they were cloned into an ectopic lacZ gene, thus preventing its transcription. The plated colonies are white. When the two sister chromatids are in very close proximity, the *loxP* sites from each sister chromatid will be able to recombine with one another, once Cre is induced with 0.1% arabinose. The products formed by this recombination are a 1*loxP* site on one sister chromatid and 3*loxP* sites on the other sister chromatid. These three *loxP* sites on one sister chromatid will recombine together, eventually leading to a 1*loxP* site. When there is only one *loxP* site in the lacZ gene, the open reading frame of *lacZ* is reconstituted and the colonies become blue when plated on Xgal. The frequency of recombination which is directly linked to the proximity between sister chromatids can be assessed by a white/blue colony count (Fig. 1). In this chapter, we also present a new method to immediately analyze recombination products formed in the minutes following Cre recombination with a semiquantitative PCR and a Bioanalyzer. The amount of 1*loxP* product versus the total amount of DNA gives the recombination frequency and thus, the amount of sister chromatid interactions (Fig. 2).

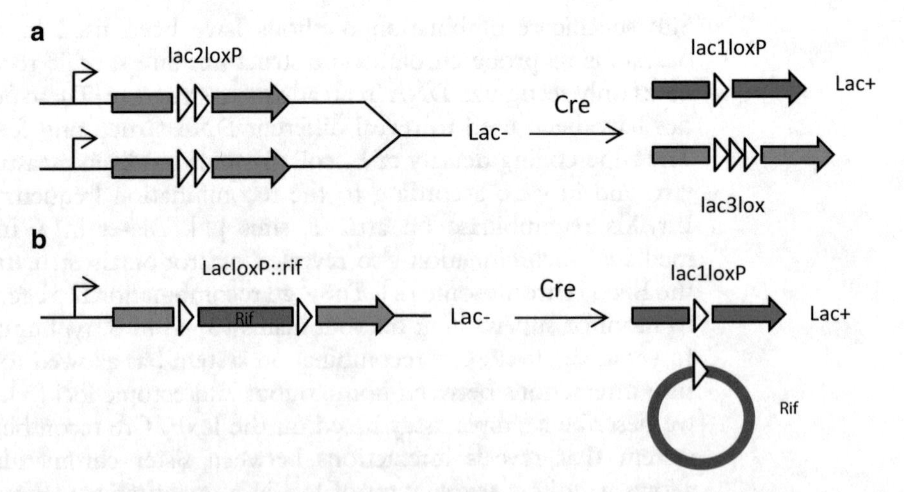

Fig. 1 (**a**) The intermolecular loxP recombination reaction occurs between two close sister chromatids. The frequency of sister chromatid interactions can be assessed by a blue/white colony count or by a semiquantitative PCR by amplifying the recombination products. (**b**) The intramolecular loxP recombination frequency is measured by the loss of the rifampicin resistance gene by recombination of the two loxP sites spaced by 1 kb

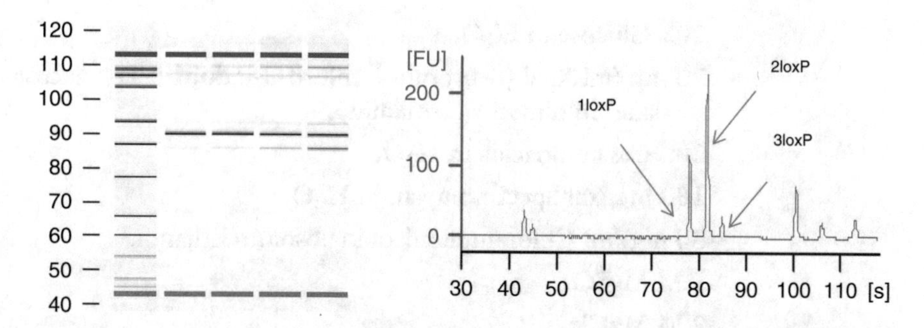

Fig. 2 (**a**) Typical Electrophoresis run of a wild-type strain upon Cre induction by arabinose. Cells were treated for 10, 20, 30, or 40 min with 0.1% arabinose and analyzed with an agilent bioanalyzer. (**b**) Typical electrophoregram of a wild-type strain upon Cre induction by arabinose. The result after 40 min of Cre induction is shown. The results are shown as Fluorescence Units (FU) in function of the time of elution

2 Materials

Store all reagents as indicated on the product. The bioanalyzer reagents must be equilibrated at room temperature for at least an hour before use.

2.1 Plasmids, Strains, and Primers

<u>*Strains*:</u>

MG1655 (*E. coli* K12).
MG1656 (*E. coli* K12 *Δlac MluI*).

<u>*Plasmids*:</u>

pGBKD3-LacloxP [6].
pGBKD3-Lacloxrif [7].
pCre [6].

<u>*Primers for insertion of loxP sites onto the chromosome*:</u>

LacloxP-ins-Forward: [50 nt of homology upstream of insertion site—GAT TGT GTA GGC TGG AGC TGC].
LacloxP-ins-Reverse: [50 nt of homology downstream of insertion site—GG TCT GCT ATG TGG TGC TAT CT].

<u>*Primers for analysis of loxP recombination products by PCR*:</u>

LacloxP-Forward: CTTCTGCTTCAATCAGCGTGCCGTC.
LacloxP-Reverse: GATCAGGATATGTGGCGGATGAGCGG.

2.2 Reagents: Stock Solutions

All solutions must be prepared using ultrapure water (by purifying deionized water, to attain a sensitivity of 18 MΩ-cm at 25 °C).

Prepare the following buffers and stock solutions. Unless otherwise specified, filter solutions using a 0.2 μm low protein binding non-pyrogenic membrane.

20% (w/v) Arabinose in H_2O.

40% Glycerol in H_2O.

20% Glucose in H_2O.

20 mg/ml Xgal (5-bromo-4-chloro-3-indolyl-β-D-galactopyranoside) in dimethylformamide.

20% Casaminoacids in H_2O.

100 mg/ml Spectinomycin in H_2O.

30 mg/ml Chloramphenicol in absolute Ethanol.

1 M $MgSO_4$.

2 M $MgCl_2$.

Dimethyl Sulfoxide (DMSO).

10× Minimum Medium A: 0.26 M KH_2PO_4, 0. 06 M K_2HPO_4, 0.01 M tri sodium citrate, 2 mM $MgSO_4$, 0.04 M $(NH_4)_2$ SO_4.

Transformation and Storage Solution (TSS): LB, 10% PEG 6000, 10 mM MgSO4, 10 mM $MgCl_2$, 5% DMSO.

LB Broth (Lennox Broth).

Select agar.

10× TBE: 1 M Tris, 1 M Borate, 0.02 M EDTA.

1.7% Agarose (molecular grade) in 1× TBE.

DNA Ladder 100 bp DNA.

2.3 Medium

For *LoxP* recombination assays in *Escherichia coli* and related bacteria, grow cells in 1× Minimum Medium A supplemented with 0.2% casaminoacids and 0.25% glycerol. pCre transformation and measure of recombination can be performed in Minimum Medium A supplemented with 0.2% glucose, 0.25% casaminoacids, 40 μg/ml Xgal and 100 μg/ml spectinomycin or on LB agar plates supplemented with 40 μg/ml Xgal and 100 μg/ml spectinomycin.

2.4 Genomic DNA Extractions

Liquid nitrogen.

Genomic DNA extraction kit.

2.5 Genomic DNA and PCR Product Quantification

DNA quantification using a Nanodrop (ThermoScientific).

2.6 PCR

Primers for strain constructions (*see* Subheading 2.1).

Insertion of LacloxP and LacloxP-rif cassettes on the chromosome:

Four 50 μl reactions were assembled using 1 ng of plasmid DNA as a matrix (pGBKD3-LacloxP or pGBKD3-LacloxP-rif), primers LacloxP-ins-Forward and LacloxP-ins-Reverse (20 μM), dNTPs (10 μM each), Mg^{2+} plus buffer and TaKaRa Ex Taq (*see* **Note 1**).

Amplification of loxP sites:

A 50 μl reaction mix was prepared using 1 μl of genomic DNA at 2 ng/μl, primers LacloxP-Forward and LacloxP-Reverse (20 μM), dNTPs (10 μM each) and Mg^{2+} plus buffer and TaKaRa Ex Taq.

2.7 PCR Product Verification on Agarose Gel

1.7% agarose gel electrophoresis in 1× TBE.

2.8 Detection of loxP Recombination Events by Bioanalyzer

Agilent DNA 1000 kit.

3 Methods

3.1 Strain Constructions

The Lac*loxP* cassette that allows measuring of intermolecular recombination was constructed by the integration of a double-stranded oligonucleotide 5'CGTAATAACTTCGTATAATGTAT GCTATACGAAGTTATGGATCCC.

CGGGTACCGAGCTCATAACTTCGTATAATGTATGCTA TACGAAGTTATCCTA-3′ into the *Cla*I restriction site of the *lacZ* gene of the pGBKD3-lacZ plasmid. We called this plasmid pGBKD3-LacloxP [6]. For the control of intramolecular recombination, a rifampicin resistance gene (*rif)* and its promoter were introduced between the 2 *loxP* sites by cloning into the *BamH*I site of pGBKD3-Laclox. We called this plasmid pGBKD3-Laclox-rif [6]. The pGBKD3-LacloxP or pGBKD3-Laclox-rif plasmids were used as matrices to insert the intermolecular and intramolecular cassettes inside intergenic regions of nonessential genes of the DY330 strain with the standard "lambda red" method [8]. These vectors contained the Lac*loxP* or the Lac*loxP*-rif cassettes adjacent to the chloramphenicol resistance gene. The Lac*loxP*::Cm or the Lac*loxP*-rif::Cm cassettes were then P1 transduced into the genome of an MG1656 strain. Expression of the Cre recombinase was driven by an arabinose-inducible promoter on a plasmid (pFX465) derived from pSC101, we called this plasmid pCre [6]. pCre also contains a PLac promoter antisens to *cre* that can be used to further repress Cre expression with IPTG in some conditions (*see* **Note 2**).

3.2 Measuring LoxP Recombination as a Function of the Number of Lac + colonies on Plate

Day 1

Transformation of the pCre plasmid in the MG1656-lacloxP and MG1656 LacloxP-rif strains.

1. Inoculate 100 ml of LB with 1 ml of overnight culture of MG1656 LacloxP and MG1656 LacloxP-rif at 37 °C. Let the cells grow to an OD600nm ≈ 0.5–0.6.

2. Centrifuge at 3400 × *g*, 10 min at 4 °C.
3. Resuspend the pellet in a 1/10 volume of cold TSS.
4. Leave the cells for 10 min on ice.
5. Mix 100 μl of cells with 2–4 ng of pCre plasmid.
6. Leave the cells with the plasmid for 10 min on ice.
7. Induce a heat shock at 37 °C for 3 min or 42 °C for 90 s.
8. Leave the sample on ice for 30 s and add 900 μl of LB.
9. Incubate for 60 min at 37 °C.
10. Plate 100 μl of the transformation on LB supplemented with 40 μg/ml Xgal and 100 μg/ml spectinomycin.
11. Incubate overnight at 37 °C (*see* **Note 3**).

Day 2

1. Select a white colony and streak on a fresh plate of Minimum Medium A supplemented with 0.2% glucose, 0.25% casaminoacids, 40 μg/ml Xgal and 100 μg/ml spectinomycin.

Day 3

1. Prepare three different 1.5 ml cultures from three different white colonies in Minimum Medium A supplemented with 0.2% glucose, 0.25% casaminoacids, and 100 μg/ml spectinomycin final concentration for both the intramolecular strain and the intermolecular strain.
2. Incubate overnight at 37 °C.

Day 4

1. Dilute the overnight cultures at 1/200 in minimum medium A supplemented with 0.2% casaminoacids and 0.2% glycerol at 37 °C, under shaking. Use a 100 ml Erlenmeyer and 25 ml of medium (*see* **Note 4**).
2. At OD600nm ≈ 0.2 take two 100 μl aliquots of each culture; dilute them at a convenient dilution to obtain about 200 colonies per 100 μl plating. This is the non induced recombination frequency reference. Add 0.1% arabinose to each culture (*see* **Notes 5** and **6**).

This step is crucial because recombination occurs as soon as the arabinose is added. Specific care must be taken for the timing.

3. For each time-point 5, 10, 15, 20 min for intermolecular recombination and between 3, 5, 10 min for intramolecular recombination (*see* **Notes 7** and **8**), dilute samples 1/100 in minimum medium A, casaminoacids and glycerol in order to dilute the arabinose and stop Cre induction. We observed that Cre recombination dramatically dropped down in the minutes following the arrest of induction [6] (*see* **Note 9**).

4. Then, dilute each sample at a convenient dilution to obtain about 200 colonies per 100 μl plating (*see* **Note 10**).
5. Plate two times 100 μl of each sample on LB plates supplemented with 40 μg/ml Xgal and 100 μg/ml spectinomycin.
6. Incubate at 37 °C overnight.
7. Count the number of white and blue colonies for each plate and each sample.
8. Quantification of recombination: (number of blue colonies/number of white colonies INTER)/(number of blue colonies/number of white colonies INTRA). The intramolecular recombination rate serves as a normalization for Cre recombination efficiency per se, which depends on the locus considered, growth conditions, and the genetic background.

3.3 LacloxP Assay by PCR Method

This alternative method relies on the quantification of the *loxP* recombination products by combining a PCR amplification of the *loxP* products with their quantification on an Agilent Bioanalyzer. This method significantly facilitates interpretation of the results when working with mutants whose viability is affected or with drugs that affect the cell cycle. In addition, we observed a higher reproducibility of the results with the PCR-Bioanalyzer method than with the plating method.

For growth conditions and Cre induction, *see* Subheading 3.2 Day 3 and Day 4 (1–2)

1. Before Cre induction and at chosen time-points after induction, take 1.5 ml of the given samples and flash freeze them immediately in liquid nitrogen in a 2 ml test tube. Frozen samples may be stored at −20 °C for several weeks.
2. Gently thaw samples and centrifuge them for 3 min at 8000 × *g*. Remove the supernatant and proceed to genomic DNA extraction according to manufacturer's instructions. Elute DNA in 100 μl of elution buffer (*see* **Note 11**).
3. Quantify DNA with the Thermo Fisher Scientific Nanodrop. Expected DNA concentration is between 20 and 80 ng/μl.
4. Dilute all samples to 2 ng/μl in molecular grade water.
5. Perform a PCR on each sample using Ex Taq enzyme from TaKaRa. Use 1 μl of diluted sample for each 50 μl reaction. Tm = 58 °C, 28 cycles. Elongation step is 30 s.
6. Verify the PCR products with a 1.7% agarose gel in TBE 1× (*see* **Note 12**). Migration should be done at no more than 10 V/cm for 1 h in order to reveal the *loxP* recombination products.
7. Take out the DNA 1000 agilent kit from 4 °C about 1 h before use to equilibrate reagents to room temperature.

8. Load chip with amplified PCR products according to the manufacturer's protocol (*see* **Note 13**).
9. Bioanalyzer software automatically performs DNA intensity peak calling and quantification. Typical results are described in Fig. 2a, b. Results are then expressed as a frequency of recombination by the following formula: (Amount of 1*loxP* + 3*loxP* products (in ng/ml))/(amount of 1*loxP* + 2*loxP* + 3*loxP* products (in ng/ml)).

4 Notes

1. We noticed that many Taq polymerases failed to properly amplify two and three successive loxP sites. The best results were obtained with TaKaRa Ex Taq in Mg^{2+} plus buffer.
2. In some genetic backgrounds or growth conditions a high level of Cre recombination of the Laclox-rif cassette can be observed immediately after pCre transformation even in the absence of induction. If required, IPTG 20 μg/ml can be used to further repress Cre expression.
3. Strains with the pCre plasmid cannot be stored at −80 °C. The pCre plasmid must be transformed fresh before each experiment.
4. The induction of pCre by arabinose in LB medium rapidly produces a high level of recombination. 100% of recombination products are observed in less than 10 min. We noticed that such a short time window does not allow to obtain highly reproducible results. We recommend performing the experiments in minimum media A or M9 supplemented with casamino acids 0.2% and glycerol 0.2% or succinate 0.2% as a carbon source. Avoid glucose to assure the proper induction of the arabinose promoter.
5. A residual Cre recombination can be observed in some cases even in the absence of induction. This background recombination should be kept below 5% and must be subtracted from the recombination samples before analysis.
6. Growth conditions are critical for reproducibility. We recommend the systematic usage of 100 ml erlenmeyers, with 8 ml of culture for your experiment. Best results were obtained when growth was carried in a water bath shaker (250 rpm).
7. Timing is very important when adding the arabinose and the drug. Indeed, recombination occurs very quickly after arabinose induction and a bias can occur between samples if not timed properly.
8. The Cre induction length should be adjusted according to the growth conditions and the genetic background.

Recombination frequency is linear as a function of time between 5% and 70%. We recommend using appropriate time points for your experiments depending on the chosen conditions. Recombination frequency is strongly affected by temperature. Intra molecular normalization is required for each temperature change.

9. When doing the experiment on plate, recombination rates are higher than with the PCR method. Time points for kinetics must be shorter than with the PCR method. We do not yet fully understand this observation. Our current hypothesis is that a significant part of the recombination reaction initiated during the induction period only gives rise to resolved recombination products after this induction period and therefore could remain undetected by the PCR method [6].
10. When using the plate method, samples have to be diluted at 1/100e directly for each time point to stop the induction of Cre as quickly as possible. When using drugs affecting cell viability, make sure to calibrate dilutions for plating in order to have a countable amount of bacteria on plate.
11. Variation between experiments can be observed when changing genomic extraction kits. We suggest you always use the same supplier for your experiments.
12. TBE 1× rather than TAE 0.5× should be used for the agarose gel verification for a better separation of the *loxP* products.
13. The DNA 1000 Agilent chip is sensitive. It is important to avoid the amplification of nonspecific products by PCR. The Tm and the number of cycles of the PCR are important factors allowing optimization of this step. When using the piko PCR machine from Thermo Scientific, optimal Tm is 58 °C and 28 cycles are enough. These parameters may need changing and optimization if using a different PCR machine.

References

1. Bliska JB, Cozzarelli NR (1987) Use of site-specific recombination as a probe of DNA structure and metabolism in vivo. J Mol Biol 194:205–218
2. Valens M, Penaud S, Rossignol M, Cornet F, Boccard F (2004) Macrodomain organization of the *Escherichia coli* chromosome. EMBO J 23:4330–4341
3. Deng S, Stein RA, Higgins NP (2004) Transcription-induced barriers to supercoil diffusion in the salmonella typhimurium chromosome. Proc Natl Acad Sci U S A 101:3398–3403
4. Staczek P, Higgins NP (1998) Gyrase and Topo IV modulate chromosome domain size in vivo. Mol Microbiol 29:1435–1448
5. Burgess SM, Kleckner N (1999) Collisions between yeast chromosomal loci in vivo are governed by three layers of organization. Genes Dev 13:1871–1883
6. Lesterlin C, Gigant E, Boccard F, Espéli O (2012) Sister chromatid interactions in bacteria revealed by a site-specific recombination assay. EMBO J 31:3468–3479
7. Espéli O et al (2012) A MatP-divisome interaction coordinates chromosome segregation with cell division in E. coli. EMBO J 31:3198–3211
8. Datsenko KA, Wanner BL (2000) One-step inactivation of chromosomal genes in *Escherichia coli* K-12 using PCR products. Proc Natl Acad Sci U S A 97:6640–6645

Chapter 4

Transposon Insertion Site Sequencing for Synthetic Lethal Screening

Yoshiharu Yamaichi and Tobias Dörr

Abstract

Transposon insertion site sequencing (TIS) permits genome-wide, quantitative fitness assessment of individual genomic loci. In addition to the identification of essential genes in given growth conditions, TIS enables the elucidation of genetic networks such as synthetic lethal or suppressor gene combinations. Therefore, TIS becomes an exceptionally powerful tool for the high-throughput determination of genotype-phenotype relationships in bacteria. Here, we describe a protocol for the generation of high-density transposon insertion libraries and subsequent preparation of DNA samples for Illumina sequencing using the Gram-negative bacterium *Vibrio cholerae* as an example.

Key words TnSeq, TIS, Mariner transposon, Transposon mutagenesis, Genomics, High-throughput sequencing, Fitness

1 Introduction

Transposon insertion sequencing (TIS) enables genome-wide, quantitative assessment of the effect of individual gene disruptions on bacterial fitness under any chosen condition. Many different approaches for TIS have been described ([1–4], with alternative acronyms HITS, INSeq, TraDIS, and TnSeq, respectively), but the underlying concept is always the same. First, a bacterial culture is subjected to saturating transposon mutagenesis, then individual transposon insertion events are identified and their abundances within mutant pools quantified via massive parallel sequencing of all transposon-chromosome junctions. In a transposon insertion mutant pool, the number of sequences derived from each insertion event is a direct measure of the abundance of the mutant carrying the insertion and thus represents its relative fitness. Since Tn insertions usually cause gene disruptions, TIS enables genome-wide identification of genes conditionally required for growth or optimal fitness. In our model organism *Vibrio cholerae*, the waterborne pathogen that causes the diarrheal disease cholera, TIS, has been

Olivier Espéli (ed.), *The Bacterial Nucleoid: Methods and Protocols*, Methods in Molecular Biology, vol. 1624,
DOI 10.1007/978-1-4939-7098-8_4,

used to identify essential housekeeping genes [5] as well as those crucial for host colonization [6, 7], growth in different environments [7, 8], and survival of antibiotic stress [9]. TIS has further been used extensively for genetic interaction mapping (e.g., synthetic lethality with a locus of interest [10–12]). Importantly, at higher levels of saturation of the library, not only the contribution to bacterial fitness of entire genes but also of gene subdomains as well as intergenic regions can be assessed [5, 13].

Conducting a TIS experiment involves the following steps: (1) Mutant library generation, (2) genomic DNA preparation, (3) adapter ligation + PCR to enrich for Tn-chromosome junctions, (4) sequencing, and (5) data analysis. In this protocol, we describe the experimental procedures to perform TIS for the purpose of synthetic lethal screening, using *V. cholerae* as an example; for other applications and general experimental design considerations, we refer the reader to recent excellent reviews on the subject [14–16].

To ensure sufficient statistical power for data analysis, mutant libraries (**step 1**) should be saturated, i.e., a transposon in each nonessential insertion site should be represented at least once in the mutant pool. To achieve this, the transposon used should not exhibit strong bias toward certain genome regions (or so-called hotspots). Here, we have employed the *Himar1* mariner transposon, which inserts into TA dinucleotide sites [17]. The genome of *V. cholerae*, a low-GC Gram-negative bacterium, contains ~190,000 TA sites [18] and we generally aim at 2–3× coverage of Tn insertion sites to ensure saturation, which is achieved by three individual experiments of selecting 200,000 colonies each after Tn mutagenesis.

We will focus here on the first four steps only, as several excellent protocols detailing statistical analyses of TIS-derived sequencing datasets are available elsewhere ([14] and references therein). In principle, the procedure described here can also be applied to conditional essentiality/fitness screens, if the comparison is between two wild-type libraries prepared from different conditions.

2 Materials

2.1 Handling Bacteria

1. Mutant and corresponding parental *V. cholerae* strains (Sm^R); "recipient" cells.
2. *E. coli* SM10 λ *pir* or MFD*pir* [19] harboring pSC189 for delivering Tn via conjugation (Amp^R, Km^R); "donor" cells.
3. LB broth and agar media (Miller).
4. Ampicillin (Amp, 100 mg/mL) *see* **Note 1**.
5. Kanamycin (Km, 50 mg/mL) *see* **Note 1.**

6. Streptomycin (Sm, 200 mg/mL) *see* **Note 1.**
7. Large square petri dish (500 cm^2).
8. Cellulose filter membrane (0.45 μm pore size, e.g., MF-Millipore HAWP02500).

2.2 Genomic DNA (gDNA) Purification

1. Cell lysis buffer: 0.1 M EDTA (pH 8.0), 2% SDS.
2. RNase solution: 4 mg/mL in H_2O. Store at −20 °C.
3. Protein precipitation solution: 7.5 M ammonium acetate.
4. 100% 2-propanol.
5. 70% ethanol.
6. 100% ethanol.
7. 9 in. Pasteur pipette, which is sealed by flaming the tip.
8. ½ × EB buffer: 5 mM Tris–HCl (pH 7.5).

2.3 Library Construction

1. Bath sonicator/alternative: fragmentase enzyme kit (NEB).
2. Thermal cycler.
3. NEB quick blunting kit.
4. Taq DNA polymerase.
5. T4 DNA ligase.
6. Phusion high-fidelity DNA polymerase.
7. Spin-column PCR purification kit.
8. Thermal cycler.
9. Qubit or equivalent fluorescent-based DNA quantification apparatus (*see* **Note 2**).

2.4 Oligo DNAs

"chain terminator": 5′-TACCACGACCA$_{NH2}$–3′.

"index fork adapter": 5′-GTGACTGGAGTTCAGACGTGT-GCTCTTCCGATCTGGTCGTGGTAT-3′.

"Himar3out primer": 5′-CGCCTTCTTGACGAGTTC-3′ (*see* **Note 3**).

"Index Rprimer": 5′-GTGACTGGAGTTCAGACGTGTG-3′.

"P5 spacer primers": equimolar mixture of the following 6 oligos (varied with underlined sequence) 5′-AATGATACGGC-GACCACCGAGATCTACACTCTTTCCCTACACGACGCTCT-TCCGATCTGACTTATCAGCCAACCTGT-3′,

5′-AATGATACGGCGACCACCGAGATCTACACTCTTTC-CCTACACGACGCTCTTCCGATCTCGACTTATCAGCCAAC-CTGT-3′,

5′-AATGATACGGCGACCACCGAGATCTACACTCTTTC-CCTACACGACGCTCTTCCGATCTATGACTTATCAGCCAA-CCTGT-3′,

5′-AATGATACGGCGACCACCGAGATCTACACTCTTTCCCTACACGACGCTCTTCCGATCTTGTCGACTTATCAGCCAACCTGT-3′,

5′-AATGATACGGCGACCACCGAGATCTACACTCTTTCCCTACACGACGCTCTTCCGATCTTCGACGACTTATCAGCCAACCTGT-3′,

5′-AATGATACGGCGACCACCGAGATCTACACTCTTTCCCTACACGACGCTCTTCCGATCTGCAGCGACGACTTATCAGCCAACCTGT-3′.

"P7 barcode primer": use one of these six primers for each sample.

5′-CAAGCAGAAGACGGCATACGAGATATTGGCGTGACTGGAGTTCAGACGTGTGCTCTTCCGATC-3′ (AD006).

5′-CAAGCAGAAGACGGCATACGAGATTACAAGGTGACTGGAGTTCAGACGTGTGCTCTTCCGATC-3′ (AD012).

5′-CAAGCAGAAGACGGCATACGAGATCACTGTGTGACTGGAGTTCAGACGTGTGCTCTTCCGATC-3′ (AD005).

5′-CAAGCAGAAGACGGCATACGAGATTGACATGTGACTGGAGTTCAGACGTGTGCTCTTCCGATC-3′ (AD015).

5′-CAAGCAGAAGACGGCATACGAGATACATCGGTGACTGGAGTTCAGACGTGTGCTCTTCCGATC-3′ (AD002).

5′-CAAGCAGAAGACGGCATACGAGATGGACGGGTGACTGGAGTTCAGACGTGTGCTCTTCCGATC-3′ (AD016).

3 Methods

3.1 Pilot Experiment (See Note 4)

1. Grow overnight culture of "donor" (in LB broth with Amp) and "recipient" (in LB broth with Sm) cells.
2. Prepare three LB agar plates without antibiotics. Let them dry well and place a 0.45 μm membrane filter on top of each plate.
3. Prepare three Eppendorf tubes and in each tube, add 500 μL of each donor and recipient culture. Mix, then pellet with tabletop centrifuge at maximum speed for 2 min.
4. Remove the supernatants and then resuspend the pellets in 1 mL of LB broth. Mix and spin again.
5. Resuspend the pellets in 50 μL LB broth and place each cell suspension onto the filter prepared in **step 2**.
6. Place these plates into a 37 °C incubator with care not to let the cell suspension spill outside the membrane (do not flip the plate).
7. Remove one plate each after 2, 4, and 6 h of incubation respectively; transfer each filter into a 50-mL centrifuge tube using forceps. Resuspend cells from the filter in 1 mL of LB broth by pipetting up/down and vortexing.

8. Make serial dilutions of resuspended cells and plate 100 μL of 1×, 1:10, and 1:100 dilutions onto LB agar plates containing Sm and Km (to select against the donor strain and for Tn insertion mutants). Incubate overnight at 30 °C.
9. Enumerate number of colonies obtained from each reaction in the given incubation time (= # of colonies × 10 × dilution factor).
10. Examine ~100 Sm^R Km^R colonies for sensitivity to Amp by patching the colonies onto an LB agar plate containing Amp and Sm (Sm^R Amp^R phenotype indicates retention of the transposon delivery plasmid) (*see* **Note 5**).
11. Calculate number of reactions required to obtain 200,000 Tn-insertion mutants. If only a small number of reactions is sufficient, use shorter incubation time, but it is essential to use identical conditions between mutant and wild-type samples.

3.2 Library Construction

The entire library construction procedure should be done in duplicate or triplicate for each sample (i.e., 2–3 independent libraries with 200,000 colonies each for WT and the same for the mutant of interest).

1. Grow overnight cultures of "donor" (in LB broth with Amp) and "recipient" (in LB broth with Sm) cells.
2. Prepare LB agar plate(s) without antibiotics. Let them dry well and place 0.45 μm membrane filters (as many as number of reactions you defined by pilot experiment to obtain ~200.000 colonies) on top of the plate(s) (as many as four membranes can be applied to each plate).
3. For each reaction, mix 500 μL of each, donor and recipient cultures in Eppendorf tube. Mix, then spin down with tabletop centrifuge at max speed for 2 min.
4. Remove the supernatant and then resuspend the pellet in 1 mL of LB broth. Mix and spin again.
5. Resuspend the pellet in 50 μL LB broth and place the cell suspension onto the filter prepared above in **step 2**.
6. Place these plates in 37 °C incubator, use caution not to let the cell suspension spill outside the membrane (do not flip the plate).
7. After the incubation time determined in the pilot experiment, remove the filter into 50-mL centrifuge tube by forceps. Resuspend cells from filters in 1 mL of LB broth by pipetting up/down and vortexing.
8. Pool cell suspensions from all reactions for each strain and adjust the volume to 6 mL. Centrifuge and resuspend cells if necessary.

9. Make 1:10 and 1:100 serial dilutions, and spread 100 μL onto LB agar plates containing Sm and Km (to provide an estimate of transposon insertion efficiency, as the actual library plates can become too crowded to count colonies).
10. Spread cell suspensions onto 500 cm^2 LB agar plate (make sure to dry these well, e.g., leave open in 37 °C incubator for 1 h prior to plating) containing Sm and Km, 2-mL each to three plates. Make sure cells are spread evenly.
11. Incubate all plates overnight at 30 °C (to minimize crowding) (*see* **Note 6**).
12. Enumerate colonies contained in resulting library (= # of colonies × 60 × dilution factor), which should be more than 200,000.
13. After the library colonies are of a good size, add 5 mL LB broth to each plate and carefully mix the cells into suspension by a glass spreader. Scrape all the cells to one corner and pipette out the cell suspension into a new 50 mL centrifuge tube.
14. Take another 5 mL LB broth and resuspend the remaining cells on the plate. Collect cells into the above-mentioned 50 mL tube.
15. Repeat **steps 11** and **12** for the remaining two library plates to collect all cells into one tube. (Usually every 30 mL LB broth used on three large square plates yields ~15 mL of cells after the scraping).
16. Vortex the collected library cells to disperse clumps and evenly mix the mutants. Transfer 3 mL of the culture, which will be used for gDNA extraction below, into 15-mL centrifuge tube, then sediment cells by centrifugation. Remove the supernatant and keep the pellet at −20 °C until you proceed with DNA extraction.
17. (optional) mix 900 μL of cells with 300 μL of 80% glycerol into cryotube to store a recoverable library at −80 °C.

3.3 gDNA Extraction

1. Thaw the cell pellet on ice. Then resuspend cells in 6 mL of cell lysis buffer.
2. Incubate the tube at 80 °C for 5 min, shaking frequently.
3. Let lysate cool down to room temperature (RT).
4. Add 15 μL of RNase solution and incubate at 37 °C for 30–60 min.
5. Let reaction cool to RT, then add 2 mL of protein precipitation solution. Mix the reaction thoroughly by shaking.
6. Incubate tube on ice for 5 min.

7. Aliquot the lysate into 2-mL Eppendorf tubes and sediment the aggregates with table-top centrifuge at max speed for 10 min.
8. Pipette out the supernatants from the tubes into a new 15-mL centrifuge tube and add 6 mL of isopropanol. Invert the tube a few times to precipitate gDNA, which will become visible as a large, white, stringy mass.
9. Spool out the gDNA mass into a new 15-mL centrifuge tube containing 10 mL of 70% ethanol, by using a sealed Pasteur pipette.
10. Invert the tube and shake several times to wash the pellet.
11. Spool out the gDNA and transfer to an Eppendorf tube containing 1 mL of 100% ethanol.
12. Invert the tube and shake several times to wash the gDNA, then pellet by centrifugation.
13. Remove the ethanol and air-dry the gDNA for 15 min.
14. Resuspend the pellet in 1.5 mL of ½ × EB (*see* **Note 7**) and incubate at 65 °C for 30–60 min or at room temperature overnight.

3.4 DNA Shearing (See Note 8)

1. Dilute gDNA in a fresh Eppendorf tube to a final concentration of 25 μg of DNA per 100 μL of H_2O.
2. Sonicate DNA to yield fragments of 200–800 bp (check efficiency by running an aliquot on an agarose gel): e.g., Power: 80%, Time: 20 min, Cycle: 30 s ON and 30 s OFF with Qsonica ultrasonic processor 800.

3.5 End Repair

1. Prepare blunting reaction, using NEB Quick Blunting Kit, in PCR tube.

   5 μg sheared gDNA.

   4 μL dNTPs (stock concentration 1 mM each).

   5.5 μL 10× reaction buffer.

   Add H_2O to 53 μL.

   2 μL Blunting enzyme mix.
2. Incubate the reaction at RT for 30 min.
3. Column-purify DNA, elute with 35 μL pre-warmed H_2O (*see* **Note 9**).

3.6 A-Tailing

1. Prepare the A-tailing reaction in a PCR tube.

   35 μL of blunted DNA.

   5 μL 10× PCR buffer.

   10 μL 10 mM dATP.

   3 μL Taq DNA polymerase.

2. Incubate the reaction at 72 °C for 45 min.
3. Column-purify DNA, elute with 50 μL pre-warmed H_2O.

3.7 Adapter Ligation

1. Prepare the following adapter mixture in a PCR tube.

   2.4 μL 100 μM "chain terminator".

   2.4 μL 100 μM "index fork adapter."

   0.2 μL 2 mM MgCl2.

2. Incubate the tube at 95 °C for 5 min, then slowly (1 °C/min) decrease the temperature to 20 °C.
3. Prepare the following ligation reaction in a PCR tube.

   1.2 μg purified DNA (from above Subheading 3.6)

   0.8 μL adapter mix (from above Subheading 3.7, **step 2**).

   1.5 μL 10× T4 DNA ligase buffer.

   Add H_2O to 14 μL.

   1 μL T4 DNA ligase.

4. Incubate at 16 °C overnight.
5. After overnight incubation, add 8 μL H_2O, 1 μL 10× T4 DNA ligase buffer and 1 μL of T4 DNA ligase to spike ligation reaction.
6. Incubate a further 2 h at 16 °C.
7. Column-purify DNA, elute with 50 μL pre-warmed (37 °C) H_2O.

3.8 Amplification of Tn-Associated gDNA

1. Prepare the following PCR reaction in Eppendorf tube.

   500 ng ligated DNA (from Subheading 3.7).

   50 μL 5× Phusion high fidelity PCR buffer.

   1.25 μL 100 μM Himar3out primer.

   1.25 μL 100 μM Index R primer.

   6.25 μL 10 mM dNTPs mix, H_2O to 250 μL.

   1.25 μL Phusion DNA polymerase.

2. Aliquot into 5 PCR tubes (50 μL each) and carry out pcr using the following cycling conditions.

   1 = 98 °C for 1 min.

   2 = 98 °C for 10 s.

   3 = 53 °C for 30 s.

   4 = 72 °C for 30 s.

   5 = go to 2, 29 times.

   6 = 72 °C for 10 min

3. Pool the 5 PCR reactions and then column-purify the PCR products, elute with 50 μL pre-warmed H_2O.

3.9 Second PCR to Add Barcodes, Illumina Attachment, and Variability Sequences

1. Prepare the following PCR reaction in Eppendorf tube.

   500 ng purified PCR product (from Subheading 3.8).

   50 μL 5× Phusion high fidelity PCR buffer.

   1.25 μL 100 μM P5 spacer primers (equimolar mixture of 6).

   1.25 μL 100 μM P7 barcode primer.

   6.25 μL 10 mM dNTPs mix.

   Add H_2O to 250 μL.

   1.25 μL Phusion DNA polymerase.

2. Aliquot into 5 PCR Tubes (50 μL each) and carry out pcr using the following cycling conditions.

   1 = 98 °C for 1 min.

   2 = 98 °C for 10 s.

   3 = 55 °C for 30 s.

   4 = 72 °C for 30 s.

   5 = goto 2, 17 times.

   6 = 72 °C for 10 min.

3. Pool the 5 PCR reactions and then column-purify the PCR products, elute with 50 μL pre-warmed (37 °C) H_2O.

3.10 Size Selection

1. Run ~2 μg of purified PCR products on a 2% agarose gel in 0.5× TAE buffer.
2. Size select by cutting out the smear between 200 and 500-bp. Make sure NOT to include the visible head-to-head primer dimers around 200 bp.
3. Column-purify DNA using a standard gel extraction procedure, elute with 50 μL pre-warmed H_2O.
4. Quantify the DNA concentration by Qubit.
5. The DNA is now ready to be run on an Illumina benchtop sequencer. We use the MiSeq v2 Reagent kit (50 cycle cartridge) with single read for 65 cycles. We routinely run 6 samples/chip in an equimolar fashion (so that total concentration of DNA/chip is 10 pM). We usually obtain ~1.5 M raw reads/experimental duplicate, of which typically ~40–50% can be mapped on the reference genome for each sample. After combining data from triplicate samples, >40% of TA sites should be covered for analysis.

4 Notes

1. Antibiotic stocks are 1000× concentrated.
2. When determining DNA concentration for Illumina sequencing (Subheading 3.10, **step 4**), it is critical to use the more

sensitive fluorescent-based DNA quantification. For other steps, a spectrometry-based DNA quantification apparatus (such as Nanodrop) can be used.

3. The Himar primer amplifies from within the transposon; if a different transposon is used, this primer sequence needs to be modified accordingly.
4. The purpose of the pilot experiment is to determine the best conditions to get the most colonies while minimizing mating time. Increasing mating times is to be avoided as the mating process by itself constitutes a selection process. Additionally, increasing the incubation time and thus permitting growth of the recipient strains may lead to an overestimation of library diversity (the number of colonies does not accurately reflect the number of insertion events due to multiplication of each insertion mutant). Alternatively, other means of transposon delivery such as electroporation and natural transformation can be used instead of conjugation.
5. Amp^R in Sm^R Km^R colonies indicates integration of the whole Tn delivery plasmid as opposed to successful transposition. A few percent of Amp^R is often seen and acceptable. However, if >10% of the cells are Amp^R, it is suggested to use an alternative Tn-delivery plasmid. When using kanamycin as the selectable marker, we often observe colonies that are not truly resistant and are able to grow in the presence of Km even without containing a transposon insertion (phenotypic resistance). When estimating library diversity, it is advisable to also restreak ~20 colonies to single colony on fresh plates containing kanamycin/streptomycin to estimate the fraction of phenotypically resistant colonies, which should be <1%.
6. The libraries can also be incubated at room temperature for better separation of colonies; however, all libraries should be treated the same way. Note that changing the incubation temperature also imposes a selection.
7. It can be very difficult to bring a large amount of dry genomic DNA into solution. We have obtained best results with half concentrated EB buffer (5 mM Tris); however, pure water can also be used. The solvent volume can also be increased to bring all DNA into solution.
8. As an alternative to sonication, the NEB fragmentase kit following the manufacturer's protocol can be used for genomic DNA fragmentation.
9. In the elution step, the column can also be left standing at 37 °C for 10 min after addition of pre-warmed H_2O to increase yield.

References

1. Gawronski JD, Wong SM, Giannoukos G et al (2009) Tracking insertion mutants within libraries by deep sequencing and a genome-wide screen for Haemophilus genes required in the lung. Proc Natl Acad Sci U S A 106:16422–16427
2. Goodman AL, McNulty NP, Zhao Y et al (2009) Identifying genetic determinants needed to establish a human gut symbiont in its habitat. Cell Host Microbe 6:279–289
3. Langridge GC, Phan MD, Turner DJ et al (2009) Simultaneous assay of every salmonella Typhi gene using one million transposon mutants. Genome Res 19:2308–2316
4. van Opijnen T, Bodi KL, Camilli A (2009) Tn-seq: high-throughput parallel sequencing for fitness and genetic interaction studies in microorganisms. Nat Methods 6:767–772
5. Chao MC, Pritchard JR, Zhang YJ et al (2013) High-resolution definition of the Vibrio cholerae essential gene set with hidden Markov model-based analyses of transposon-insertion sequencing data. Nucleic Acids Res 41:9033–9048
6. Fu Y, Waldor MK, Mekalanos JJ (2013) Tn-Seq analysis of Vibrio cholerae intestinal colonization reveals a role for T6SS-mediated antibacterial activity in the host. Cell Host Microbe 14:652–663
7. Kamp HD, Patimalla-Dipali B, Lazinski DW et al (2013) Gene fitness landscapes of Vibrio cholerae at important stages of its life cycle. PLoS Pathog 9:e1003800
8. Möll A, Dörr T, Alvarez L et al (2015) A D, D-carboxypeptidase is required for Vibrio cholerae halotolerance. Environ Microbiol 17:527–540
9. Dörr T, Alvarez L, Delgado F et al (2016) A cell wall damage response mediated by a sensor kinase/response regulator pair enables beta-lactam tolerance. Proc Natl Acad Sci U S A 113:404–409
10. Dörr T, Möll A, Chao MC et al (2014) Differential requirement for PBP1a and PBP1b in in vivo and in vitro fitness of Vibrio cholerae. Infect Immun 82:2115–2124
11. Möll A, Dörr T, Alvarez L et al (2014) Cell separation in Vibrio cholerae is mediated by a single amidase whose action is modulated by two nonredundant activators. J Bacteriol 196:3937–3948
12. Wang Q, Millet YA, Chao MC et al (2015) A genome-wide screen reveals that the vibrio cholerae phosphoenolpyruvate phosphotransferase system modulates virulence Gene expression. Infect Immun 83:3381–3395
13. Yamaichi Y, Chao MC, Sasabe J et al (2015) High-resolution genetic analysis of the requirements for horizontal transmission of the ESBL plasmid from *Escherichia coli* O104:H4. Nucleic Acids Res 43:348–360
14. Chao MC, Abel S, Davis BM, Waldor MK (2016) The design and analysis of transposon insertion sequencing experiments. Nat Rev Microbiol 14:119–128
15. van Opijnen T, Camilli A (2013) Transposon insertion sequencing: a new tool for systems-level analysis of microorganisms. Nat Rev Microbiol 11:435–442
16. Barquist L, Boinett CJ, Cain AK (2013) Approaches to querying bacterial genomes with transposon-insertion sequencing. RNA Biol 10:1161–1169
17. Chiang SL, Rubin EJ (2002) Construction of a mariner-based transposon for epitope-tagging and genomic targeting. Gene 296:179–185
18. Heidelberg JF, Eisen JA, Nelson WC et al (2000) DNA sequence of both chromosomes of the cholera pathogen *Vibrio cholerae*. Nature 406:477–483
19. Ferrières L, Hémery G, Nham T et al (2010) Silent mischief: bacteriophage mu insertions contaminate products of *Escherichia coli* random mutagenesis performed using suicidal transposon delivery plasmids mobilized by broad-host-range RP4 conjugative machinery. J Bacteriol 192:6418–6427

Part II

Study of Bacterial Nucleoid with Whole Genome Analysis Method

Chapter 5

WGADseq: Whole Genome Affinity Determination of Protein-DNA Binding Sites

Mickaël Poidevin, Elisa Galli, Yoshiharu Yamaichi, and François-Xavier Barre

Abstract

We present a method through which one may monitor the relative binding affinity of a given protein to DNA motifs on the scale of a whole genome. Briefly, the protein of interest is incubated with fragmented genomic DNA and then affixed to a column. Washes with buffers containing low salt concentrations will remove nonbound DNA fragments, while stepwise washes with increasing salt concentrations will elute more specifically bound fragments. Massive sequencing is used to identify eluted DNA fragments and map them on the genome, which permits us to classify the different binding sites according to their affinity and determine corresponding consensus motifs (if any).

Key words Next-generation sequencing (NGS), Genomics, High-throughput sequencing, Site-specific DNA binding protein

1 Introduction

DNA binding proteins participate in a wide range of DNA-specific processes, such as replication, recombination, repair, chromosome segregation, and transcription. In addition, they play a role in global cellular processes, such as the positioning and timing of formation of the cell division apparatus in bacteria. A number of classical methods have been developed to study the interaction of a protein with DNA: Electrophoretic Mobility Shift Assay (EMSA) and DNA pull down experiments are used to determine its binding strength [1, 2]; chemical or nuclease footprints are used to determine the exact region of DNA it covers [1, 2]. However, they can only be used to study the binding to a single DNA sequence motif at a time, making it tedious to determine the range of DNA sequences a given protein might bind to. With the development

Authors "Mickaël Poidevin and Elisa Galli" contributed equally to this work.

Olivier Espéli (ed.), *The Bacterial Nucleoid: Methods and Protocols*, Methods in Molecular Biology, vol. 1624,
DOI 10.1007/978-1-4939-7098-8_5, © Springer Science+Business Media LLC 2017

of high-throughput sequencing and data analysis methods, Chromatin Immunoprecipitation (ChIP) has been used to determine which genomic positions a given DNA binding protein might be bound to [1, 2]. However, the relative affinity of the protein for each genomic position cannot be deduced from the data because other proteins within the cell might alter its DNA binding properties or mask some of its target sites.

Here, we present a highly adaptable method that permits us to determine if a given DNA binding protein targets a specific DNA sequence motif and when this is the case, monitor its relative affinity to each of the genomic positions in which such a motif is found. The method, WGADseq, is based on standard protein purification techniques and high-throughput sequencing. Our laboratory has used the method described herein to identify the SlmA-DNA binding sites (SBSs) on the whole genome of *Vibrio cholerae* [3]. SlmA is a Nucleoid Occlusion factor, it prevents placement of the cell division machinery over unsegregated chromosomes [4]. *Escherichia coli* SlmA binding sites were identified using ChIP and shown to be distributed over the entire chromosome except the replication terminus region [5, 6]. The genome of *V. cholerae* is divided into two circular chromosomes, chr1 and chr2. Chr1 emanates from the mono-chromosomal ancestor of *V. cholerae*, whereas chr2 derives from a horizontally acquired plasmid. WGADseq results permitted us to explain how the cell division was coordinated with the replication and segregation cycle of chr2.

In brief, WGDAseq consists in incubating the purified protein of interest with sheared genomic DNA (gDNA) (Fig. 1a, i) and purifying the protein-DNA complex by standard biochemical procedures such as affinity purification (Fig. 1a, ii). As an example, we present the use of a recombinant peptide tagged at its N-terminus with 6 histidine residues. As an alternative, chemical properties of the purified protein or an antibody against it can be used. Unbound DNA fragments are washed away and protein-bound DNA fragments are recovered in a step-wise elution with buffers of increasing salt concentration (Fig. 1a, iii). DNA fragments recovered in the flow through (unbound) and in the elution fractions are subjected to High-Throughput Sequencing (Fig. 1a, iv) and their relative frequency is determined by informatics mapping (Fig. 1b, c).

2 Materials

2.1 DNA Shearing

1. TE 1× (10 mM Tris–HCl pH 7.5, 1 mM EDTA).
2. MilliTUBE 1 mL AFA Fiber (Covaris).
3. Sonicator S220 (Covaris).

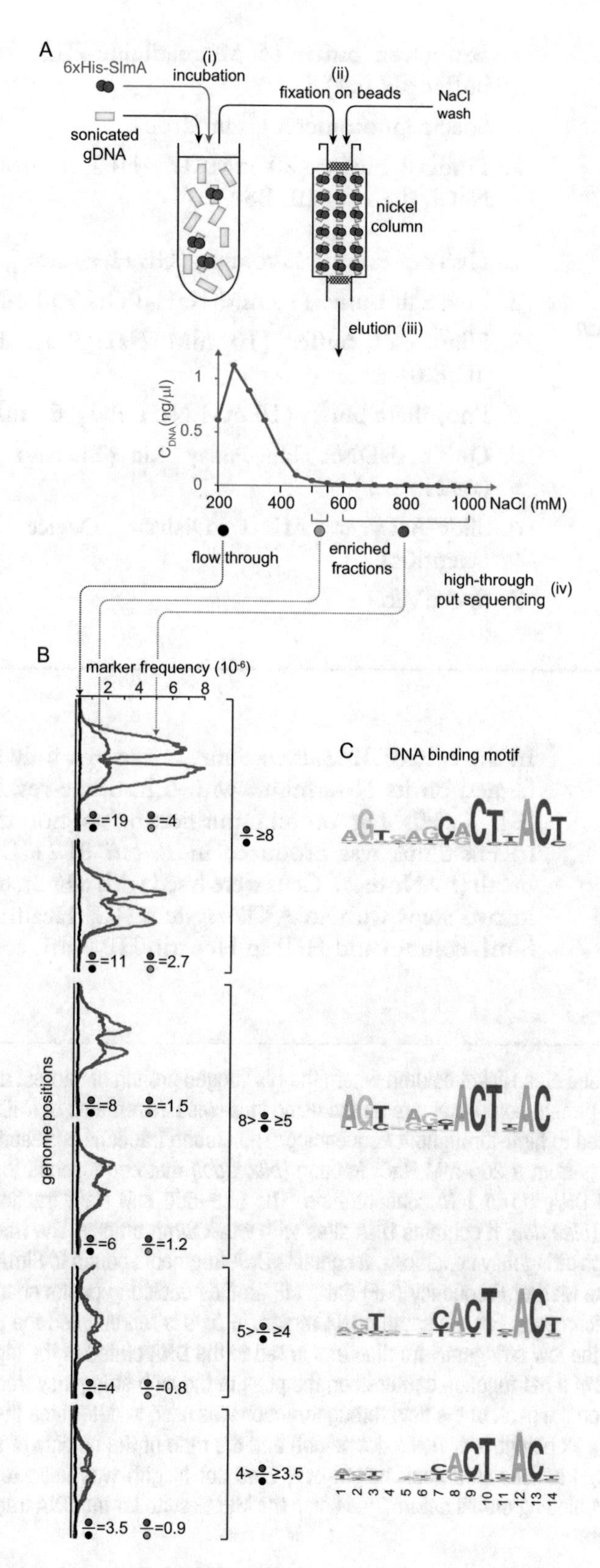

Fig. 1 Step-by-step protocol overview (**a**) Schematic representation of the protocol used to purify the DNA target sequences. (*i*) 6×His-SlmA and sheared gDNA are incubated together, (*ii*) the DNA-protein mixture is

2.2 Protein and Genomic DNA (gDNA) Binding

1. Extinction buffer (6 M guadinine-HCl, 20 mM phosphate buffer pH 6.4).
2. Spectrophotometer (Nanodrop).
3. Binding buffer (20 mM Tris–HCl, 1 mM EDTA, 80 mM NaCl, 0.1 mg/mL BSA).

2.3 Loading of Protein-DNA Complexes and Elution of Bound DNA

1. HisTrap HP 1 mL column (GE Healthcare).
2. Low Salt buffer (10 mM NaH_2PO_4, 100 mM NaCl, pH 8.0).
3. High Salt buffer (10 mM NaH_2PO_4, 1000 mM NaCl, pH 8.0).
4. Phosphate buffer (10 mM NaH_2PO_4, 50 mM NaCl, pH 8.0).
5. Qubit dsDNA HS Assay Kit (Thermo Fisher Scientific) (*see* **Note 1**).
6. Slide-A-Lyzer MINI Dialysis Device (Thermo Fisher Scientific).
7. SpeedVac.

3 Methods

3.1 Protein Purification

In the case of *V. cholerae* SlmA, we used a fully functional peptide tagged on its N-terminus with 6 histidine residues, 6×His-SlmA [3]. 6×His tag protein purification is not developed. Briefly, 6×His-SlmA was produced in *E. coli* BL21 DE3 Δ*slmA* in LB broth (*see* **Note 2**). Cells were lysed with a French press and purified in two steps with an AKTA system (GE Healthcare, HisTrap HP 5 mL column and HiTrap Heparin HP 5 mL column).

Fig. 1 (continued) loaded on a Nickel column where the His-tagged protein of interest and the bound DNA are retained, (*iii*) the DNA-protein complexes are eluted using step-wise increments of NaCl and (*iv*) the fractions of interest are subjected to high-throughput sequencing. The elution fractions correspond to increasing NaCl concentrations: it starts from a 200 mM NaCl fraction (*black dot*) that corresponds to the flow through and contains the unbound DNA, up to 1 M concentration. The 500–550 mM NaCl fraction corresponds to low stringency conditions (*blue dot*). It contains DNA sites with weak SlmA affinity. The fraction from 600 mM to 1 M corresponds to high stringency conditions. It contains DNA fragments bound to SlmA with high affinity (*red dot*). (**b**) Analysis of the Marker Frequency (MF) data. MF profiles depicting peaks characteristic around SBS with different fold enrichments. For the same gDNA region, in *blue* is represented the peak corresponding to the DNA contained in the low stringency fraction and in red to the DNA eluted in the high stringency fraction. The ratio of the height of a hat function centered on the peak of the high stringency fraction to the height of a hat function centered on the peak of the flow through fraction was used to determine the fold of enrichment of any gDNA locus (*red* peak height over *black* dot height) and the ratio of the heights of the hat function of the high to low stringency fractions (*red* peak height over *blue* dot height) was used to classify the different binding motifs. (**c**) DNA binding motifs determined using the MEME suite for the DNA fragments of the different classes described in (**b**)

3.2 DNA Shearing (See Note 3)

1. Dilute gDNA in 15 mL tube to a final concentration of 500 μg/mL in TE 1×. Do not exceed 500 μg/mL concentration or gDNA shearing efficiency will be impacted negatively.
2. Fill milliTUBE AFA Fiber with 1 mL of gDNA at 500 μg/mL.
3. Sonicate with Covaris S220 (parameters: Duty cycle: 10%, Cycle per burst: 200, Peak incident power: 140 W, Time: 20 min).
4. Repeat four times to prepare 2 mg of sheared gDNA.
5. Confirm gDNA shearing efficiency by running 1–2 μL aliquot on 1.5% agarose gel. gDNA sheared profile should show yield fragments of 50–300 bp with an average ~150 bp.

3.3 Protein and gDNA Binding

We routinely used 1:20 molar ratio for protein-gDNA binding reaction. Classically, 1 nmol of protein is mixed with 2 mg of sheared gDNA (which corresponds to 20 nmol of DNA fragment with an average size of 150 bp).

1. Spin the purified protein 10 min at 18,000 × *g*.
2. Dilute 1 μL of protein in 10 μL of Extinction buffer.
3. Measure OD at 280 nm with spectrophotometer.
4. Look for your protein extinction coefficient on website (http://www.biomol.net/en/tools/proteinextinction.htm).
5. Apply the formula:

   Protein concentration (M) = (OD 280 nm × 10)/extinction coefficient.
6. Mix 1 nmole of protein with 20 nmole of sheared gDNA to 5 mL of 1× Binding buffer.
7. Incubate gently under rotative shaking at 4 °C from 2 h to overnight.

3.4 Protein-Sheared gDNA Complex Purification

1. Load slowly 5 mL mix on 1 mL HisTrap column (GE Healthcare, Ref# 17–5247-01) with 5 mL syringe.
2. Wash column with 10 V (10 mL) of Low Salt buffer.
3. Elute gDNA fragments with a range of 50 mM NaCl concentration from 200 to 1000 mM. 1 mL of each concentration is added to column with 1 mL syringe.
4. Recover each fraction in 1.5 mL eppendorf tube.
5. Quantify DNA in each fraction with Qubit dsDNA HS Assay Kit.
6. Dialyze fractions against 0.1× TE with Slide-A-Lyzer MINI Dialysis Device. Contiguous fractions can be combined (*see* Fig. 1, blue and red fractions).
7. Concentrate fractions with SpeedVac from ~1 mL to 0.1 mL.

3.5 High-Throughput Sequencing

gDNA fragments recovered in the flow through and the different elution fractions can be analyzed with any standard High-Throughput Sequencing technique. In the *V. cholerae* SlmA example presented here, the flow through (Fig. 1b, black dot), a fraction eluted with low stringency buffers (Fig. 1b, blue dot) and a fraction eluted with high stringency buffers (Fig. 1b, red dot) were analyzed. DNA libraries were prepared with a Beckman Coulter SPRIworks according to the supplier's recommendations (A84803). They were sequenced on Illumina Hiseq 2000 using HiSeq SBS 50 Cycles Kit (FC-401-4001) with single read for 50 cycles.

3.6 Data Analysis

1. MF profiles calculation: MF profiles were created by counting the total number of reads that start at a given genome position. Because the total number of reads that is obtained can significantly differ between samples, we divided the MF profile of each sample by its total number of reads for quantitative comparison of the MF profiles of different fractions. The MF profiles were smoothed by attributing to each position the sum of the relative reads obtained over a fixed window around this region. In the example we present, we sequenced about 20 million reads for each sample for a total genome length of about 4 million bp. The MF data was smoothed over a sliding window of 40 bp. Examples of typical marker frequency profiles are shown in Fig. 1b.
2. Determination of the genomic DNA sequences corresponding to putative binding sites: as the DNA fragments are in average 150 bp long, MF of positions in the 150 bp region surrounding a motif bound by the protein of interest should be higher in the high salt elution fractions than in the flow through. Based on this assumption, regions of interests were searched in the following recursive manner: (1) the genomic position with the highest ratio between the high salt MF profile and the flow through MF profile is selected; (2) the center, width, and height of a hat function that best fit the MF ratio within a 200 bp around this point are calculated; (3) the center, width, and height of the hat function are stored and the MF ratio data is set to zero in a 200 bp region surrounding its center before proceeding to step (1). The loop is stopped once no more positions with a MF ratio higher than a given threshold are found. In the example presented, we used a threshold of 3.5.
3. The recovered DNA sequences are submitted to the Multiple EM for Motif Elicitation (MEME; [7]) on line server to determine any common DNA binding consensus motif. The distances between the center points of each peak (as calculated in **step 2**) and its corresponding motif are calculated.

4. Strength of the binding motif: the height of the hat function on each peak reports the strength of the binding of the protein to the motif found in the region of the peak. In the example, peaks were determined for low stringency (blue MF profile) and high stringency MF profile (red MF profile). Peak positions were identical, demonstrating the validity of the procedure. However, their height differed. In some, the height was higher in the red MF profile. They correspond to the highest affinity DNA motif. In others, the height was higher in the blue MF profile. They correspond to motifs with weaker affinity. In the example we present, the peaks could be classified into four categories based on the ratio of the heights of the hat functions obtained with the red and flow through profiles (red dot height over black dot height). In the two categories in which the ratio between the heights of the hat functions in the red and blue MF profiles (red dot height over blue dot height) was higher than 1, we found a palindromic consensus. In the two other categories (in which the ratio between the heights of the hat functions in the red and blue MF profiles was lower than 1), we found half of the palindromic motif, demonstrating that the method could serve to monitor the relative affinity of a given protein to all the DNA motifs. These results were confirmed by classical EMSA experiments [3].

4 Notes

1. When determining DNA concentration for Illumina sequencing, it is critical to use the more sensitive fluorescent-based DNA quantification. For other steps, a spectrometry-based DNA quantification apparatus (such as Nanodrop) can be used.
2. To avoid potential contamination of endogeneous (*E. coli*) SlmA through conformation of heterologous dimer formation, for the protein expression and purification of *V. cholerae* SlmA we used bEYY1029, a strain derived from BL21 DE3 but in which the *slmA* gene was deleted.
3. Alternatively, conventional sonication apparatus could be used, but appropriate settings must be determined to obtain DNA fragments with comparable sizes.

Acknowledgments

This work was supported by the European Research Council under the European Community's Seventh Framework Programme (FP7/2007-2013 grant agreement no. 281590) to F.-X.B.

References

1. Mahony S, Pugh BF (2015) Protein-DNA binding in high-resolution. Crit Rev Biochem Mol Biol 50:269–283
2. Dey B, Thukral S, Krishnan S et al (2012) DNA-protein interactions: methods for detection and analysis. Mol Cell Biochem 365:279–299
3. Galli E, Poidevin M, Le Bars R et al (2016) Cell division licensing in the multi-chromosomal *Vibrio cholerae* bacterium. Nat Microbiol 1: 16094
4. Bernhardt TG, de Boer PA (2005) SlmA, a nucleoid-associated, FtsZ binding protein required for blocking septal ring assembly over chromosomes in E coli. Mol Cell 18:555–564
5. Cho H, McManus HR, Dove SL, Bernhardt TG (2011) Nucleoid occlusion factor SlmA is a DNA-activated FtsZ polymerization antagonist. Proc Natl Acad Sci U S A 108:3773–3778
6. Tonthat NK, Arold ST, Pickering BF et al (2011) Molecular mechanism by which the nucleoid occlusion factor, SlmA, keeps cytokinesis in check. EMBO J 30:154–164
7. Bailey TL, Boden M, Buske FA et al (2009) MEME SUITE: tools for motif discovery and searching. Nucleic Acids Res 37:W202–W208

Chapter 6

High-Resolution Chromatin Immunoprecipitation: ChIP-Sequencing

Roxanne E. Diaz, Aurore Sanchez, Véronique Anton Le Berre, and Jean-Yves Bouet

Abstract

Chromatin immunoprecipitation (ChIP) coupled with next-generation sequencing (NGS) is widely used for studying the nucleoprotein components that are involved in the various cellular processes required for shaping the bacterial nucleoid. This methodology, termed ChIP-sequencing (ChIP-seq), enables the identification of the DNA targets of DNA binding proteins across genome-wide maps. Here, we describe the steps necessary to obtain short, specific, high-quality immunoprecipitated DNA prior to DNA library construction for NGS and high-resolution ChIP-seq data.

Key words Chromatin immunoprecipitation, Next-generation sequencing, Genome-wide maps, DNA target, DNA fragmentation, DNA sonication, Affinity-purified antibody, Bacterial nucleoid

1 Introduction

ChIP-sequencing (ChIP-seq), which combines two methods, chromatin immunoprecipitation and next-generation sequencing, has made a tremendous impact in many biological research fields from eukaryotes to prokaryotes. Knowing the locations where proteins interact with DNA is essential for understanding their functionality in a system. Many methods are available to study their binding sites and specificity of interaction. However, only ChIP-seq can give high-resolution in vivo data mapped across the entirety of a genome. Since its first use in eukaryotes [1–4], to its use in prokaryotes [5, 6], ChIP-seq technology has continued to advance, while the cost of sequencing has decreased. Consequently, ChIP-seq is becoming an increasingly accessible tool. Of note, ChIP-seq has been successfully utilized to investigate nucleoprotein complexes that shape the bacterial nucleoid, giving insights into a range of processes such as local and global organization,

Olivier Espéli (ed.), *The Bacterial Nucleoid: Methods and Protocols*, Methods in Molecular Biology, vol. 1624,
DOI 10.1007/978-1-4939-7098-8_6,

replication, segregation, as well as global regulation of gene expression by nucleoid-associated proteins.

Current and widely used methods to capture protein-DNA interactions in a live cell population couple ChIP with the use of covalent and reversible formaldehyde cross-linking, namely X-ChIP. Following crosslinking, the cells are lysed and DNA is extensively fragmented, through enzymatic digestion or sonication. In this protocol, we describe DNA fragmentation through sonication using an automated rotating water bath system. We have found this method to be the most efficient and reproducible that results in a uniform size of DNA among samples with an average of ~200 bp required for high-resolution ChIP-seq data. Another key step involves the selective immunoprecipitation of the fragmented protein-DNA complexes using protein-specific antibodies. The quality and specificity of antibodies used for this assay are of utmost importance to prevent nonspecific pulldowns and datasets with a low signal-to-noise ratio. Here, we also describe a method for membrane strip affinity purification of antibodies using rabbit polyclonal sera raised against the protein of interest. Following immunoprecipitation, proteins are eliminated using a proteinase K digestion and the samples are reverse cross-linked prior to DNA purification. We have found that using a simple chloroform and isoamyl alcohol purification followed by an isopropanol precipitation, is a cost-efficient method that results in a high recovery of quality DNA; however, other options such as bead and DNA purification kits are effective alternatives.

The ChIP-seq protocol described here is adapted from Cho et al. (2011) and has contributed to important achievements related to bacterial nucleoids, such as the essential involvement of SlmA in *Escherichia coli* in regulating FtsZ ring assembly [7]. More recently, Sanchez et al. (2015) used high-resolution ChIP-seq data that allowed for the physico-mathematical modeling of the $ParB_F$ propagation along the DNA and the proposal of a new model of stochastic self-assembly for the F plasmid partition complex [8].

2 Materials

Prepare all solutions using Milli-Q® or any form of ultra-pure water with a sensitivity of 18.2 MΩ.cm and molecular grade reagents. Unless otherwise specified, filter solutions using a 0.45 μm low protein binding non-pyrogenic membranes and store them at room temperature.

2.1 Antibody Purification

Prepare the following buffers:

1. 2× sample buffer: 100 mM Tris pH 6.8, 5% SDS, 20% glycerol, 0.05% Bromo-phenol blue. Add β-mercaptoethanol (50 μl.ml^{-1}) just before mixing with protein samples.

2. Ponceau S: Mix 50 mg Ponceau S (Sigma-Aldrich), 2.5 ml acetic acid ($\geq$99%) with 47.5 ml of water.
3. 10× Tris-Buffered Saline (TBS): Mix 20 ml of 1 M Tris pH 7.6, 20 ml of 5 M NaCl with 960 ml of water.
4. TBS-Tween20: Dissolve 0.1% Tween20® in 1× TBS.
5. 0.1 M Glycine-HCl pH 2.2: Dissolve 1.5 g of glycine in water, adjust to pH 2.2 with 5 M HCl, and add water up to a volume of 200 ml.
6. 10 mM NaPhosphate buffer pH 7.2: for 100 ml, mix 72 ml of 0.1 M Na_2PO_4 with 28 ml of 0.1 M NaH_2PO_4.

2.2 Bacterial Culture and Cross-Linking

1. For *Escherichia coli* and related bacteria, grow cells in LB medium or other appropriate media specific of the tested growth condition.
2. 36% formaldehyde commercial solution.
3. 2.5 M Glycine: Dissolve 92.83 g of glycine powder in 400 ml of water. Transfer solution to a graduated cylinder and add water up to a volume of 500 ml.
4. Cold TBS pH 7.6: See antibody purification (Subheading 2.1, **item 3**).

2.3 Cell Lysis and Sonication

1. 1.5 ml tubes with a low DNA binding grade.
2. Rotating water bath sonicator: We recommend the Bioruptor® plus (Diagenode). The M220 Focused-ultrasonicator (Covaris) also provides reproducible and accurate results.
3. Lysis buffer: 10 mM Tris–HCl pH 7.8, 100 mM NaCl, 10 mM EDTA, 20% sucrose, 1 $mg.ml^{-1}$ lysozyme (*see* **Note 1**).
4. 2× IP buffer: 50 mM Hepes-KOH pH 7.5, 150 mM NaCl, 1 mM EDTA, 1% Triton X-100, 0.1% sodium deoxycholate, 0.1% SDS, 1 mM PMSF (*see* **Note 2**).

2.4 Immuno-precipitation

1. Magnetic rack.
2. Magnetic protein A beads (Ademtech).
3. Rocking tube agitator for use at 4 °C.
4. Blocking buffer: 0.1 $\mu g.\mu l^{-1}$ BSA, 1 $\mu g.\mu l^{-1}$ tRNA.
5. Wash buffer 2: 50 mM Hepes-KOH pH 7.5, 500 mM NaCl, 1 mM EDTA, 1% Triton X-100, 0.1% sodium deoxycholate, 0.1% SDS, 1 mM PMSF (*see* **Note 2**).
6. Wash buffer 3: 10 mM Tris–HCl pH 7.8, 250 mM LiCl, 1 mM EDTA, 0.5% IGEPAL CA-630 (Sigma-Aldrich), 0.1% sodium deoxycholate, 1 mM PMSF (*see* **Note 2**).
7. TE Buffer: 10 mM Tris pH 7.5, 1 mM EDTA.

8. Elution buffer: 50 mM Tris–HCL pH 7.5, 10 mM EDTA, 1% SDS.
9. 4× reducing buffer (Invitrogen).

2.5 Reverse Crosslink

1. RNase: 10 $mg.ml^{-1}$.
2. Proteinase K: 10 $mg.ml^{-1}$.

2.6 DNA Purification

1. 5 M NaCl.
2. Pure chloroform ($CHCl_3$) and Isoamyl alcohol solutions.
3. 20 $mg.ml^{-1}$ Glycogen.
4. 70% ethanol and 100% Isopropanol.
5. 10 mM Tris–Cl pH 8.5.

3 Methods

Carry out all procedures at room temperature (23 °C) unless otherwise specified.

3.1 Antibody Purification

For high-quality ChIP-sequencing data, we recommend using affinity-purified antibodies raised against the proteins of interest.

1. Recuperate 100 μg of the purified protein of interest and mix with 300 μl of 2× sample buffer (*see* **Note 3**).
2. Load the protein sample on an appropriate 1-well SDS-PAGE. Run the gel and transfer subsequently by Western blotting on nitrocellulose membrane.
3. Rinse the membrane twice in 1× TBS for 5 min.
4. Cut a vertical strip at one extremity of the membrane. Rinse it with 500 μl of Ponceau S to identify the location of the protein band (*see* **Note 4**).
5. Align the colored strip to the full membrane and excise a horizontal strip containing the protein band, as small as possible. Discard the colored strip.
6. Rinse the protein strip in 20 ml of 1× TBS for 10 min at room temperature with gentle rocking.
7. Rinse with 20 ml of 1× TBS-0.1% Tween20® with 10% milk for 1 h at room temperature with gentle rocking.
8. Rinse with 6 ml of 1× TBS-tween supplemented with 500 μl of antibody serum, incubate for 3 h at room temperature on a rocking platform with gentle rocking, or overnight in a cold room with gentle rocking.
9. Wash the strip five times for 10 min in 1× TBS-0.1% Tween20®.
10. Wash the strip two times for 5 min with 10 mM Na-phosphate buffer pH 7.2.

11. Place the strip on saran wrap and add 200 μl of 0.1 M Glycine pH 2.2.
12. Recover and place the eluate in a 1.5 ml tube that contains 40 μl 1 M Na_2HPO_4.
13. Repeat **step 13** and recuperate the second eluate in the same tube.
14. Rinse the strip with 10 mM Na-Phosphate buffer pH 7.2 for 10 min.
15. Store the affinity-purified antibody at 4 °C or at −20 °C for short- or long-term storage, respectively.
16. Place the strip on a paper towel and allow drying for several minutes.
17. The strip can be reused, place in a sterile tube and store at 4 °C (*see* **Note 5**).

3.2 ChIP-Sequencing

All 1.5 ml tubes used are DNA low binding grade.

3.2.1 Bacterial Growth Culture

1. Inoculate a 10 ml LB preculture, containing necessary antibiotics, with an isolated colony on LB agar of the bacterial strain to be studied and incubate at the appropriate temperature under agitation overnight.
2. Dilute the preculture 200-fold in 100 ml of fresh LB medium, and incubate at 37 °C under agitation, until OD_{600} ~ 0.6.

3.2.2 Cross-Link

1. Aliquot 80 ml of the culture into an Erlenmeyer flask.
2. Add 2.2 ml of 36% formaldehyde solution (1% final concentration) per sample and incubate at room temperature (23 °C) for 30 min with gentle agitation (90 rpm).
3. Add 16 ml of 2.5 M glycine (0.5 M final concentration) to quench the cross-linking reaction and incubate at room temperature for 15 min with gentle agitation (90 rpm).
4. Transfer the entire cross-linked samples into appropriate centrifuge tubes.
5. Centrifuge for 10 min 6,000 × *g* at 4 °C.
6. Discard the supernatant and resuspend the pellet with 96 ml of cold TBS pH 7.6.
7. Centrifuge for 10 min 6,000 × *g* at 4 °C.
8. Discard the supernatant and resuspend the pellet in 1 ml of cold TBS (*see* **Note 6**), and aliquot 500 μl into two separate 1.5 ml tubes.
9. Centrifuge for 10 min at 6,000 × *g* at 4 °C.
10. Carefully discard supernatants, no liquid should remain in the bacterial pellets (*see* **Note 7**).

3.2.3 Cell Lysis

1. Resuspend pellet in 500 μl of lysis buffer (*see* **Note 8**).
2. Add 50 μl of 10 $mg.ml^{-1}$ lysozyme and incubate at 37 °C for 30 min.
3. Add 500 μl of IP buffer and 10 μl of 100 mM PMSF.

*3.2.4 Sonication (See **Note 9**)*

1. Aliquot 130 μl of the sample into six 1.5 ml DNA low binding grade tubes (six tubes per sample).
2. Set aside 100 μl of sample to serve as a non-sonicated control and place on ice.
3. Sonicate samples using the Bioruptor® plus (Diagenode) or equivalent apparatus. Sonication conditions for Bioruptor: 3 rounds of 28 cycles of 30 s on, 30 s off (*see* **Note 10**).
4. Pool the six sonicated samples tubes into one tube.
5. Centrifuge the sonicated and the non-sonicated sample for 30 min at 18,000 × *g* at 4 °C.
6. Recuperate the supernatant into a new 1.5 ml tube.
7. For the sonicated sample, aliquot 500 μl into 1.5 ml tube to serve as the sample to be immunoprecipitated (IP), and aliquot 100 μl into 1.5 ml tube to serve as the input (*see* **Note 11**).
8. Test the efficiency of DNA fragmentation by either analyzing the DNA size using a bioanalyzer (e.g., Agilent 2100) or agarose gel electrophoresis (*see* Fig. 1).

3.2.5 Immuno-precipitation

1. Combine 500 μl of IP sample with the application-specific antibody at the required concentration (*see* **Note 12**), and incubate overnight at 4 °C with gentle agitation (90 rpm).
2. In a separate tube, add 25 μl of Protein A magnetic beads to 225 μl of blocking buffer and incubate overnight at 4 °C with gentle agitation (90 rpm).
3. For an IP negative control, prepare a tube containing 500 μl of sonicated IP sample that will receive IgG antibodies and will go through the same immunoprecipitation process as the IP sample.
4. Recover the blocked Protein A magnetic beads by pelleting using a magnetic rack and discard the blocking buffer.
5. Add the IP sample and antibody mixture to tube containing pre-blocked beads and incubate for 2 h at 4 °C with gentle agitation (90 rpm).
6. Perform consecutive washes of the beads for 10 min with gentle agitation at 4 °C, precipitating the beads for at least 5 min between each wash (*see* **Notes 13–15**).
   (a) 1 wash with 300 μl 2× IP buffer.
   (b) 2 washes with 300 μl buffer 2.
   (c) 1 washes with 300 μl buffer 3.
   (d) 1 wash with 300 μl TE buffer.

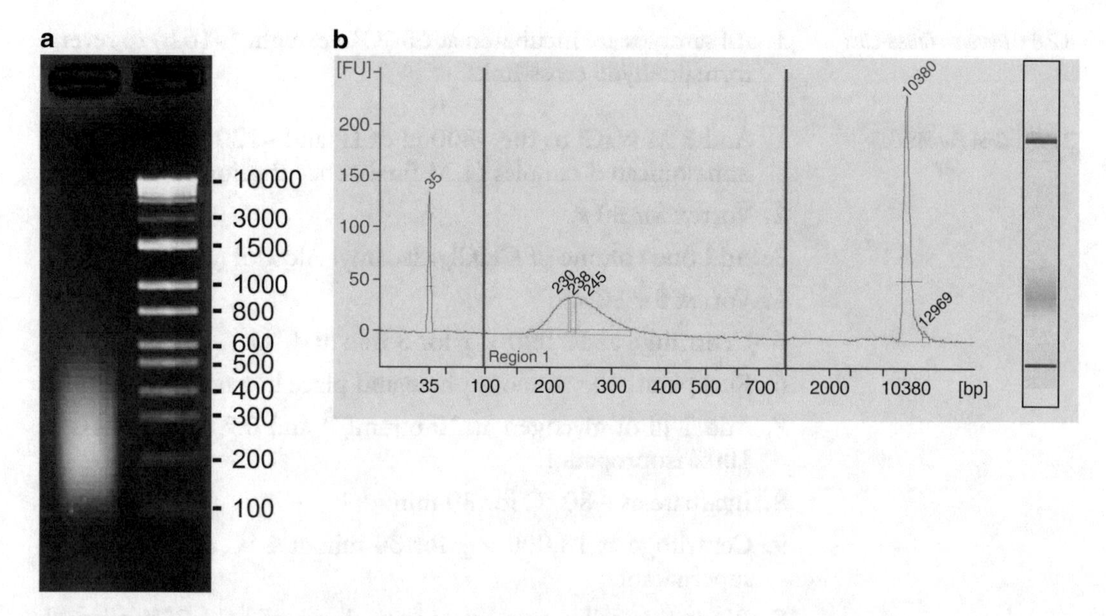

Fig. 1 DNA profiles of sonicated formaldehyde-treated cells prior (**a**) and after (**b**) library preparation. (**a**) Sonicated and purified DNA fragments from the ChIP procedure prior to library preparation were analyzed by 1.8% agarose gel electrophoresis showing fragment sizes between 100 and 400 bp. Molecular weight (base pair) markers are indicated on the *right*. The brightness and contrast of the image has been modified by linear scaling for amplification of the signal. (**b**) Following library preparation, the size of DNA fragments, containing an 80 bp barcode and sequencing adaptor, was controlled by Agilent 2100 Bioanalyzer using Agilent High sensitivity DNA kit. Low- and high-DNA markers, noted 35-bp and 10,380-bp, respectively, were added in each sample prior to electrophoresis. The average size of the sample after adaptor subtraction was ~170 bp

3.2.6 IP Samples: Elution, RNase, and Proteinase K Digestion

1. After the final wash, resuspend the beads in 300 μl of elution buffer (*see* **Notes 16** and **17**).
2. Add 1 μl of RNase A at 10 mg.ml^{-1} (33.3 μg.ml^{-1} final concentration) and incubate for 1 h at 37 °C.
3. Add 3 μl of proteinase K at 10 mg.ml^{-1} (0.1 mg.ml^{-1} final concentration), and incubate for 2 h at 37 °C.
4. Pellet the magnetic beads using the magnetic rack and recover the supernatant in a new 1.5 ml tube (*see* **Note 18**).

3.2.7 Input Samples: RNase and Proteinase K Digestion

1. On ice, thaw the 100 μl of input and non-sonicated samples that had been previously frozen (**step 7** from Subheading 3.2.4; *see* **Note 19**).
2. Add 3.3 μl of 1 mg.ml^{-1} RNase A (dilute stock tenfold; 33.3 μg.ml^{-1} final concentration) and incubate for 1 h at 37 °C.
3. Add 1 μl of 10 mg.ml^{-1} proteinase K (0.1 mg.ml^{-1} final concentration) and 10 μl of 10% SDS (final 1%), and incubate for 2 h at 37 °C.

3.2.8 Reverse Cross-Link

1. All samples are incubated at 65 °C overnight (~16 h) to reverse formaldehyde cross-links.

3.2.9 DNA Purification

1. Add 5 M NaCl to the ~300 μl of IP and ~120 μl of input and non-sonicated samples (1 M final concentration).
2. Vortex for 30 s.
3. Add one volume of $CHCl_3$/Isoamyl Alcohol (24:1) mix.
4. Vortex for 30 s.
5. Centrifuge at 18,000 × *g* for 5 min at 4 °C.
6. Recuperate the aqueous phase and place in a new 1.5 ml tube.
7. Add 1 μl of glycogen at 20 mg.ml^{-1} and 0.7 volume of cold 100% isopropanol.
8. Incubate at −80 °C for 30 min.
9. Centrifuge at 18,000 × *g* for 30 min at 4 °C and remove the supernatant.
10. Rinse the pellet two times with 1 ml of cold 70% ethanol, centrifuging at 18,000 × *g* for 10 min at 4 °C between each wash.
11. Resuspend the DNA pellet in 100 μl of 10 mM Tris–HCl, pH 8.5. Check the DNA profile (*see* **Note 20**) and quantify the amount of DNA recovered (*see* **Note 21**).

3.2.10 DNA Library Preparation

1. Both immunoprecipitated (ChIP-DNA) and non-immunoprecipitated (input DNA) are used to prepare the DNA library for next-generation sequencing. Many new sequencing machines can be used for the sequencing step (Illumina, Ion Torrent, etc.), but the DNA Library preparation always involves the following steps:
   (a) Repair of 3′ and 5′ ends.
   (b) Preparation of adaptor ligated DNA.
   (c) Size selection.
   (d) Amplification via Polymerase Chain Reaction (PCR) of adaptor ligated DNA.
   (e) Cleanup of amplified library.
2. Prepare the libraries according to the manufacturers' instructions.

3.2.11 Next-Generation Sequencing

1. Prepare templates for sequencing using the DNA libraries. For this step the sequencing workflow differs depending on the sequencing technology used. Briefly,

(a) For the Ion Torrent technology, the library is clonally amplified by emulsion PCR (emPCR) onto Ion Sphere™ particles. In emulsion PCR methods, the surface of the spheres contains oligonucleotide probes with sequences that are complementary to the adaptors binding the DNA fragments. The spheres are then compartmentalized into water-oil emulsion droplets. Each of the droplets capturing one bead is a PCR microreactor that produces amplified copies of a single DNA template. Then, spheres are inserted into the individual sensor wells by spinning the chip in a centrifuge. Ion Torrent™ Technology directly translates chemically encoded information (A, C, G, T) into digital information (0, 1) on a semiconductor chip to provide the sequences of individual DNA fragments.

(b) For the Illumina technology, the DNA templates are bridge amplified to form clonal clusters inside a flow cell. The library is loaded into a flow cell, where fragments are captured on a lawn of surface-bound primers complementary to the library adapters. Each fragment is then amplified into distinct, clonal clusters through bridge amplification. Several million dense clusters of DNA are generated in each channel of the flow cell. Illumina technology utilizes a labeled reversible terminator–based method that detects single bases as they are incorporated into DNA template strands. Successive sequencing cycles provide the sequence of individual DNA fragments.

2. Perform high-throughput sequencing. We use single-ends sequencing as paired-ends do not provide advantages in ChIP-seq (for most cases). We recommend obtaining over ten million reads for high-resolution ChIP-seq with bacterial genomes, especially when more information in the ChIP-seq pattern than just binding sites is sought after.

3.2.12 Software Available for ChIP-Seq Data Analysis and Publication

1. For peak visualization, convert ChIP-seq data file to the BEDGRAPH format. This format allows for its use in Integrated Genome Viewer, which displays the entirety of ChIP-seq data with the possibility to view several datasets simultaneously. Assess the reads quality using FastQC.
2. Map the reads to the reference genome. We used TMAP (Torrent Suite Software) but other software packages are available (BowTie, BWA). Reads count was determined using Genomecov bedtool. Alternatively, the Galaxy platform (https://galaxyproject.org/) also offers a full workflow and data integration system.

3. Use CLC sequence viewer to determine the exact sequence of peaks on genome with the capability to select and copy sequences. Reference genome of ChIP-seq experiments must be in FASTA format.
4. To analyze specific sections of ChIP-seq data, we have found that Excel (Microsoft) provides a user-friendly interface to analyze in detail, portions of ChIP-seq data. The limitation of Excel is that it is unable to show the entirety of data, as the file size becomes too large. ChIP-seq data must be in.txt format.
5. R also works as a powerful tool to analyze ChIP-seq data, but requires prior knowledge on the usage of the interface.
6. The Gene Expression Omnibus (GEO) database is a major repository that stores high-throughput functional genomics datasets that are generated using both microarray-based and sequence-based technologies [9]. Datasets are submitted to GEO to be made freely available to the reviewing process of publication and subsequently for further analysis by the scientific community. In addition to serving as a public archive for these data, GEO has a suite of tools that allow users to identify, analyze, and visualize data relevant to their specific interests.

4 Notes

1. Lysozyme must be stored at −20 °C; therefore, the addition of lysozyme to the lysis buffer is done at the time of the experiment and is added to each sample individually.
2. PMSF must be stored at −20 °C; therefore, the addition of PMSF to the 2× IP buffer is done at the time of the experiment and is added to each sample individually.
3. A high level (~100 μg) of the protein of interest is required to purify a large amount of specific, polyclonal antibodies from the serum. Proteins can either be produced as a his-tagged recombinant, purified in a one-step procedure using Nickel-affinity chromatography purification, or through a native purification procedure. In this latter case, enriched side fractions that are usually discarded could be used to avoid to wasting the protein of interest. Crude extracts are not suitable for use in this antibody purification procedure.
4. When rinsing the strip with Ponceau S, protein band may be difficult to distinguish. To eliminate the staining of Ponceau S on surrounding areas, rinse with sterile water.

5. Protein strips can be stored at 4 °C and reused for antibody purification up to seven times. To reuse a protein strip, start antibody purification protocol at **step** 7.
6. Each sample is separated into two tubes. Tube 1 serves as the experimental sample, and tube 2 serves as a safety sample.
7. This is a freeze point. Flash freezing of pellets must be done with liquid nitrogen and stored at −80 °C until needed.
8. Some proteins may be sensitive to proteolysis during the lysis step. In this case, we recommend the addition of Protease Inhibitor Cocktail Tablets (cOmplete™, Mini, EDTA-free Protease Inhibitor Cocktail (Roche)) in the lysis buffer.
9. We recommend using the Bioruptor® Plus. The described sonication conditions that ensure reliable DNA fragmentation are adapted to this instrument (*see* sonicator's instructions) and result in a similar distribution of the DNA fragments size. Importantly, if using another instrument it is imperative to properly adapt the sonication conditions. We recommend verifying the size of the DNA fragments of each sample after sonication as variations may bias subsequent analyses. Set the cooling device of the sonicator Bioruptor® Plus at 4 °C 30 min before use.
10. During sonication, the sonicator water bath must be maintained at 4 °C, as it is critical to keep the samples at a low temperature. In between sonication rounds let the sonicator cool down for 10 min, as Bioruptor sonicator will shut off in the middle of sonicating if overheated.
11. This is a freeze point. Samples can be stored at −20 °C or −80 °C. Input and non-sonicated samples will rejoin the protocol at the DNA purification step.
12. Antibody concentration used for immunoprecipitation varies based on protein of interest. It is recommended to perform immunoprecipitations using purified protein and several concentrations of antibody to determine the amount needed for ChIP (*see* Fig. 2).
13. When resuspending magnetic beads between washes, avoid creating bubbles as this will result in inefficient bead pelleting and bead loss.
14. Filter or low retention tips are not required, but can reduce sample loss during resuspension steps.
15. Bead pelleting can be performed at room temperature.
16. From this point, it is possible to finish the procedure with the IPure kit v2 (Diagenode) to complete the elution, reverse cross-link, and purify DNA.

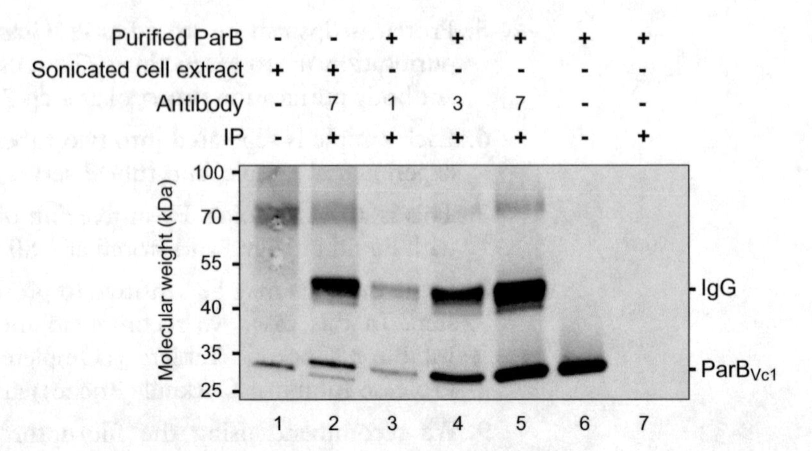

Fig. 2 Optimization step and control of immunoprecipitation. Western blot analysis of $ParB_{Vc1}$ proteins from crude cell extracts of *Vibrio cholerae* (*lanes 1–2*) or purified $ParB_{Vc1}$ protein (0.38 μg; *lanes 3–7*) after SDS-PAGE (4–16%). $ParB_{Vc1}$ proteins are from the ParABS system present on chromosome 1 of *V. cholerae*. Protein samples were subjected (+) or not (−) to immunoprecipitation (IP) using the indicated amount (μg) of purified anti-$ParB_{Vc1}$ antibody. The optimal antibody concentration for immunoprecipitation was determined with increasing amount of antibody. The total amount of $ParB_{Vc1}$ contained in the sample was recovered using 7 μg of anti-$ParB_{Vc1}$ antibody (compare *lanes 5* and *6*). Molecular weights (kDa) are indicated on the *left*

17. To control immunoprecipitation efficiency, recover 5 μl of IP elution sample and add 20 μl of elution buffer. Heat at 65 °C for 15 min and then place the sample on magnetic rack, and recuperate cleared sample. Mix with sample buffer and use 10 μl to analyze through Western blotting.
18. Reverse cross-link is performed in the absence of the magnetic beads.
19. As a control of the protein baseline to test for immunoprecipitation efficiency, recuperate 5 μl of input sample and add 20 μl of 2× IP buffer. Heat at 65 °C for 15 min and then place the sample on magnetic rack, and recuperate cleared sample. Mix with sample buffer and use 10 μl to analyze through Western blotting.
20. To determine the sonication profile of the input and non-sonicated samples, purified DNA can be run on a 0.5× TBE, 1.8% agarose gel to observe a DNA smear indicating the average fragment size per sample. In addition, the use of a bioanalyzer is highly recommended for automated sizing with a more precise and digital format of DNA sample fragment sizes.
21. For precise quantification of DNA quantity, we strongly recommend the use of the Qubit® 3.0 Fluorometer for Input and IP samples prior to library preparation. DNA recovery for IP samples can vary significantly, from 2 ng up to 200 ng, as it

depends on the proteins of interest and the quality/specificity of the corresponding antibodies. We usually construct DNA libraries using 2 ng of IP or Input DNA samples.

Acknowledgments

We thank C. Turlan and C. Guynet for careful reading and suggestions concerning this manuscript, and J. Rech for his technical support provided while adapting the procedures described in this protocol. This work was supported by Agence National pour la Recherche (ANR-14-CE09-0025-01) and by the ComUE Université de Toulouse (APR2014).

References

1. Barski A, Cuddapah S, Cui K, Roh T-Y, Schones DE, Wang Z et al (2007) High-resolution profiling of histone methylations in the human genome. Cell 129:823–837
2. Robertson G, Hirst M, Bainbridge M, Bilenky M, Zhao Y, Zeng T et al (2007) Genome-wide profiles of STAT1 DNA association using chromatin immunoprecipitation and massively parallel sequencing. Nat Methods 4:651–657
3. Johnson DS, Mortazavi A, Myers RM, Wold B (2007) Genome-wide mapping of in vivo protein-DNA interactions. Science 316:1497–1502
4. Mikkelsen TS, Ku M, Jaffe DB, Issac B, Lieberman E, Giannoukos G et al (2007) Genome-wide maps of chromatin state in pluripotent and lineage-committed cells. Nature 448:553–560
5. Lun DS, Sherrid A, Weiner B, Sherman DR, Galagan JE (2009) A blind deconvolution approach to high-resolution mapping of transcription factor binding sites from ChIP-seq data. Genome Biol 10:R142
6. Kahramanoglou C, Seshasayee ASN, Prieto AI, Ibberson D, Schmidt S, Zimmermann J et al (2011) Direct and indirect effects of H-NS and Fis on global gene expression control in *Escherichia coli*. Nucleic Acids Res 39:2073–2091
7. Cho H, McManus HR, Dove SL, Bernhardt TG (2011) Nucleoid occlusion factor SlmA is a DNA-activated FtsZ polymerization antagonist. Proc Natl Acad Sci U S A 108:3773–3778
8. Sanchez A, Cattoni DI, Walter J-C, Rech J, Parmeggiani A, Nollmann M, Bouet J-Y (2015) Stochastic self-assembly of ParB proteins builds the bacterial DNA segregation apparatus. Cell Syst 1:163–173
9. Clough E, Barrett T (2016) The gene expression omnibus database. In: Mathé E, Davis S (eds) Statistical genomics. Springer, New York, pp 93–110. doi:10.1007/978-1-4939-3578-9_5. Accessed 19 Apr 2016

Chapter 7

Generation and Analysis of Chromosomal Contact Maps of Bacteria

Martial Marbouty and Romain Koszul

Abstract

This methods article described a protocol aiming at generating chromosome contact maps of bacterial species using a genome-wide derivative of the chromosome conformation capture (3C) technique. The approach is readily applicable on a broad variety of gram + and gram-bacterial species. It describes and addresses known caveats and technicalities associated with the technique, and should be of interest to any laboratory interested to perform a multiscale analysis of the genome structure of its species of interest.

Key words Chromosome conformation capture, Hi-C, 3C, Genome organization, Nucleoid

1 Introduction

This method aims at characterizing the average tridimensional (3D) organization of bacterial genomes from cell populations (*see* [1]). Using 3C–seq, a derivative of chromosome conformation capture (3C; [2]), genome-wide contact maps of any bacterial species can be generated [1, 3–7] (*see* Fig. 1). The immediate interest of the 3C–seq approach, compared to other 3C derivatives such as Hi-C [8], is that it does not require a ligation products enrichment step, limiting the experimental manipulations and increasing the chances to recover exploitable contact data from any species with minimal tries and errors. The drawback resulting from the absence of this enrichment step is that more reads have to be sequenced to reach at the appropriate number of contact events. However, given the limited sizes of most bacterial genomes, this approach remains amenable and convenient for teams aiming at exploring the 3D organization of their favorite species. Once this protocol is mastered, the biotin enrichment step described originally in Lieberman-Aiden et al. [8] can theoretically be easily introduced through tries and error approach.

Olivier Espéli (ed.), *The Bacterial Nucleoid: Methods and Protocols*, Methods in Molecular Biology, vol. 1624,
DOI 10.1007/978-1-4939-7098-8_7, © Springer Science+Business Media LLC 2017

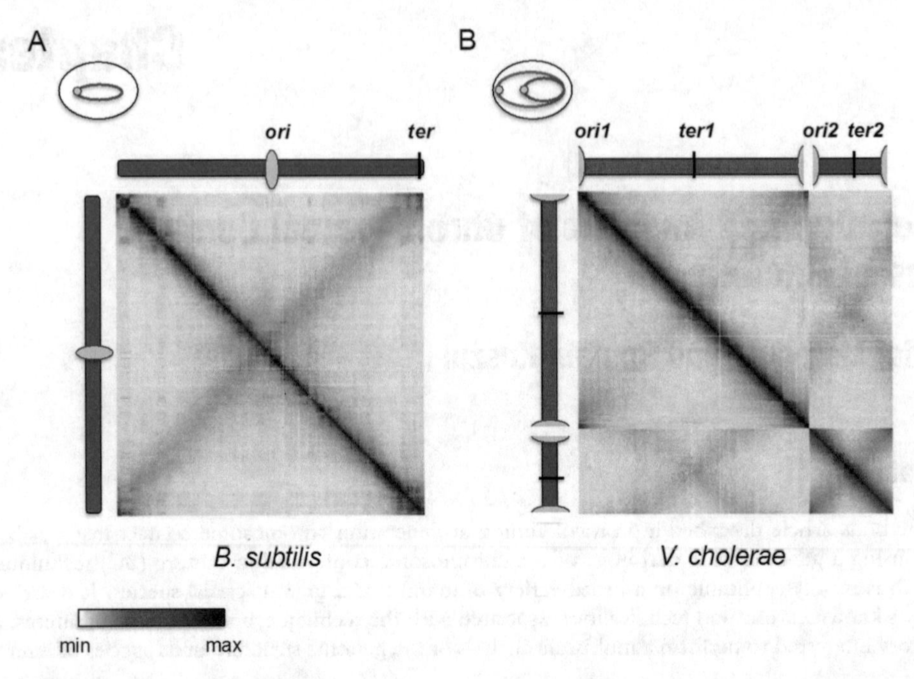

Fig. 1 Normalized genomic contact maps obtained from asynchronous population of *Bacillus subtilis* (**a**) and *Vibrio cholera* (**b**) in rich medium (data from [5, 6]). *X* and *Y* axes represent genomic coordinates. Origins and terminus of replication of the different chromosomes are indicated. The colorscale reflects the frequency of contacts between two regions of the genome (a.u.), from white (rare contacts) to dark red (frequent contacts)

Briefly, 3C quantifies the frequencies of collisions between restriction fragments (RFs) along (and between) the DNA molecule(s) of a genome [2]. These frequencies of contact reflect presumably the relative distances between the loci monitored, hence their average organization within the cellular compartment that contains the DNA (i.e., the nucleus for eukaryotes, and the cell for bacteria). In bacteria, the nucleoid 3D organization presents some links with transcription [3, 5, 6, 9], and also with DNA replication [5–7].

The 3C protocol starts with fixation of a cellular culture using a cross-linking agent (typically formaldehyde), resulting in covalent bounds between proteins and DNA and the generation of complexes of proteins and DNA. The DNA molecules will therefore "freeze" predominantly in the dispositions displayed within the majority of the cell of the population, reflecting their physiological configuration. To quantify the contacts between the different DNA regions of the genome frozen in the vicinity of each other's, two steps are necessary. First, the cells are lysed and the crosslinked chromatin is digested with a carefully chosen restriction enzyme. The insoluble part of the raw chromatin extract is then isolated through centrifugation, diluted, and subjected to a ligation reaction. Using the insoluble fraction diminishes the background by removing small DNA molecules that were not cross-linked in large complexes [10]. Performing ligation in diluted conditions aims at alleviating the ligation events involving molecules trapped in different crosslinked complexes. After ligation the crosslink is reversed and the DNA purified. The resulting 3C library consists of a mix of different ligation products whose relative abundance reflects their average spatial proximity within the cell population at the time of the fixation step. The different religation events within a 3C library are typically and conveniently quantified using pair-end (PE) sequencing and genomic contacts maps generated through a variety of protocols [11–13].

This section describes the experimental protocol for generating a 3C library of a bacterial species. Protocol to prepare the 3C library for PE sequencing can be found elsewhere [14, 15]. Construction and analysis of 3C libraries do not require special equipment besides a sequencer apparatus. However, the preparation of the assay requires careful planning. The choice of the restriction enzyme and of the crosslinking conditions are critical for the success of the experiment, and must be carefully thought through before starting (*see* **Notes 1** and **2**). Numerous programs, toolboxes, algorithms, and routines are available to analyze the data generated by this protocol, and will not be discussed here [11, 14–16] (https://github.com/koszullab/HiC-Box, http://mirnylab.bitbucket.org/hiclib/).

2 Materials

2.1 3C Library Components

1. 15 mL disposable conical tubes.
2. 50 mL disposable conical tubes.
3. Filtration unit 0.22 μm.
4. 1.5 and 2 mL lo-binding microcentrifuge tubes (Eppendorf, Hambourg, Germany).
5. Ready to lyze lysozyme (Ready-Lyze, Lysozyme solution, Epicentre, Madison, USA) (*see* **Note 3**).
6. Restriction enzyme and corresponding restriction enzyme buffer (*see* **Note 2**).
7. 5 U/μL T4 DNA ligase (Weiss Units).
8. 20 mg/mL Proteinase K in water.
9. 10 mg/mL DNAse-free RNAse A in water.
10. 37% Formaldehyde solution (v/v) (Sigma-Aldrich, Saint Louis, Missouri, USA).
11. 2.5 M Glycine: weigh 75.07 g of glycine and transfer to a 1 L cylinder. Add water to a volume of 400 mL and dissolve glycine using a magnetic stirrer and a stir bar (*see* **Note 4**). Filtrate on a 0.22 μm filtering unit and store at room temperature (RT).

12. 10% Sodium dodecyl sulfate (w/v) (SDS) in water. Add 20 mL of 20% SDS (*see* **Note 5**) in a 50 mL disposable conical tube. Add 20 mL of water. Mix gently by returning tube several times. Store at RT.
13. 10% Triton X-100 (v/v) in water. Add 5 mL of Triton X-100 in a 50 mL falcon. Add 45 mL of water and incubate in a 37 °C water bath until complete dissolution (it can take several hours). Store at RT.
14. 10× ligation buffer (without ATP): 500 mM Tris–HCl pH 7.4, 100 mM $MgCl_2$, 100 mM DTT. Add 100 mL of Tris–HCl pH 7.5, 20 mL of $MgCl_2$ 1 M and 10 mL of DTT 2 M to a 500 mL cylinder. Add water to reach 200 mL, mix and filtrate on 0.22 μm filtering unit. Split as 10 mL aliquot and store at −20 °C.
15. 10 mg/mL bovine serum albumin (BSA) in water. Store as 1 mL aliquots at −20 °C.
16. 100 mM Adenosine triphosphate (ATP) pH 7.0 in water. Weigh 1 g of ATP and transfer to a 50 mL falcon. Add 14 mL of water. Add 1.6 mL of NaOH 1 M. Complete to 16.7 mL with water. Check that the pH is around 7.0. Filtrate on 0.22 μm filtering unit. Store as 1 mL aliquots at −20 °C (*see* **Note 6**).
17. 500 mM EDTA in water, pH 8.0.
18. 3 M sodium acetate in water, pH 5.2. Weigh 204.12 g of sodium acetate and transfer to a 1 L cylinder. Complete with water to 400 mL, and adjust pH to 5.2 with acid acetic 100%. Complete to 500 mL with water. Filtrate on a 0.22 μm filtering unit and store at RT.
19. Isopropanol.
20. 10:9:1 phenol:chloroform:isoamylalcohol pH 8.2.
21. 100% Ethanol.
22. TE buffer, pH 8.0. Add 5 mL of TE 10× to a 50 mL falcon. Add 45 mL of water and filtrate on a 0.22 μm filtering unit. Store at RT.
23. 16 °C water bath.
24. 65 °C oven.
25. Magnetic stirrer and stir bar.
26. Variable temperature incubator (25 °C, 30 °C, and 37 °C).
27. Dry bath at 65 °C.
28. Refrigerated tabletopcentrifuge (for 50 mL falcon tubes).

3 Methods

3.1 Generation of a 3C Library of Bacteria

The generation of the 3C library takes 2 days, and the generation of the sequencing library an additional 2–3 days. 3C libraries can be stored at −20 °C and therefore the two processes can be easily separated. Whereas it remains difficult to prepare more than four libraries at a time, processing the samples for sequencing can be performed at a larger scale, the limiting step being then, to some extent, the purification of molecules of a size appropriate for sequencing (*see* **Note 7**). Therefore, timing is important criteria when planning to do the experiment.

3.1.1 Culture Fixation

1. Grow bacteria in your favorite medium (volume :100 mL) until reaching a concentration of approximately 1×10^7 cells/mL (1×10^9 total cells) (*see* **Note 8**).
2. Add 8.5 mL of the fresh formaldehyde solution (i.e., 37%) to the culture (final concentration of 3%) (*see* **Note 2**).
3. Incubate the cells for 30 min at room temperature (RT) under gentle agitation with a magnetic stirrer.
4. Transfer the cell culture at 4 °C for another 30 min under gentle agitation.
5. Move the culture at RT and add 25 mL of Glycine 2.5 M (final concentration: 470 mM) to quench the remaining formaldehyde; incubate under agitation for 5 min at RT.
6. Relocate the culture at 4 °C and keep the cells under gentle agitation for an extra 15 min.
7. Pellet the fixed cells at 4 °C (3500 × g—10 min).
8. Wash the cells with 10 mL of the initial medium.
9. Pellet the fixed cells at 4 °C (3500 × g—10 min).
10. Suspend the cells into 1 mL of medium and transfer them into one 1.5 mL microcentrifuge tube.
11. Pellet the cells at 4 °C (3500 × g—10 min).
12. Remove the supernatant and flash freeze the pellet (i.e., in liquid nitrogen or dry ice + ethanol).
13. Store pellet at −80 °C until use.

NB: Do not store the pellet for more than 6 months (*see* **Note 9**).

3.1.2 3C Library Construction

Day 1

1. Thaw the pellet on ice during half an hour.
2. Resuspend the cells in 500 μL of 1× TE buffer.
3. Add 5 μL of ready-to-lyze lysozyme.
4. Incubate at 37 °C during 30 min.

5. Add 25 μL of 10% SDS per tube (final concentration: 0.5%).
6. Incubate at room temperature for 10 min.
7. Transfer 100 μL of lysate into 4 × 1.5 mL microcentrifuge tube (100 μL per tube) containing 400 μL of a restriction reaction mix (10× restriction buffer 50 μL, Triton X-100 10% 50 μL, 100 U restriction enzyme, water).
8. Transfer 50 μL of remaining lysate into 2 × 1.5 mL microcentrifuge tube (50 μL per tube) containing 200 μL of a control reaction mix (control ND—nondigested: 10× restriction buffer 25 μL, Triton X-100 10% 25 μL, water; control D—digested: 10× restriction buffer 25 μL, Triton X-100 10% 25 μL, 50 U restriction enzyme, water).
9. Incubate for 3 h at the appropriate temperature for the chosen restriction enzyme.
10. Take the two control tubes (nondigested and digested controls, respectively). Add 15 μL of SDS 10% and 15 μL of proteinase K to each tube and incubate them at 65 °C overnight (these controls will then be furthered processed at **step 24**).
11. Centrifuge the four remaining tubes at 16,000 × *g* during 20 min at room temperature in order to isolate the insoluble fraction of the crosslinked chromatin [10].
12. Remove the supernatant and suspend each pellet in 500 μL of H_2O.
13. Dilute the samples in 4 × 7.5 mL of a precooled (4 °C—on ice) ligation reaction mix (10× ligation buffer 800 μL, BSA 10 mg/mL 80 μL, ATP 100 mM 80 μL, water) in 15 mL conical tubes.
14. Add 25 units of T4 DNA ligase.
15. Homogenize the reaction by inverting the tubes two to three times.
16. Incubate for 4 h in a 16 °C waterbath.
17. Add 100 μL of EDTA 500 mM per tube to stop the reaction.
18. Add 100 μL of proteinase K (20 mg/mL), 100 μL of SDS 10% and incubate the tube overnight at 65 °C.

Day 2

19. The next morning, cool down the tubes at room temperature and transfer the solution to two 50 mL conical tubes (16 mL per tube).
20. Add 1.6 mL of 3 M Na Acetate pH 5.0 and 16 mL isopropanol and incubate at −80 °C for 1 h in order to precipitate DNA (*see* **Note 10**).

21. Centrifuge the tube in an appropriate centrifuge at 10,000 × g during 20 min at 4 °C.
22. Remove the supernatant and dry the pellet on the bench (*see* **Note 11**).
23. Suspend each pellet in 900 μL of TE buffer 1× and transfer them in 2 × 2.0 mL microtube.
24. Perform a DNA extraction for each tube using 900 μL of phenol:chloroform. Also extract the DNA from control samples from **step 15** using 300 μL of phenol:chloroform: isoamylalcohol.
25. Recover 2 × 400 μL of the aqueous phase (upper phase) for each tube (800 μL per tube in total) (and 1 × 250 μL for control tubes) and transfer them into 1.5 mL microcentrifuge tube. Adjust the volume in the control tubes to reach 400 μL.
26. Add 40 μL of 3 M Na Acetate pH 5.0 and 1 mL of cold ethanol to each tube.
27. Vortex the tubes and incubate at −80 °C for 30 min.
28. Centrifuge the tubes at 16,000 × *g* for 20 min; discard the supernatants.
29. Wash each DNA pellet with 500 μL of cold 70% ethanol.
30. Centrifuge tubes at 16,000 × *g* for 20 min and remove the supernatant.
31. Dry pellet by incubating them on a 37 °C dry bath.
32. Suspend each pellet in 30 μL TE buffer 1× supplemented with RNAse A (1 mg/mL final concentration).
33. Incubate at 37 °C for 45 min.
34. Pool the tubes containing the 3C libraries.
35. Estimation of the quality and quantity on a 1% agarose gel (Fig. 2; *see* **Note 12**).
36. *Optional: store at − 80 °C as ~ 6 μg DNA aliquots.*

4 Notes

1. The choice of the restriction enzyme and buffer is an important parameter of the experiment. Many restriction enzymes become inactive under the experimental conditions described in the protocol (i.e., brut cellular extract). The cheapest enzymes—usually the best characterized—provide the best candidates to generate a 3C library. Consequently, we highly recommend choosing "classical" restriction enzymes when designing the experiment. However, it is still possible that the enzyme selected is not active. Restriction buffer used to

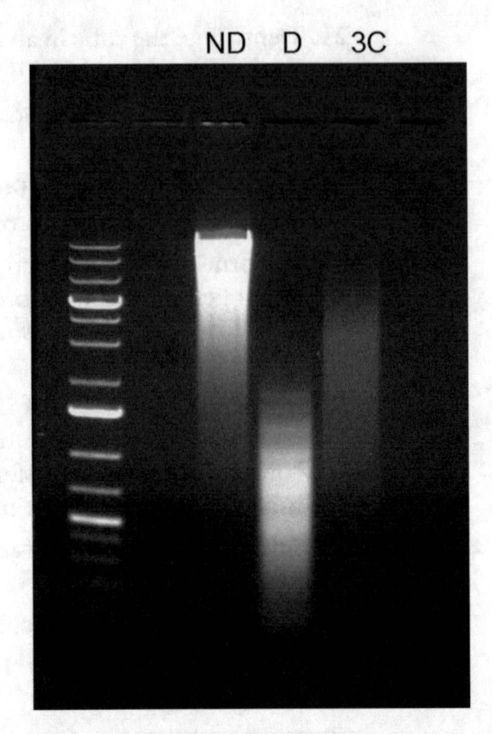

Fig. 2 Photography of gel electrophoresis migration of DNA at various steps of a 3C library construction. *ND:* nondigested control; *D:* restriction control; *3C:* expected migration on a gel of a 3C library

construct 3C libraries provides better results if they contain DTT. Consequently, we strongly suggest avoiding NEB buffers that don't contain DTT (NEBuffer 1.1, 2.1, 3.1 and CutSmart).

2. The experimental conditions for the fixation step have also to be carefully interpreted. It was shown that the likelihood for a RF to be crosslinked is dependent on the probability for 1 bp to be crosslinked, and thus on the incubation parameters in presence of a fixative reagent [11]. For instance, a 4-cutter (restriction enzyme recognizing a 4 pb site) will require a higher concentration of crosslinking reagent than a 6-cutter. The protocol described in this article is designed for enzymes that generate RFs with a distribution average lower than 500 bp (+/− 200 bp). We do not recommend enzymes that generate RF with a distribution average lower than 300 bp.
3. Bacteria lysis is typically achieved by lyzozyme treatment. However this treatment sometimes appears less efficient as a result of fixation. Gram + are also more difficult to lyse. In such cases, lysis can be improved through mechanical disruption (e.g., French Press or glass beads [4, 14]).

4. Dissolving glycine at this concentration can take hours. The process can be accelerated by gently warming the solution (40–50 °C).
5. SDS treatment is a critical step. We noticed an important drop in the quality of the library when the SDS begins to precipitate (warming prior use does not solve the problem). Change the SDS solution immediately upon signs of precipitation.
6. ATP is a critical cofactor of the ligase reaction. In order to avoid any problem due to ATP degradation, discard thawed aliquots after use.
7. For size selection, we routinely use a Pippinprep apparatus (Sage Science), though other methods such as gel purification work well.
8. The mixed culture of 100 mL with a concentration of 1×10^7 cells of genome sizes ~3–4 Mb is sufficient to generate one library in the conditions described in the protocol. For other conditions, the crosslinking step will have to be adapted, as it will change the ratio DNA-Protein-formaldehyde (*see* **Notes 1** and **2**).
9. We have noticed a quality decreased after storage of more than 6 months.
10. After 1 h at −80 °C, solution will froze. You should also notice the apparition of white precipitate. Prompt freezing is necessary for a good recovery of libraries.
11. The pellet does not have to be entirely dry, since DNA will be subsequently processed through a phenol:chloroform extraction.
12. Quantify the libraries on a gel using an image quantification software (such as Image J or Quantity One) is a better solution as large DNA fragments and impurities prevent use of Nanodrop or Qbit quantification.

Acknowledgments

This research was supported by funding to R.K. from the European Research Council under the seventh Framework Program (FP7/2007-2013, ERC grant agreement 260822) and from Agence Nationale pour la Recherche (HiResBaCS ANR-15-CE11-0023).

References

1. Umbarger MA, Toro E, Wright MA, Porreca GJ, Baù D, Hong S-H et al (2011) The three-dimensional architecture of a bacterial genome and its alteration by genetic perturbation. Mol Cell 44(2):252–264
2. Dekker J, Rippe K, Dekker M, Kleckner N (2002) Capturing chromosome conformation. Science 295(5558):1306–1311
3. Le TBK, Imakaev MV, Mirny LA, Laub MT (2013) High-resolution mapping of the spatial organization of a bacterial chromosome. Science 342(6159):731–734
4. Marbouty M, Cournac A, Flot J-F, Marie-Nelly H, Mozziconacci J, Koszul R (2014) Metagenomic chromosome conformation capture (meta3C) unveils the diversity of chromosome

organization in microorganisms. eLife 3: e03318
5. Marbouty M, Le Gall A, Cattoni DI, Cournac A, Koh A, Fiche J-B et al (2015) Condensin- and replication-mediated bacterial chromosome folding and origin condensation revealed by hi-C and super-resolution imaging. Mol Cell 59(4):588–602
6. Val M-E, Marbouty M, de Lemos MF, Kennedy SP, Kemble H, Bland MJ et al (2016) A checkpoint control orchestrates the replication of the two chromosomes of Vibrio cholerae. Sci Adv 2:e150194. Available from: http://www.ncbi.nlm.nih.gov/pmc/articles/PMC4846446/
7. Wang X, Le TBK, Lajoie BR, Dekker J, Laub MT, Rudner DZ (2015) Condensin promotes the juxtaposition of DNA flanking its loading site in Bacillus Subtilis. Genes Dev 29(15):1661–1675
8. Lieberman-Aiden E, van Berkum NL, Williams L, Imakaev M, Ragoczy T, Telling A et al (2009) Comprehensive mapping of long-range interactions reveals folding principles of the human genome. Science 326 (5950):289–293
9. Le TB, Laub MT (2016) Transcription rate and transcript length drive formation of chromosomal interaction domain boundaries. EMBO J 35(14):1582–1595
10. Gavrilov AA, Gushchanskaya ES, Strelkova O, Zhironkina O, Kireev II, Iarovaia OV et al (2013) Disclosure of a structural milieu for the proximity ligation reveals the elusive nature of an active chromatin hub. Nucleic Acids Res 41(6):3563–3575
11. Cournac A, Marie-Nelly H, Marbouty M, Koszul R, Mozziconacci J (2012) Normalization of a chromosomal contact map. BMC Genomics 13:436
12. Imakaev M, Fudenberg G, McCord RP, Naumova N, Goloborodko A, Lajoie BR et al (2012) Iterative correction of Hi-C data reveals hallmarks of chromosome organization. Nat Methods 9(10):999–1003
13. Yaffe E, Tanay A (2011) Probabilistic modeling of Hi-C contact maps eliminates systematic biases to characterize global chromosomal architecture. Nat Genet 43 (11):1059–1065
14. Cournac A, Marbouty M, Mozziconacci J, Koszul R (2016) Generation and analysis of chromosomal contact maps of yeast species. Methods Mol Biol 1361:227–245
15. Lajoie BR, Dekker J, Kaplan N (2015) The Hitchhiker's guide to hi-C analysis: practical guidelines. Methods 72:65–75
16. Wingett S, Ewels P, Furlan-Magaril M, Nagano T, Schoenfelder S, Fraser P et al (2015) HiCUP: pipeline for mapping and processing Hi-C data. F1000Res 4:1310. Available from: http://f1000research.com/articles/4-1310/v1

Chapter 8

Nucleoid-Associated Proteins: Genome Level Occupancy and Expression Analysis

Parul Singh and Aswin Sai Narain Seshasayee

Abstract

The advent of Chromatin Immunoprecipitation sequencing (ChIP-Seq) has allowed the identification of genomic regions bound by a DNA binding protein in-vivo on a genome-wide scale. The impact of the DNA binding protein on gene expression can be addressed using transcriptome experiments in appropriate genetic settings. Overlaying the above two sources of data enables us to dissect the direct and indirect effects of a DNA binding protein on gene expression. Application of these techniques to Nucleoid Associated Proteins (NAPs) and Global Transcription Factors (GTFs) has underscored the complex relationship between DNA-protein interactions and gene expression change, highlighting the role of combinatorial control. Here, we demonstrate the usage of ChIP-Seq to infer binding properties and transcriptional effects of NAPs such as Fis and HNS, and the GTF CRP in the model organism *Escherichia coli* K12 MG1655 (*E. coli*).

Key words NAP, GTF, Regulation of transcription initiation, ChIP-Seq, Z-score, MACS

1 Introduction

Chromatin ImmunoPrecipitation (ChIP) sequencing has become one of the most important methods for discovering the binding sites of NAPs and TFs on the DNA in vivo. In a ChIP experiment DNA and proteins are first cross-linked to strengthen protein-DNA interactions. The cross-linked chromatin is then sheared within a size range of 200–500 base pairs (bp). Next, the protein of interest is immuno-precipitated using an appropriate antibody. The cross-links are reversed and the DNA obtained is used either for sequencing (ChIP-seq) or for hybridization on a microarray-based platform (ChIP-chip).

ChIP studies have been used to understand developmental processes and disease associations in eukaryotes [1]. The roles of DNA binding proteins in bacterial chromosome maintenance and gene regulation have also been uncovered using this method. One of the first uses of the ChIP method for bacteria was the analysis of

Olivier Espéli (ed.), *The Bacterial Nucleoid: Methods and Protocols*, Methods in Molecular Biology, vol. 1624,
DOI 10.1007/978-1-4939-7098-8_8,

the genome-wide distribution of cAMP-receptor protein (CRP) on *E. coli* chromosome, which resulted in the suggestion that this GTF might in fact be a NAP [2]. Since then various groups have carried out experiments to determine the genome wide binding patterns of various NAPs and GTFs including, but not limited to, HNS, Fis, HU, IHF, FNR, Fur, and LRP [3–7].

A certain degree of care must be taken while performing a ChIP experiment. The control usually is categorized broadly into two categories: (a) *Input*: the fragmented genomic sample extracted before immuno-precipitation; (b) *mock-IP*: the sample treated without the antibody or with a nonspecific antibody such as IgG (Immunoglobulin G).

This article shares our experience performing ChIP-seq experiments with *E. coli* NAPs and GTFs, exploring the computational aspects of such studies.

2 Materials

Hardware: Computer with installed UNIX, Linux or MAC OSX (with xcode installed separately), with a minimum of 4 GB of RAM. *Software*: All software listed below are open source tools. Whereas some of these procedures make use of sophisticated algorithms including the Burrows-Wheeler procedure for rapidly aligning millions of reads to a reference sequence, many others can also be implemented efficiently using easy-to-write scripts in programming languages such as PERL or PYTHON.

Install the following software: FastQC [8], Cutadapt [9], Burrows wheeler aligner (BWA) [10], SAMtools [11], Bedtools [12], and MACS [13].

Install R [14] (check the newest stable version) and bioconductor packages such as Genefilter [15].

UCSC archaeal genome browser [16] for visualization (web only).

MEME-ChIP [17] (web only).

Sample names:

The filenames below assume paired end sequencing.

- ChIP biological replicate 1—ChIP1.read1.fastq and ChIP1.read2.fastq.
- ChIP biological replicate 2—ChIP2.read1.fastq and ChIP2.read2.fastq.
- Input control replicate 1—Input1.read1.fastq and Input1.read2.fastq.
- Input control replicate 2—Input2.read1.fastq and Input2.read2.fastq.
- Reference genome sequence—reference.fasta (*see* **Note 1**).

- Sequences of putative gene regulatory regions upstream of operons – regulatory.bed.
- Differentially expressed lead genes (lead gene from each operon) list obtained from RNA-Seq experiment of the relevant mutant NAP/GTF—rnaseq.diffexp.leadgenes.txt.

3 Methods

Install the software and packages mentioned in Subheading 2. All of these are open source and the installation is straightforward.

3.1 Quality Check

After obtaining the reads, check the quality using FastQC software. This tool gives the output in html format where you can see the sequence quality of the reads, sequence duplication, %GC content, and adapter contamination.

3.2 Alignment to Reference Genome

Reads are aligned to the reference genome using BWA (*see* **Note 2**).

1. The first step is the indexing of the reference fasta file; it only needs to be done once.

```
$ bwa index reference.fasta
```

2. The alignment of the reads in the *.fastq* files to the indexed reference genome is generally reported in the *.sai* or *.sam* file formats (*see* **Note 3**).

   The *.sai* file is a machine-readable binary file. In contrast, the *.sam* file is a tab-delimited text file, which contains positions of reads mapping to the regions of genome and the flag column for the mapped reads. This file can be opened, subject to its size, in any text viewer.

```
$ bwa aln -q 30 reference.fasta ChIP1.read1.fastq > ChIP1.read1.sai
$ bwa aln -q 30 reference.fasta ChIP1.read2.fastq > ChIP1.read2.sai
$ bwa aln -q 30 reference.fasta ChIP2.read1.fastq > ChIP2.read1.sai
$ bwa aln -q 30 reference.fasta ChIP2.read2.fastq > ChIP2.read2.sai
$ bwa aln -q 30 reference.fasta Input1.read1.fastq > Input1.read1.sai
$ bwa aln -q 30 reference.fasta Input1.read2.fastq > Input1.read2.sai
$ bwa aln -q 30 reference.fasta Input2.read1.fastq > Input2.read1.sai
$ bwa aln -q 30 reference.fasta Input2.read2.fastq > Input2.read2.sai
```

-q command followed by the integer trims the reads which are below the *Q* value of 30. The user also can set a lesser threshold of *Q* 15/20 and compare the output with the more stringent threshold to decide which value to proceed with.

3. At the next step you can integrate both the read 1 and read 2 file of each sample.

```
$ bwa sampe reference.fasta ChIP1.read1.sai ChIP1.read2.
sai ChIP1.read1.fastq ChIP1.read2.fastq > ChIP1.sam
$ bwa sampe reference.fasta ChIP2.read1.sai ChIP2.read2.
sai ChIP2.read1.fastq ChIP2.read2.fastq > ChIP2.sam
$ bwa sampe reference.fasta Input1.read1.sai Input1.read2.
sai Input1.read1.fastq Input1.read2.fastq > Input1.sam
$ bwa sampe reference.fasta Input2.read1.sai Input2.read2.
sai Input2.read1.fastq Input2.read2.fastq > Input2.sam
```

*In single end sequencing, results are in one read file per sample, in which case use the "samse" alignment command shown below –

```
$ bwa aln -q 30 reference.fasta ChIP1.fastq > ChIP1.sai
$ bwa samse reference.fasta ChIP1.sai ChIP1.fastq > ChIP1.
sam
```

4. Reads that do not map to the reference genome, and those that map to multiple sites on the chromosome, are discarded. For single end sequencing, the integers in the flag field column (column number 2) in the *.sam* file have the following properties: 4 means read unmapped whereas 0 and 16 flags are for read mapped in the forward and reverse strands respectively. Likewise, there are different sets of flags for paired end sequencing—83,99,147, and 163, which denote mapped and proper paired reads.
5. At this stage calculate the fraction of reads mapped to the reference genome. The fraction of unmapped reads should ideally be less than 20%. For bacterial samples, most reads should have been mapped uniquely, which can be ascertained by an inspection of the alignment score (column number 2), where a score of 0 indicates unmapped or multiply mapped reads.
6. The *.sam* files are bulky files to work with and they can be converted to *.bam* files that are the binary version of the same and occupy smaller disk space and are faster to work with (*see* **Note 4**).

   The conversion can be done using SAMtools package with the following.

```
$ samtools view -bS ChIP1.sam ChIP1.bam
$ samtools view -bS ChIP2.sam ChIP2.bam
$ samtools view -bS Input1.sam Input1.bam
$ samtools view -bS Input2.sam Input2.bam
```

To check total number of reads -

```
$ samtools view ChIP1.bam | wc -l
```

To check only the mapped reads for further downstream analysis -

```
$ samtools view -c -F 0x40 ChIP1.bam
```

-c - count the number of occurrences, –F - to remove, 0x40 flag -unmapped reads.

After checking for the number of reads, user can use the following command to work only with reads that are mapped.

```
$ samtools view -h -F 0x40 ChIP1.bam > ChIP1.mapped.bam
```

-h to include header in the output file.

*To count mapped reads in single end dataset.

```
$ samtools view -c -F 4 ChIP.bam
```

-c - count the number of occurrences, –F 4 - remove unmapped reads.

*To extract for mapped reads in single end dataset -

```
$ samtools view -h -F 4 ChIP.bam > ChIP.mapped.bam
```

-h to include header in the output file.

To sort the bam file according to the read name -

```
$ samtools sort -o ChIP1_sort -n ChIP1.mapped.bam
$ samtools sort -o ChIP2_sort -n ChIP2.mapped.bam
$ samtools sort -o Input1_sort -n Input1.mapped.bam
$ samtools sort -o Input2_sort -n Input2.mapped.bam
```

sort command sorts the output according to the user given option, –o for output filename, –n option to sort it according to the read name.

7. Next is the conversion of the BAM file to the read depth per genomic interval tab delimited BED file. This file can be used as

input for visualization of ChIP-Seq peaks even before the analysis. The user needs to install bedtools and type the following on the terminal.

```
$ bedtools genomecov -d -ibam ChIP1_sort.bam > ChIP1.cov
$ bedtools genomecov -d -ibam ChIP2_sort.bam > ChIP2.cov
$ bedtools genomecov -d -ibam Input1_sort.bam > Input1.cov
$ bedtools genomecov -d -ibam Input2_sort.bam > Input2.cov
```

where –d option is for computing the coverage per base and ibam stands for input bam file. This step might take longer time to run.

3.3 Analysis of Read Count Distribution and Replicate Correlation

The *.cov* output file has three columns in which two columns are of interest: the second column with the base position and third with the coverage computed for that specific position.

1. Using this information plot the density of the read count distribution in R environment. In the R environment first step is to import the tab-delimited input bed file. Below input command assumes the user has a tab-delimited text file without the header row.

```
> mat1 <- read.delim ("ChIP1.cov", header=F, quote=" ")
> plot(density (mat1$V3))
```

The typical density plot of a GTF/NAP with sequence-specific binding has a heavy right end tail (Fig. 1).

2. Plot the corresponding distribution for the input control. This will show a severely curtailed right tail, when compared to the ChIP sample in Fig. 1.

```
> mat2 <- read.delim ("Input1.cov", header=F, quote=" ")
> lines(density (mat2F$V3), lty= 2, lwd = 2)
```

3. Normalize each ChIP and input sample internally using the following procedure. Assume that the background distribution of read counts is normal. Please note that this assumption remains to be rigorously tested. Calculate Z score for each base position, where μ—mode of the read distribution calculated using the shorth function, s—standard deviation calculated using Mean Absolute Deviation (MAD) function in R. Repeat for the following for each replicate including input.

```
> library(genefilter)
> F_1 <- read.table("ChIP1.cov")$V3
> F_1_m <- shorth(F_1, tie.limit=1)
```

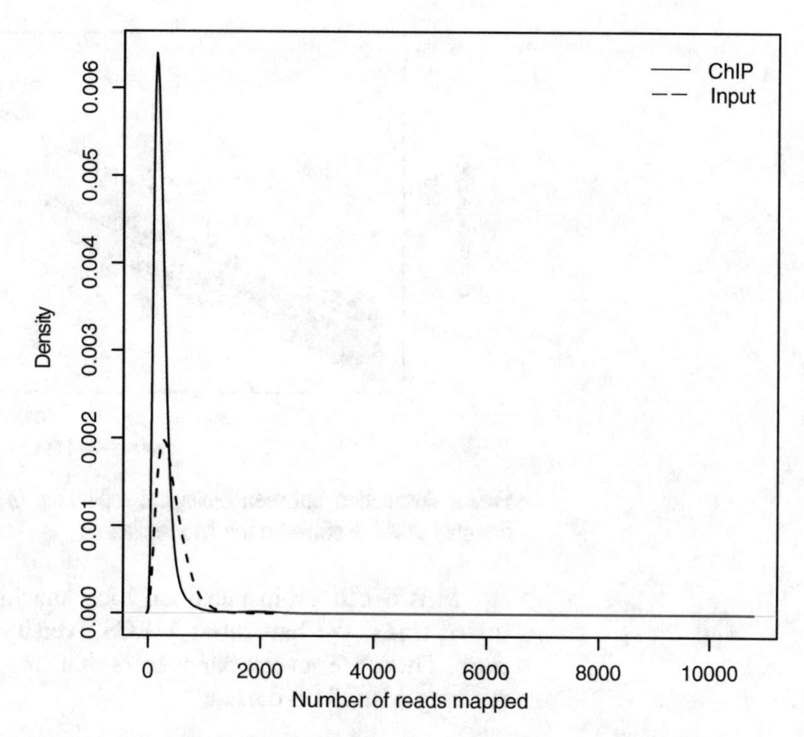

Fig. 1 Density plot of the read count distribution for ChIP (*solid line*) and Input (dashed line) of an example NAP/GTF

```
> F_1_s <- mad(F_1[which(F_1<=mean(F_1))])
> F_1_z <- (F_1 - F_1_m)/(F_1_s)
```

4. Calculate the correlation between the Z score values between the biological replicates. Good correlation between the biological replicates is an important estimator for a successful ChIP experiment (*see* **Note 5**). Since the Z-score is a normalized measure of the signal, not only should correlation coefficient be high, the scatter plot should lie along the 45° line. It is possible that a good correlation is obtained only at Z-score values above a threshold, say $Z > 3$ (Fig. 2).

```
> cor.test(F_1_z, F_2_z)
> smoothScatter(F_1_z, F_3_z)
> abline (0,1)
```

3.4 Peak Calling with MACS

Model-based Analysis of ChIP-Sequencing (MACS) identifies regions bound by a NAP/GTF/Histone modification. The model assumes the read distribution to be Poisson and then performs three key steps to find enrichment—removal of redundant reads, adjustment of read position based on fragment size distribution, and calculation of peak enrichment using local background normalization [13, 18].

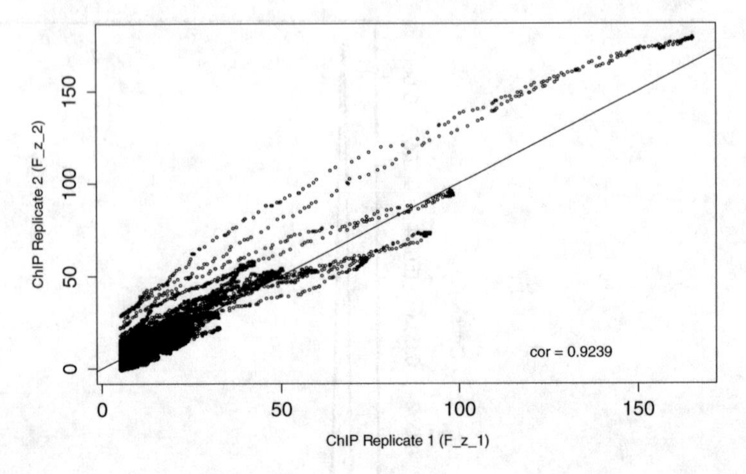

Fig. 2 Correlation between biological replicates of ChIP sample. *Cor* depicts correlation value between the two replicates

MACS can be installed on local machine using the author's instructions. We have used MACS2 version for our analysis purpose. There are several parameters that one has to consider before running MACS on dataset.

```
$ macs2 callpeak -t ChIP1.bam ChIP2.bam -c Input1.bam Input2.
bam -f BAMPE -g 4.6e7 -n output
```

-t for ChIP-Seq treatment file.

-c for input/mock data control. MACS can also work without this dataset.

-f for the format of the input files. MACS takes several read formats including SAM, BAM, BED, ELAND. For paired end reads BAM and ELAND formats can be used by specifying it as BAMPE and ELANDMULTIPLET. If this option is not specified MACS by default will decide the format automatically (*see* **Note 6**).

-g is the parameter for genome size.

-p is value cutoff. If you don't set this default will be 1e-5.

The output contains several files named ChIP1_peaks.bed, ChIP1_peaks.xls, ChIP1_summits.bed etc. ChIP1_peaks.bed has the start and end of the genomic coordinates of the putative binding sites. The fourth column corresponds to the name of the file and fifth is the -log10 (q value) also seen in the ChIP1_peaks.bed. The log2 fold-change cutoff is 1.2 and greater (*see* **Note 7**).

3.5 Peak Visualization

Peak visualization in the UCSC genome browser gives detailed information on whether peaks are clustered in specific regions of chromosome, evolutionary conservation with other organisms, gene annotation tracks (refseq) to name a few (Fig. 3). One can

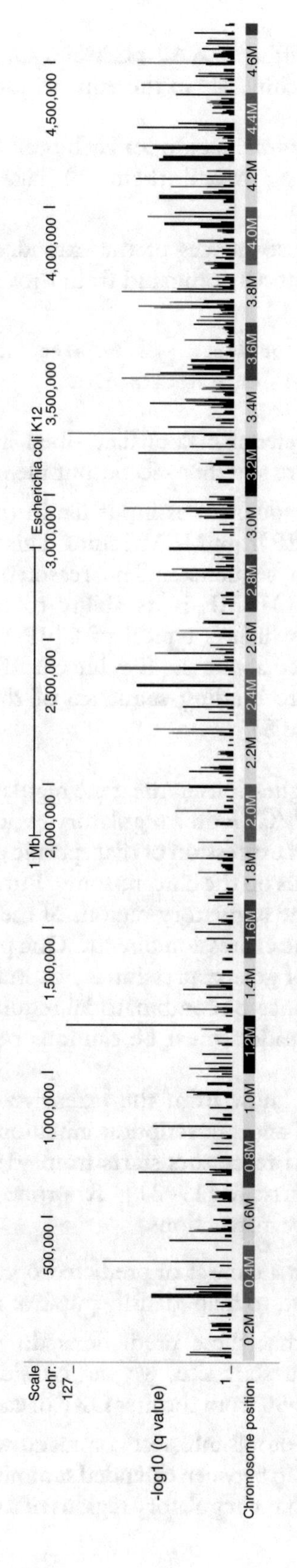

Fig. 3 Peaks as visualized in UCSC archaeal genome browser. *E. coli* K12 MG1655 chromosome positions are on *x*-axis, −log 10 (*q* value) of the peaks (as computed by MACS) on the *y*-axis

also combine different NAP peak files into one file and view the differences and similarity in the same window.

3.6 Motif Finding

1. Take the summit region for each peak from the MACS output summits file, extend them 20 bases both upstream and downstream.
2. Extract the sequences of the extended summit region using bedtools getfasta command from those genomic regions.

```
$ bedtools getfasta -fi reference.sequence -bed ChIP1_-
peaks.bed -fo chip.motifseq.txt
```

   -fi –input reference fasta file, –bed input file for which the sequences are sought, –fo output file name
3. Use these sequences as input file in the web-based platform, MEME-ChIP from MEME suite. This program will take equal length fasta sequences. The reason why use MEME-ChIP rather than MEME is its ability to take a large number of sequences, which is typical of ChIP-Seq datasets. The motif thus obtained above the E-value cutoff can then be considered the signature binding sequence of the NAP/GTF of interest (*see* **Note 8**).

3.7 Integrating ChIP-Seq and Gene Expression Study

One of the key questions in the gene regulation field is whether the binding of NAP/GTF on a regulatory region of a gene can explain the regulation of expression of that specific gene. GTFs/NAPs bind to various regions on the chromosome. But, only those peaks which are present in the regulatory regions of the chromosome are likely to influence gene expression directly. One point to note here is that the regulation of gene expression is not straightforward, as there is increasing evidence of combinatorial regulation by several GTFs/NAPs; hence, readers must be cautious before interpreting these results.

We already know from the extensive gene-centric studies of gene regulation and transcription initiation in *E. coli* that binding of activators and repressors starts from ~150 bp upstream till the transcription start site [19–21]. To probe the role of NAP/GTF follow the below instructions.

1. First, obtain a dataset of predicted operons for the genome of your interest, to help identify putative regulatory regions.
2. Assuming that these predictions do not always include the transcription start site, we can consider the region between −300 and +50 from the first ORF of each operon as regulatory.
3. Next use bedtools intersect command which calculates the fraction of overlap between extended summit regions (peaks) of ChIP datasets with the regulatory regions of the operons (*see* **Note 9**).

```
$ bedtools intersect -a ChIP1_peaks.bed -b regulatory.bed
> ChIP1_regulatory.intersect.bed
```

4. Next look for the peaks which lie in the regulatory regions and overlap them with the list of differentially expressed operon list (at least one gene in the operon is differentially expressed) from the RNA-Sequencing of the mutant GTF/NAP experiment.

```
$ awk 'NR==FNR{a[$1];next} ($1 in a)' ChIP1_regulatory.
intersect.bed rnaseq.diffexp.leadgenes.txt > leadgenes.
overlap.chip.rnaseq.txt
```

provided that first column of both input files has the name of lead gene of the operon.

Following this, the user obtains a tab-delimited file with genomic regions (list of operons) which are bound in their respective regulatory region by the NAP/GTF and are differentially expressed in mutant NAP/GTF background. This indicates whether the binding effect of the NAP/TF on the gene expression is direct or indirect. Based on the position of binding from the transcription start site, user can also predict whether the GTF is an activator or repressor; for this one will presumably require a more precise binding site identification than is permitted by the resolution of the ChIP, something that can be obtained by combining ChIP peaks with motif identification, or by using higher-resolution experimental techniques such as ChIP-exo.

4 Notes

1. The version of genome being used as the reference should be the same throughout the analysis. Different web-based platforms might use their own version of a given reference genome. Therefore, before using a web server such as the UCSC genome browser, the user should ensure that version of genome being used is appropriate.
2. There are certain compatibility issues within different versions of softwares, for the analysis of paired end sequences use BWA version 0.6 (older) compared to 0.7.5 (newer release) which works well with the single end alignment. Also, do not update the version of software if not necessary, there are small changes that might affect the output.
3. Due to large size of *.fastq* files, it is recommended to compress them by using gzip command on the terminal.

```
$ gzip filename
```

BWA alignment commands can be run on these compressed files in the same way as the uncompressed ones.

4. Processing of *.sam* files might take longer time. Also, they tend to occupy large space, once converted to *.bam*, compress the *.sam* files in a similar way as *.fastq*. Also, sort both *.bam* and *.bed* files, this makes processing of these files more speed efficient and less error prone.
5. Only those genomic regions that are enriched in both the biological replicates should be considered for downstream analysis.
6. MACS by default filters duplicate reads, and keeps one tag at one location. In case the mapped reads file shows error while being processed, user can use the sorted *.bam* file derived from the original *.bam* file (total reads) for further MACS analysis.
7. The overlap between different methods predicting the GTF/NAP binding is crucial. In our datasets we see ~90% overlap between the Z-score-based method of Kahramanoglou et al. and MACS-based peak calls. At this point, we do not know where errors lie, but it may be prudent to take peaks called by multiple methods into consideration.
8. NAP/GTF will not always have very simple motif, sometimes the motif pattern changes drastically due to other factors such as growth phase and presence/absence of other GTFs/NAPs. A complex motif in these conditions does not mean ChIP experiment has not worked, it means that the particular sequence might have more than one NAP/GTF binding site.
9. There are several open-source software alternatives available for each step in this process. Many of these steps, including counting reads, assigning peaks to genes of interest and finding overlaps in peak positions across datasets can be easily coded in programming languages such as PERL or PYTHON, which allows complete control over how these steps are performed. This approach is advisable if your experiment involves special conditions, which may not be covered by standard software.

References

1. Park PJ (2011) ChIP–seq: advantages and challenges of a maturing technology. Nat Rev Genet 10:669–680
2. Grainger DC, Hurd D, Harrison M, Holdstock J, Busby SJW (2005) Studies of the distribution of *Escherichia coli* cAMP-receptor protein and RNA polymerase along the E.coli chromosome. Proc Natl Acad Sci U S A 102:17693–17698
3. Kahramanoglou C et al (2011) Direct and indirect effects of H-NS and Fis on global gene expression control in *Escherichia coli*. Nucleic Acids Res 39:2073–2091
4. Prieto AI et al (2012) Genomic analysis of DNA binding and gene regulation by homologous nucleoid-associated proteins IHF and HU in *Escherichia coli* K12. Nucleic Acids Res 40:3524–3537
5. Myers KS et al (2013) Genome-scale analysis of *Escherichia coli* FNR reveals complex features of transcription factor binding. PLoS Genet 9:11–13
6. Seo SW et al (2014) Deciphering fur transcriptional regulatory network highlights its

complex role beyond iron metabolism in *Escherichia coli*. Nat Commun 5:4910

7. Cho B et al (2008) Genome-scale reconstruction of the Lrp regulatory network in *Escherichia coli*. Proc Natl Acad Sci 105:19462–19467
8. Andrews S (2010) FastQC: a quality control tool for high throughput sequence data. Available online at : http://www.bioinformatics.babraham.ac.uk/projects/fastqc
9. Martin M (2011) Cutadapt removes adapter sequences from high-throughput sequencing reads. EMBnet Journal 17(1), pp.10–12. doi: http://dx.doi.org/10.14806/ej.17.1.200
10. Li H, Durbin R (2009) Fast and accurate short read alignment with Burrows-wheeler transform. Bioinformatics 25:1754–1760
11. Li H et al (2009) The sequence alignment/map format and SAMtools. Bioinformatics 25:2078–2079
12. Quinlan AR, Hall IM (2010) BEDTools: a flexible suite of utilities for comparing genomic features. Bioinformatics 26:841–842
13. Zhang Y et al (2008) Model-based analysis of ChIP-Seq (MACS). Genome Biol 9:R137
14. R Core Team (2016) R: A language and environment for statistical computing. R Foundation for Statistical Computing, Vienna, Austria. URL https://www.R-project.org/
15. Gentleman R, Carey V, Huber W and Hahne F (2016) *genefilter: genefilter: methods for filtering genes from high-throughput experiments*. R package version 1.52.1.
16. Schneider KL, Pollard KS, Baertsch R, Pohl A, Lowe TM (2006) The UCSC Archaeal genome browser. Nucleic Acids Res 34:D407–D410
17. Bailey TL et al (2009) MEME SUITE: tools for motif discovery and searching. Nucleic Acids Res 37:202–208
18. Feng J, Liu T, Qin B, Zhang Y, Liu XS (2012) Identifying ChIP-seq enrichment using MACS. Nat Protoc 7:1728–1740
19. Nasser W, Schneider R, Travers A, Muskhelishvili G (2001) CRP modulates fis transcription by alternate formation of activating and repressing nucleoprotein complexes. J Biol Chem 276:17878–17886
20. Appleman JA, Ross W, Salomon J, Richard L (1998) Activation of *Escherichia coli* rRNA Transcription by FIS during a Growth Cycle. J Bacteriol 180:1525–1532
21. Browning DF, Busby SJ (2004) The regulation of bacterial transcription initiation. Nat Rev Microbiol 2:57–65

Part III

Molecular Biology Methods to Study Nucleoid Structuring Factors

Chapter 9

Isolation and Analysis of RNA Polymerase Supramolecular Complex with Associated Proteins

Sanja Mehandziska, Alexander M. Petrescu, and Georgi Muskhelishvili

Abstract

Transcription machinery plays a central role in both the gene expression and nucleoid compaction. In this chapter we elaborate on the optimization of RNA polymerase purification protocol using a mild procedure with the purpose of preserving its native composition. This protocol combines protein extraction under non-denaturing conditions, heparin based affinity purification, and consequent BN-PAGE–SDS-PAGE separation. The outcome is an experimental procedure for screening RNA polymerase composition with associated proteins, in various bacterial strains or mutant backgrounds. With modifications in the column purification step, this procedure can be applied for isolation and identification of the components of other multi-protein complexes.

Key words RNA polymerase, Multi-subunit complexes, Heparin affinity purification, Blue native polyacrylamide gel electrophoresis (BN-PAGE), Mass Spectrometry (MS)

1 Introduction

Understanding of the complexity of cellular processes and their regulation requires development of methodologies maximally preserving the functional communications between the interacting components in the native multi-subunit protein assemblies. This is especially relevant for the multi-subunit RNA polymerase (RNAP) complex, the major transcribing machine of the bacterial cell interacting not only with transcription initiation, elongation, and termination factors but also other DNA modifying proteins and metabolic enzymes yet to be discovered. The rationale of our experimental setup aiming at isolation of RNAP holoenzyme together with associated proteins (dubbed hereafter the RNAP supramolecular complex) is that this enzymatic complex plays a central role in organizing both the gene expression program and the dynamic nucleoid structure [1, 2]. The highly purified multi-subunit RNA polymerase preparations do not reflect the actual composition of the enzymatic complex within the cell, whereas

Olivier Espéli (ed.), *The Bacterial Nucleoid: Methods and Protocols*, Methods in Molecular Biology, vol. 1624,
DOI 10.1007/978-1-4939-7098-8_9,

knowing the latter is pivotal for understanding the relationships between transcription, translation, and metabolism. In this study we elaborate on the optimization of RNAP purification protocol using a mild procedure with the purpose of preserving the native composition of the RNAP enzymatic assembly. This protocol combines protein extraction under non-denaturing conditions, heparin-based affinity purification, and consequent Blue native polyacrylamide gel electrophoresis (BN-PAGE) followed by SDS-PAGE separation [3, 4].

Heparin has a highly dense negative charge, which mimics the DNA in the cell and attracts the basic proteins. The heparin-based purification of RNA polymerase relies on the affinity of RNAP to heparin. Using increasing salt gradient, the RNAP can easily be eluted from the column. This step provides an intact RNAP complex mixed with some other proteins and protein complexes. To get the pure RNAP supramolecular complex, the heparin extracts are subjected to BN-PAGE gel electrophoresis originally described by Schägger and von Jagow [5] for the separation of enzymatically active membrane protein complexes under mild conditions. BN-PAGE has several important advantages for studying multi-protein complexes in general, especially because it provides information about the size, number, protein composition, and relative abundance of different multi-protein complexes [3, 5, 6]. The buffering capacity of the Bis-Tris buffer (pH 7.5–7.7) used for BN-PAGE is optimal for native protein separation compared to native Tris-glycine gels (pH 9.3–9.5) [7]. At neutral pH the protein modifications are brought to a minimum, whereas at basic pH, modifications of cysteine and lysine are likely to occur, due to the nucleophilic character of their side-chains [8]. In addition, the Coomassie Brilliant Blue G-250 present in the cathode and sample buffer binds to the surface of the proteins and protein complexes including exposed hydrophobic regions, providing overall negative charge [9]. The Coomassie does not act as a detergent, while the proteins negatively charged at the surface are able to move through the gel independently from their isoelectric point and therefore their size can be estimated based on their migration [3].

Next, the RNAP supramolecular complex is subjected to denaturing SDS-PAGE to separate its individual components. Finally these components are excised from the gel, in-gel trypsin digested and identified by Mass Spectrometry (MS) or tandem MS/MS. The outcome is an experimental procedure for screening the RNA polymerase composition in various strains or mutant backgrounds. In our preparations derived from growing *Escherichia coli* cells we observe ribosomal proteins, DNA gyrase, and several key metabolic enzymes associated with RNA polymerase, suggesting that the central processes of transcription, translation, and metabolism are tightly interconnected in the cell.

2 Materials

2.1 Protein Purification

1. PBS: 137 mM NaCl (8 g of NaCl), 2.7 mM KCl (0.2 g of KCl), 10 mM Na_2HPO_4 (0.57 g anhydrous or 1.44 g $Na_2HPO_4 \times 12$ H_2O), 1.8 mM KH_2PO_4 (0.24 g of KH_2PO_4). Dissolve the chemicals in ddH_2O, then adjust pH to 7.4 and the final volume to 1 l.
2. Buffer 0: 10 mM Tris-HCl pH 8.0 (20 ml of 1 M Tris-HCl), 10 mM $MgCl_2$ (20 ml of 1 M $MgCl_2$), 0.1 mM EDTA (400 μl of 0.5 M EDTA), 0.1 mM DTT (200 μl of 1 M DTT), 5% (v/v) Glycerol (116.3 ml of 86% Glycerol). First prepare stocks of 1 M Tris-HCl pH 8.0, 1 M $MgCl_2$, 0.5 M EDTA pH 8.0 and 1 M DTT, then mix appropriate volumes to get the desired final concentrations (amounts given are for 2 l final volume). Vacuum filter the buffer with 0.22 μm filter and degas for 10 min before use. DTT should always be freshly prepared before use.
3. Buffer B: 10 mM Tris-HCl pH 8.0, 10 mM $MgCl_2$, 0.1 mM EDTA, 0.1 mM DTT, 5% (v/v) Glycerol, 1 M NaCl (400 ml of 5 M NaCl) (amounts given are for 2 l final volume). Prepare in the same manner as Buffer 0 above. DTT should always be freshly prepared before use.
4. Sonication/Equilibration Buffer = Buffer A: 10 mM Tris-HCl pH 8.0, 10 mM $MgCl_2$, 0.1 mM EDTA, 0.1 mM DTT, 5% (v/v) Glycerol, 230 mM NaCl. Prepare Buffer 0 and Buffer B, then mix 230 ml of Buffer B with 770 ml of Buffer 0 to make Buffer A. Then filter and degas as Buffer 0 (*see* **Note 1**).
5. RNAP storage buffer: 20 mM Tris-HCl pH 7.5 (2 ml of 1 M Tris-HCl), 100 mM NaCl (2 ml of 5 M NaCl), 0.1 mM EDTA (20 μl of 0.5 M EDTA), 1 mM DTT (100 μl of 1 M DTT), 50% Glycerol v/v (58.14 ml of 86% Glycerol) (amounts given are for 100 ml final volume). Filter buffer with 0.22 μm filter using a syringe. This buffer also allows for preservation of RNAP activity, recipe adapted from the commercial New England Biolabs *E. coli* RNA polymerase holoenzyme storage buffer.
6. Protease Inhibitor Mix HP PLUS (SERVA Electrophoresis, Heidelberg, Germany). Mix pellet with provided 1 ml DMSO and aliquot before freezing.
7. Benzonase® Nuclease ≥ 250 units/μl, ≥ 90% (SDS-PAGE), recombinant, expressed in *E. coli*, buffered aqueous glycerol solution (Sigma-Aldrich Chemie GmbH, Steinheim, Germany).
8. 0.25–0.50 glass beads (Carl Roth GmbH, Karlsruhe, Germany).
9. Ultrasonic processor UP100H (100 W, 30 kHz) (Hielscher Ultrasonics GmbH, Teltow, Germany).

10. Amicon® Ultra Centrifugation Filters, Molecular Weight Cut-off 50 kDa, 15 ml (Merck Millipore, a part of Merck KGaA, Darmstadt, Germany).
11. Pierce® BCA Protein Assay Kit (Thermo Scientific, Darmstadt, Germany).
12. Microplate reader GENius (Tecan Instruments, Salzburg, Austria) and Magellan software v6.5.
13. HiTrap Heparin 1 ml column (GE Healthcare, Munich, Germany).
14. ÄKTApurifier 10 system (GE Healthcare, Munich, Germany).
15. Sample Loop 2.0 ml, INV-907 (GE Healthcare, Munich, Germany).

2.2 BN-PAGE

The BN-PAGE buffers were taken and adjusted from refs. [3, 4]. The Coomassie solution recipes were taken from ref. [10].

1. 3× BN-Gel Buffer: 150 mM Bis-tris (15.69 g of Bis-tris). Adjust pH to 7.0 with HCl. Fill up with ddH_2O to a final volume of 500 ml. Store at 4 °C.
2. Acrylamide-Bisacrylamide Mix: Mix 17.3 ml of 40% 19:1 acrylamide-bisacrylamide with 82.7 ml of 40% 37.5:1 acrylamide-bisacrylamide. This will result in a solution with acrylamide-bisacrylamide ratio of 32:1, with 40% (w/v) total monomer concentration (acrylamide plus crosslinker). Store at 4 °C.
3. 4% Separating Gel (BN-PAGE): 6.67 ml of 3× BN-Gel Buffer, 2 ml Acrylamide-Bisacrylamide Mix, 11.33 ml ddH_2O, 72 μl of 10% APS, 7.2 μl TEMED. Add APS and TEMED immediately before pouring the gel. This recipe is sufficient to cast a 20-ml gel.
4. 15% Separating Gel (BN-PAGE): 6.67 ml of 3× BN-Gel Buffer, 7.15 ml Acrylamide-Bisacrylamide Mix, 1.36 ml ddH_2O, 4.67 ml of 86% Glycerol, 56 μl of 10% APS, 5.6 μl TEMED. Add APS and TEMED immediately before pouring the gel. This recipe is sufficient to cast a 20-ml gel.
5. 3.2% Stacking Gel (BN-PAGE): 3.00 ml of 3× BN-gel Buffer, 0.72 ml Acrylamide-Bisacrylamide Mix, 5.28 ml ddH_2O, 120 μl of 10% APS, 12 μl TEMED. Add APS and TEMED immediately before pouring the gel.
6. Cathode Buffer: 15 mM Bis-tris (3.14 g of Bis-tris), 50 mM Tricine (8.96 g of Tricine), 0.02% Coomassie blue G-250 (0.2 g of Coomassie blue G-250). Prepare 1 l as a 10× stock, adjust pH to 7.0 with HCl, and store at 4 °C. Dilute 1:10 with ddH_2O before use. Do not substitute with other types of Coomassie dye.

7. Anode Buffer: 50 mM Bis-tris (10.46 g of Bis-tris). Prepare 1 l as a 10× stock, adjust pH to 7.0 with HCl, and store at 4 °C. Dilute 1:10 with ddH_2O before use.
8. Coomassie Staining Solution: 5% (w/v) Aluminum sulfate (100 g of Aluminum sulfate 14–18 hydrate), ethanol 10% (v/v) (192 ml of ≥ 99.9% ethanol), 0.2% (w/v) Coomassie blue G-250 (0.4 g of Coomassie blue G-250), 2% (v/v) orthophosphoric acid (47 ml of concentrated acid). Dissolve first the Aluminum sulfate in ddH_2O, then add the ethanol and homogenize. Next dissolve the Coomassie blue G-250. At the end add the orthophosphoric acid (adjust to a final volume of 2 l with ddH_2O) (*see* **Note 2**).
9. Coomassie Destaining Solution: 10% (v/v) ethanol (192 ml of ≥ 99.9% ethanol), 2% (v/v) orthophosphoric acid (47 ml of concentrated acid) (adjust to a final volume of 2 l with ddH_2O).
10. Bis-tris PUFFERAN ≥ 99%.
11. NativeMark™ Unstained Protein Standard (Invitrogen, Life Technologies GmbH, Frankfurt, Germany).
12. NativePAGE™ Sample Buffer (4×) (Invitrogen, Life Technologies GmbH, Frankfurt, Germany).
13. Peristaltic Pump P-1 (GE Healthcare, Munich, Germany).
14. Eco-Maxi System EBC (017-402) (Biometra, Göttingen, Germany).
15. Prolite basic light box (Kaiser Fototechnik GmbH & Co.KG, Buchen, Germany).

2.3 SDS-PAGE

1. SDS Sample Buffer: 12.5 mM Tris-HCl (250 μl of 0.5 M Tris-HCl, pH 6.8), 4% (w/v) SDS (0.4 g of SDS powder), 20% Glycerol (2.33 ml of 86% Glycerol), 0.02% (w/v) Bromophenol blue (0.002 g of Bromophenol blue). Adjust pH to 6.8. To reduce disulfide bonds, add 9 ml β-mercaptoethanol (to a total volume of 100 ml buffer). β-mercaptoethanol should always be added fresh, therefore pre-aliquot the sample buffer and keep it as such, adding the β-mercaptoethanol directly before use. SDS as a powder and β-mercaptoethanol are toxic. Use gloves and work under a hood.
2. 4× SDS Running buffer: 96 mM Tris (23.26 g of Tris), 800 mM Glycine (120.11 g of Glycine), 0.4% (w/v) SDS (8 g of SDS powder) (adjust to a final volume of 2 l with ddH_2O). Dilute to 1× in ddH_2O before use.
3. 8% Separating Gel (SDS-PAGE): 7.5 ml of 1.5 M Tris-HCl, pH 8.0, 6 ml Acrylamide-Bisacrylamide Mix, 300 μl of 10% SDS, 16.1 ml ddH_2O, 125 μl of 10% APS, 12.5 μl TEMED.

Add APS and TEMED immediately before pouring the gel. This recipe is sufficient to cast a 30-ml gel.

4. 15% Separating Gel (SDS-PAGE): 7.5 ml of 1.5 M Tris-HCl, pH 8.0, 11.25 ml Acrylamide-Bisacrylamide Mix, 300 μl of 10% SDS, 3.75 ml ddH_2O, 7.1 ml of 86% Glycerol, 110 μl of 10% APS, 11 μl TEMED. Add APS and TEMED immediately before pouring the gel. This recipe is sufficient to cast a 30-ml gel.
5. 3.5% Stacking Gel (SDS-PAGE): 5 ml of 0.5 M Tris-HCl, pH 6.8, 1.95 ml Acrylamide-Bisacrylamide Mix, 12.65 ml ddH_2O, 200 μl of 10% APS, 20 μl TEMED. Add APS and TEMED immediately before pouring the gel.
6. ColorPlus^{TM} Prestained Protein Marker (New England Biolabs GmbH, Frankfurt, Germany).
7. Low-melting agarose (Biozym, Hessisch Oldendorf, Germany).

3 Methods

3.1 Preparation of Whole Protein Extract

1. Harvest 2×10^{11} cells (approx. 500 ml at $OD_{600} = 0.5$) and pellet by centrifugation at 4470×*g* for 30 min at 4 °C.
2. Resuspend the cell pellet in 40 ml of 1× ice-cold PBS, then transfer the suspension to a 50 ml falcon tube, and centrifuge at 10,304×*g* for 10 min at 4 °C. Wash once more with 40 ml ice-cold PBS. At this point the pellet can be frozen and kept at − 20 °C or − 80 °C for several weeks.
3. Resuspend the pellet in 2 ml of ice-cold sonication/equilibration buffer. Then add 20 μl of a 100× stock protease inhibitor mix and 0.6 μl benzonase (calculated to have approx. final concentration of 70 U/ml). In addition, add 2 g of glass beads (diameter = 0.25–0.50 mm).
4. Disrupt the cells by sonication and bead vortexing. Set the sonicator power to 90%, 0.5 s ON–0.5 s OFF cycle. Then repeat the process five times: 30 s sonication → 1 min on ice → 30 s vortexing (*see* **Note 3**).
5. Incubate sample on ice for 1 h. This step is required for the benzonase to degrade the cellular DNA and RNA, which would otherwise interfere with the purification process.
6. Transfer supernatant to one or more 2 ml microtubes, centrifuge at 16,060×*g* for 30 min at 4 °C to get rid of cell debris. Transfer the supernatant to a new tube and centrifuge again for 10 min. Transfer the protein extract to a new tube and keep on ice until purification (have the microtubes pre-chilled on ice before each transfer).

3.2 Heparin Based Purification (Using HPLC)

1. Equilibrate the HiTrap Heparin HP column 1 ml with 10 CV (= Column Volumes) of equilibration buffer, flow rate 1 ml/min.
2. Wash the injection loop with 5 ml equilibration buffer. Then slowly load 2 ml of the protein sample using a 3 ml syringe. Flicker the syringe to get rid of small bubbles. Be careful not to introduce any bubbles into the loop. Upon injection into the column they could disrupt the gel bed and affect the column efficiency and reproducibility (*see* **Note 4**).
3. Start a gradient purification method comprising of the following steps:
   (a) Equilibration step: 5 CV Buffer A (flow 1 ml/min)
   (b) Injection of sample (volume 2 ml, flow 0.5 ml/min)
   (c) 5 CV Buffer A (= equilibration buffer) (flow 0.5 ml/min)
   (d) 6 CV linear gradient 0–70% target B (Buffer B = Buffer A with 1 M NaCl)
   (e) 0 CV step 100% B
   (f) 1 CV 100% B (used to wash off any remaining sample in the column)
4. Collect 1 ml fractions.
5. Pull together the peak fractions eluting at approx. 40% B (*see* **Note 5**) (with these settings, peak elutes at fractions 8 and 9) (*see* **Note 6**).
6. Transfer the fractions to an ultrafiltration column (Molecular Weight Cutoff 50 kDa, 15 ml). Fill up to 15 ml with Buffer 0.
7. Concentrate by centrifugation in a swing rotor centrifuge at 3,990 × *g*, for 1 h, at 4 °C. Remove flow-through and add 5 ml of Buffer 0 to the column. The final salt concentration in the sample should be ≤ 50 mM NaCl. Higher salt concentrations could affect the sample run in the BN-PAGE.
8. Mix the sample by inverting the column three to five times.
9. Centrifuge as in **step 7** until the final volume of the concentrated sample is approx. 200 μl.
10. Transfer the protein sample into a chilled 1.5 ml microtube and keep on ice.
11. Measure the protein concentration of the now concentrated and desalted sample.
12. At this point the sample can be frozen and stored at − 20 °C.
13. If multiple use of the sample is required and longer storage is intended (up to several months), the buffer of the sample from **step 9** can be exchanged to RNAP storage buffer. This buffer also allows for preservation of RNAP activity.

14. Simply add 2× 1 ml of the RNAP storage buffer to the sample from **step 9** and centrifuge as in **step** 7 to get to a final sample volume of approximately 100–200 μl.
15. Measure the protein concentration of the concentrated samples using the Pierce® BCA Protein Assay Kit, according to the Microplate Procedure. Measure absorbance at 562 nm with a microplate reader.

3.3 Pouring of BN-PAGE Gel (0.75 mm Thick)

This part of the protocol was performed similarly to the protocol in refs. [3, 4] (*see* **Note** 7).

1. For loading large gels when no gradient mixer is available, use a bottle-tube-bottle system, as described below (for clarity also check Fig. 1):
2. Place 2× 50 ml bottles on the same level (on same magnetic stir plates). Put a small stirrer in the bottle close to the pump. Connect tubing from the bottle to the pump, and from the pump to the gel system.
3. Prepare the 4 and 15% separating gel solutions. Add the APS and TEMED immediately before use.
4. Pour 13.5 ml from each of the gel solutions into the corresponding bottles (4% into the bottle not connected to the pump and 15% into the bottle connected to the pump via tubing and having a small magnetic stirrer inside).
5. Connect a flexible silicone or PVC tubing (inner diameter = 3.2 mm) to a syringe and insert the tubing into the bottle with low percent gel (4%). Prime the tubing with the 4% gel solution. Squeeze the tubing tightly with your fingers such that the gel solution does not flow and insert the end that was connected to the syringe into the high percent gel solution (15%), while slightly lifting the bottle with low percent gel solution (4%) until there are no bubbles left in the tubing and the two gel solutions are in contact.
6. Switch on the magnetic stirrer, making sure no bubbles are created. Switch on the peristaltic pump, set to a flow rate of 10 ml/min, and cast the gel.
7. Allow the whole liquid to enter the gel apparatus. If small bubbles form in the gel, gently tap the glass plates to release them to the surface. Leave space for approx. 1 cm of stacking gel from the bottom of the gel wells. When casted, overlay the gel gently with *n*-butanol (*see* **Note 8**).
8. Clean the pouring apparatus with ddH_2O immediately after use.
9. Allow the gel to polymerize for at least 30 min at room temperature.

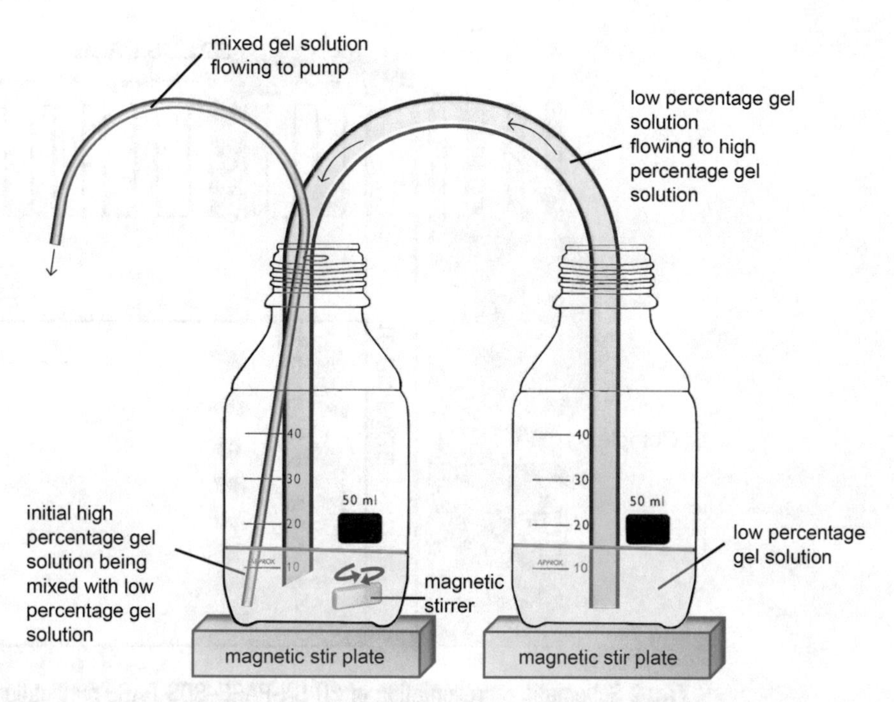

Fig. 1 Schematic representation of bottle-tube-bottle system for preparing gradient gels without use of gradient mixer. Two 50 ml bottles are placed on the same level (on same magnetic stir plates). In each of them equal volumes of two gels solutions with different acrylamide concentrations (%) are added. The lower percentage (%) gel solution is in contact with the higher percentage (%) gel solution via silicone or PVC tubing. The higher % gel solution is placed closer to the pump (*left panel*) and the lower one further away from it (*right panel*). The higher % gel solution is connected to the pump via thin silicone tubing. In addition, a magnetic stirrer is added to this bottle to mix the two gel solutions, creating a gradient of concentrations as the mixed gel solution flows away to the pump and is being casted continually into the gel plates, and the lower % gel solution flows into the mix

10. Remove the *n*-butanol, wash with ddH_2O, pour out the ddH_2O and dry the surface with a piece of thin Whatman filter paper.
11. Prepare 3.2% stacking gel, adding APS and TEMED immediately before use.
12. Pour the stacking gel on top of the separating gel and introduce the comb between the glass plates, avoiding bubbles. Best done with placing it under an angle, from one end of the comb first, allowing the bubbles to exit from the other end.
13. Immediately before sample loading, slowly remove the comb. Twist it back and forth while pulling straight upward. In this way it is easily separated from the glass slides, and the gel wells are not disturbed. If disturbance happens, straighten disturbed wells with a thin spatula that fits between the glass plates.

3.4 Separation of Concentrated Heparin Fractions by BN-PAGE

1. Wash the gel wells with cathode buffer using syringe and needle to get rid of acrylamide pieces.

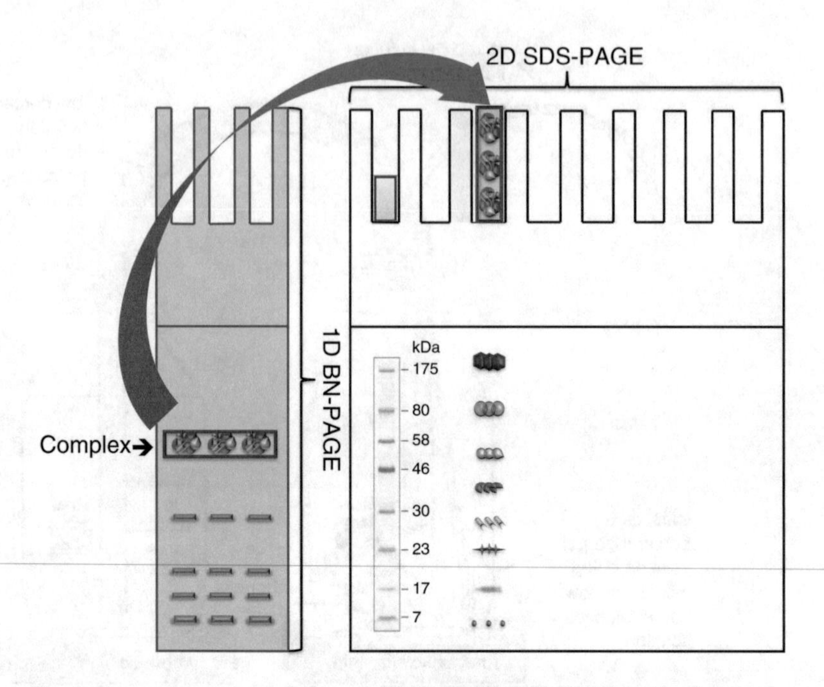

Fig. 2 Schematic representation of 2D BN-PAGE–SDS-PAGE purification procedure of RNAP supramolecular complex. First the Heparin-purified sample is loaded in triplicate in the first dimension BN-PAGE. The band containing RNAP is separated from contaminant proteins and complexes, cut out and placed in a second dimension SDS-PAGE to separate the different components of the assembly

2. Then load 10 μl Native protein marker in two control wells (best on both sides of the gel, easier for later cutting) and load 3× 50 μg of each sample mixed with 4× Native Sample Buffer in three consecutive wells (*see* **Note 9**).
3. Fill the inner chamber with Cathode Buffer and the outer/lower chamber with Anode Buffer.
4. Apply 150 V and run until the dye front reaches the bottom of the gel (8 h, for 12 cm long, 4–15% gel) (*see* **Notes 10** and **11**).
5. The BN-PAGE–SDS-PAGE procedure for determining the RNAP composition is presented in a schematic way in Fig. 2. First the Heparin-purified sample is loaded in triplicate in the first dimension BN-PAGE. The band containing RNAP is separated from contaminant proteins and complexes, cut out and placed in a second dimension SDS-PAGE to separate the different components of the assembly.
6. Real example of the first dimension BN-PAGE, with wild type sample and purified RNAP control is shown in Fig. 3a.

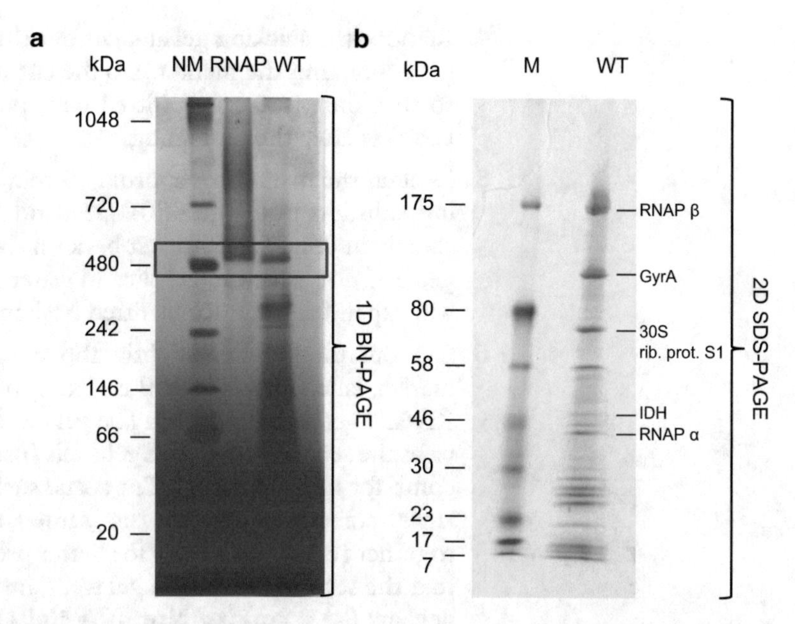

Fig. 3 (**a**) Real example of the first dimension BN-PAGE, with wild type RNAP supramolecular complex purified from exponentially growing CSH50 *Escherichia coli* cells and purified RNAP control, running just above the 480-kDa band. (**b**) Second dimension SDS-PAGE with the RNAP supramolecular complex denatured into its individual components. The polymerase components β and α (RNAP β and RNAP α, respectively), DNA gyrase subunit A (GyrA), small ribosomal proteins, e.g., 30S ribosomal protein S1 (30S rib. prot. S1) and metabolic enzymes such as isocitrate dehydrogenase (IDH) as identified by mass spectrometry, are indicated on the gel

7. For loading large gels when no gradient mixer is available, use a bottle-tube-bottle system, as described below (for clarity also check Fig. 1).

3.5 Second Dimension SDS-PAGE Gel (1 mm Thick)

1. Prepare 1 mm thick, 8–15% gradient SDS-PAGE gel with acrylamide-bisacrylamide ratio of 32:1 (*see* **Note 12**). It is very important that this gel is thicker than the first dimension BN-PAGE for the gel fitting to work smoothly. The BN-PAGE gel pieces swell during the processing procedure (*see* **Note 13**). The stacking gel should be at least 3 cm long, measuring from the bottom of the wells, to allow the proteins to exit the first dimension gel and stack nicely, before entering the separating gel of the second dimension.
2. Remove the BN-PAGE gel in the plates from the electrophoresis apparatus and gently pry up one plate.
3. For better visualization during the processing procedure, place the glass slide with the gel lying on top of a prolite basic light box.

4. Remove the stacking gel and cut out the lane of the BN-PAGE gel containing the marker. Do the cutting in a zig-zag manner so that the piece can be fitted back perfectly into its original position after the destaining.
5. Destain the marker for approx. 10 min in a coomassie destaining solution until the 480-kDa band becomes visible. Rinse shortly in dH_2O and place it back on the glass slide next to the gel. Do not leave the gel slice in water for too long because it will expand and cannot be fitted back into its original position.
6. Cut out the gel area slightly above and below the 480-kDa marker lane (in this area the RNAP band runs—use purified RNAP as control, Fig. 3a). The gel piece width should be exact to fit the second dimension gel wells (use the second dimension comb for measurement). Cut a lane such that the three sample lanes corresponding to the same biological sample stay together (cut along a ruler for better precision). It is important that the second dimension gel wells are long enough to fit this gel lane (≥ 2 cm) (*see* **Notes 14** and **15**).
7. Place the BN-PAGE gel slices in 2 ml microtubes with 500 μl SDS Sample Buffer containing freshly added β-mercaptoethanol and incubate for 10 min at room temperature. Perform **steps 7–13** under the hood to prevent inhalation of β-mercaptoethanol.
8. Boil the BN-PAGE gel slices briefly (2× 10 s) in a microwave. Do not extend boiling since the microtubes might burst.
9. Incubate the BN-PAGE gel slice in the hot SDS Sample Buffer for another 15 min at room temperature.
10. Load the BN-page slices in the wells of the second dimension gel and overlay all of the gel wells with pre-melted 0.5% low melting agarose in 1× SDS running buffer (0.1 g low melting agarose in 20 ml buffer is sufficient). This step is needed to keep the first dimension gel pieces in place during the gel run (otherwise they float away from the gel wells).
11. Apply 15 μl protein marker on a piece of Whatman filter paper and insert it in one of the wells (*see* **Note 16**).
12. Allow the low melting agarose to solidify for a few minutes.
13. Place the gel in the chamber, fill with 1× SDS running buffer. Apply 150 V and run until the dye front reaches the bottom of the gel (approx. 12–14 h, for 12 cm long, 8–15% gel) (*see* **Note 17**).
14. As a final representative result, we show the RNAP complex purified from exponentially growing CSH50 *Escherichia coli* cells. The second dimension SDS-PAGE with the RNAP complex denatured into its individual components is shown in Fig. 3b. The polymerase components β, α, and other associated proteins were identified using MS or MS/MS analyses

(as in ref. [11]). In total by this method we identified 17 different proteins associated with RNAP, and some of them are indicated in the figure.

4 Notes

The most critical troubleshooting points are mentioned in the protocol itself. However here we present a list with more detailed explanations for some of them.

1. The NaCl concentration of the sonication/equilibration buffer might need to be empirically optimized if there is no binding of the RNA polymerase to the column. It is always best to compare the conductivity of the buffer (the ability to transmit an electrical charge) because it reflects best the amount of ions in the buffer.
2. The mentioned order in the preparation procedure is mandatory in order to form the coomassie colloids, which greatly increase the stain sensitivity.
3. If you have several samples for sonication, you can first sonicate one, leave it to incubate on ice and continue with sonicating the other samples, always making sure to clean the sonicator with ethanol and clean tissue in between. Once this cycle is over, proceed with vortexing each sample and sonicating again. In this manner the incubation cycles will be the same for all samples although slightly longer than 1 min. Do sonication of maximum four to five samples at a time. If you have more samples, make a separate batch and repeat the whole process again.
4. To load the sample into the sample loop of HPLC it is always better to use bigger sample volume than the volume of the loop, since some of the sample through capillary force runs out into the waste. The sample loop should always be properly primed with sonication/equilibration buffer using a syringe, carefully without inserting any bubbles. When only a limited sample volume is available, small amount of sonication/equilibration buffer is taken from a clean microtube into the sample-containing syringe so that it primes the syringe needle, preventing big sample loss. In addition, sample loading should be very slow and continuous, pressing millimeter by millimeter. This prevents bubble introduction and sample loss due to capillary forces.
5. Theoretical concentration calculated from the software, due to gradient delay of approx. 2 CV, the actual salt concentrations is approx. 530 mM NaCl.

6. The peak elution might change, depending on the gradient settings. It is recommendable to check in which fractions the RNAP β or β′ subunit elutes by running a simple SDS-PAGE gel and pull those fractions together for further concentration.
7. The BN-PAGE solutions are prepared without addition of ε-aminocaproic acid (as given in ref. [3]). It is not needed if protease inhibitors are added to the samples.
8. For the gel system that we used, it was recommended to use *n*-butanol instead of isopropanol. Otherwise, both can be used.
9. Loading of a smaller protein amount into an individual well prevents sample aggregation during the gel run.
10. We perform the BN-PAGE run at room temperature, since it shortens the running time and gives higher reproducibility. We did not see changes in the RNAP composition between the samples run at 4 °C and at room temperature.
11. The 4–15% BN-PAGE is suitable for separation of proteins with a size range of 15,000–10,000 kDa.
12. The 32:1 Acrylamide-bisacrylamide ratio is crucial for both the BN-PAGE and SDS-PAGE. Previous to the BN-PAGE we tested other gel systems with standard Acrylamide-bisacrylamide ratio of 37.5:1. Proteins could not exit the first and enter the second dimension gel, although they were separating in the first dimension. In the SDS-PAGE this ratio helps during peptide extraction. In addition, the usage of BN-PAGE produced clearer bands than other gel systems, due to the presence of Coomassie G-250 in the cathode buffer.
13. It is important for the second dimension gel to be slightly thicker than the first dimension gel, since the gel pieces swell during the processing procedure. We use 0.75 mm gel for the first and 1 mm gel for the second dimension. In previous trials 1 and 1.5 mm gels were used. However, the 1.5 mm gel is not optimal for the in-gel trypsin digestion and peptide extraction. The gel matrix is tremendously increased for the same amount of peptides present, compared to the 1 mm gel.
14. The cutting of the RNAP band from the BN-PAGE might need optimization too. Prior to proceeding with this procedure, make a control BN-PAGE (*see* **Note 15**) that can be stained and de-stained to visualize the level at which the RNAP band runs. At different growth stages or in different cell types the size and conformation of the RNAP complex might vary, therefore it might run at different levels. Note that the marker bands in a native gel do not reflect the exact protein size, since the protein run is dependent not only on the size but also on the conformation. The marker bands in this case are used for visual control and approximate orientation.

15. The control BN-PAGE for destaining is prepared in exactly the same manner as described before, with Coomassie G-250 still present in the cathode buffer. However the gel needs to stain for approx. an hour with Coomassie staining solution, to stain and fix the proteins nicely, since the coomassie colloids have higher staining sensitivity. Destaining of the BN-PAGE is possible with long incubations of at least 3 h in a coomassie destaining solution, even though the top part of the gel remains slightly blue. To enhance color contrast, incubate destained gel for 10 min in ddH_2O before imaging. For longer storage, soak gels in 5% acetic acid O/N and then store at 4 °C wrapped in plastic foil.
16. The protein marker is applied on a piece of Whatman filter paper and then inserted into a well, such that it can get inserted into a gel well that is already pre-filled with low melting agarose (before it solidifies), as mentioned in Subheading 3.5, **step 11**. Low melting agarose is applied everywhere to keep running conditions across the gel wells more uniform.
17. The 8–15% SDS-PAGE is suitable for nice separation of proteins with a size range of 7–250 kDa.

5 Disclosures

No conflicts of interest declared.

Acknowledgements

This work was supported by the Deutsche Forschungsgemeinschaft (DFG MU 1549/7-1). The authors thank Steffi Teresa Jimmy and Zharko Daniloski (respectively carrying out their MSc and BSc thesis work in the lab) who participated in the development of the method and used it to generate results.

References

1. Geertz M, Travers A, Mehandziska S, Janga SC, Shimamoto N, Muskhelishvili G (2011) Structural coupling between RNA polymerase composition and DNA supercoiling in coordinating transcription: a global role for the omega subunit? MBio 2(4):e00034-11
2. Jin DJ, Cagliero C, Zhou YN (2013) Role of RNA polymerase and transcription in the organization of the bacterial nucleoid. Chem Rev 113(11):8662–8682. doi:10.1021/cr4001429
3. Swamy M, Siegers GM et al (2006) Blue native polyacrylamide gel electrophoresis (BN-PAGE) for the identification and analysis of multiprotein complexes. Sci STKE 2006(345):pl4
4. Fiala GJ, Schamel WWA, Blumenthal B (2011) Blue Native Polyacrylamide Gel Electrophoresis (BN-PAGE) for analysis of multiprotein complexes from cellular lysates. J Vis Exp (48): e2164. doi:10.3791/2164
5. Schägger H, von Jagow G (1991) Blue native electrophoresis for isolation of membrane

protein complexes in enzymatically active form. Anal Biochem 199:223–231
6. Schägger H, Cramer WA, von Jagow G (1994) Analysis of molecular masses and 20 oligomeric states of protein complexes by blue native electrophoresis and isolation of membrane protein complexes by two-dimensional native electrophoresis. Anal Biochem 217:220–230
7. Hachmann JP, Amshey JW (2005) Models of protein modification in Tris–glycine and neutral pH Bis–Tris gels during electrophoresis: effect of gel pH. Anal Biochem 342:237–245
8. Baslé E, Joubert N, Pucheault M (2010) Protein chemical modification on endogenous amino acids. Cell Chem Biol 17(3):213–227
9. Schägger H (2001) Blue native gels to isolate protein complexes from mitochondria. Methods Cell Biol 65:231–244
10. Dyballa N, Metzger S (2009) Fast and sensitive colloidal coomassie G-250 staining for proteins in polyacrylamide gels. J Vis Exp (30):e1431. doi:10.3791/1431
11. Trypsin gold, mass spectrometry grade. In-gel digestion protocol. Promega Technical Bulletin, Part # TB309 4–5 (revised 3/13)

Chapter 10

A Chromosome Co-Entrapment Assay to Study Topological Protein–DNA Interactions

Larissa Wilhelm and Stephan Gruber

Abstract

Chromosome organization, DNA replication, and transcription are only some of the processes relying on dynamic and highly regulated protein–DNA interactions. Here, we describe a biochemical assay to study the molecular details of associations between ring-shaped protein complexes and chromosomes in the context of living cells. Any protein complex embracing chromosomal DNA can be enriched by this method, allowing for the underlying loading mechanisms to be investigated.

Key words Agarose plug, Cysteine cross-linking, HaloTag labeling, DNA entrapment, Chromosome organization, Beta-clamp, Smc-ScpAB, BMOE

1 Introduction

Many essential processes depend on large protein assemblies acting on chromosomal DNA, such as DNA replication mediated by the replisome sliding along chromosomal DNA. To understand how these processes work at a mechanistic level, it is elementary to study the nature of association between the involved proteins and chromosomes. From a general perspective, there are two ways for proteins to associate with DNA: via a direct physical contact between protein and DNA, and by a topological interaction, i.e., by ring-shaped protein complexes encircling one or more DNA double helices. Examples for the latter case include DNA translocases (helicase acting in DNA replication and repair, FtsK in chromosome segregation), replicative sliding clamps, DNA topoisomerases, DNA repair proteins (such as MutS), and SMC protein complexes [1–8]. Often the establishment of topological associations with DNA relies on sophisticated and tightly regulated loading processes requiring loading factors and dedicated mechanisms for protein ring opening and closure. To study such processes occurring on the native substrate, i.e., the replicating chromosome, we have established an assay based on the co-entrapment of

Olivier Espéli (ed.), *The Bacterial Nucleoid: Methods and Protocols*, Methods in Molecular Biology, vol. 1624,
DOI 10.1007/978-1-4939-7098-8_10, © Springer Science+Business Media LLC 2017

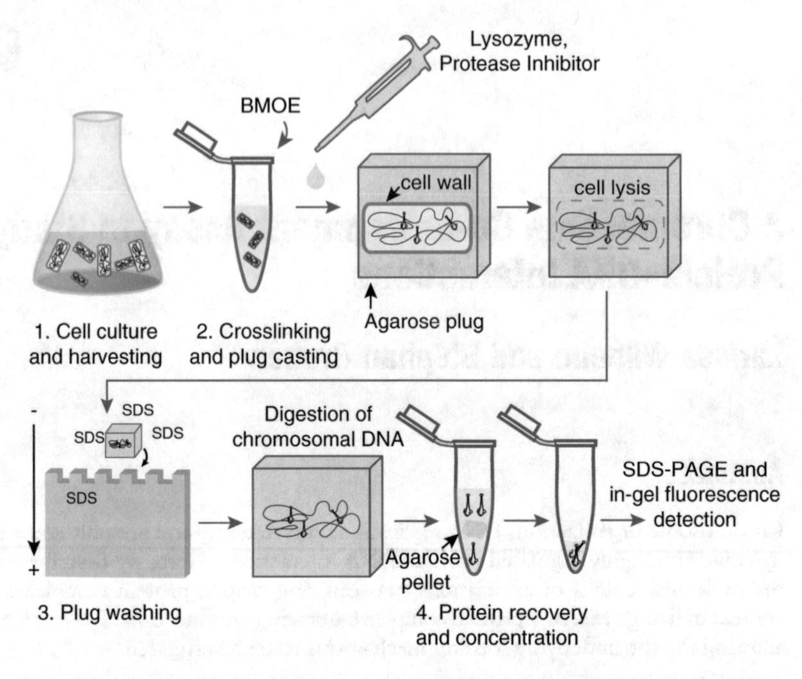

Fig. 1 A scheme for the chromosome co-entrapment assay. After harvesting, cells are cross-linked with BMOE, cast into agarose plugs and subsequently lysed. The agarose plug is loaded onto a SDS-PAGE gel for protein removal. Thereafter, chromosomal DNA is digested and all remaining proteins are recovered. The proteins are concentrated and analyzed by SDS-PAGE

proteins with intact chromosomal DNA in agarose plugs [9]. The assay principle is simple (*see* Fig. 1): Proteins are cross-linked into covalently closed rings in vivo, chromosomal DNA is isolated in agarose plugs under stringent conditions, and stably co-entrapped proteins are recovered and analyzed.

During the co-entrapment assay, topological protein–DNA interactions are preserved via the formation of covalent protein rings by site-specific protein cross-linking. To do so, pairs of cysteine residues are introduced at all protein–protein ring interfaces so that they can be specifically cross-linked by a thiol-specific compound (such as BMOE). The design of appropriate pairs of cysteines is relatively straightforward when structural information is available; in the absence of detailed information, cysteine pairs have also been identified by screening approaches [10]. Ideally, cysteine mutants are incorporated into the endogenous locus by allelic replacement; alternatively, they can be expressed ectopically. After cross-linking, bacterial cells are immobilized in agarose plugs—protecting the chromosomal DNA from any shearing forces—and enzymatically lysed, similar to the preparation of genomic DNA in agarose plugs for pulsed-field electrophoresis [11]. The agarose plugs are then exposed to harsh protein denaturing conditions (i.e., incubation with SDS in an electric field). During

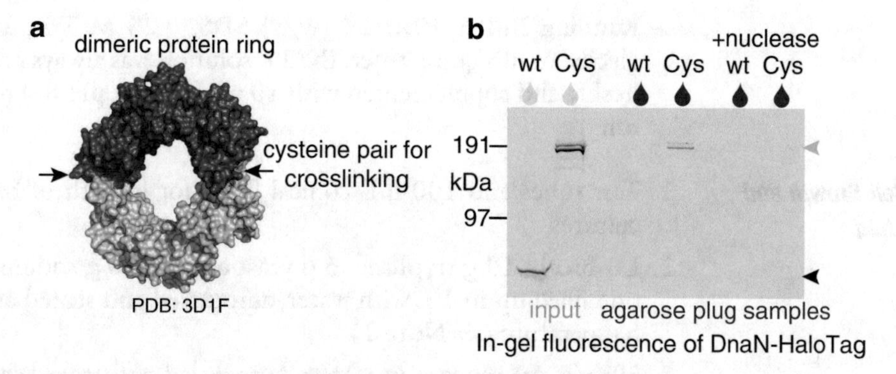

Fig. 2 (**a**) Isolation of the DNA sliding clamp by chromosome co-entrapment. Crystal structure of DnaN (DnaN monomers in *light* and *dark blue*, respectively). A pair of cysteine residues at the dimerization interface allows cross-linking of DnaN into a covalently closed ring (panel **a** modified from (**9**)). (**b**) SDS-PAGE analysis of chromosome co-entrapment samples of DnaN. On the *left* side protein extracts from cross-linked cells are shown ("input," *light blue droplet*). Wild-type protein results in monomer species only (*black arrowhead*). DnaN (Cys) protein usually produces a weak monomer band (*black arrowhead*) as well as single and double cross-linked dimer species of higher intensities (*blue arrowhead*). On the right side the agarose plug samples are depicted (*dark blue droplet*). Only cysteine cross-linked species are recovered from agarose plugs. The co-entrapment is lost when samples are treated with nuclease during cell lysis

the final step, stably entrapped proteins are released from chromosomes by DNA digestion and recovered from the agarose (*see* Fig. 2). The proteins are then ready for analysis by SDS-PAGE.

So far, the co-entrapment assay was successfully applied to two protein complexes in *B. subtilis*: Smc-ScpAB and the β-sliding clamp DnaN [9]. The assay should be readily applicable to other protein structures (including DNA helicases, topoisomerases and polymerases) in *B. subtilis* and other bacteria. In principle, similar approaches should also be feasible for eukaryotic chromosomes.

2 Materials

All solutions used for this assay are prepared using ultrapure water (PURELAB Ultra, 18.2 MΩ cm).

2.1 SDS-PAGE

1. SDS-PAGE gel (Bio-Rad Mini PROTEAN Tetra Cell System): with 1.5 mm spacer plates and 9-well combs for 1.5-mm gels.
2. Resolving gel buffer (10 mL per gel): 6% (w/v) acrylamide–bis solution; 0.39 M Tris–HCl, pH 8.8; 0.1% (w/v) SDS; 0.1% (w/v) APS; 0.1% (v/v) TEMED in ultrapure water.
3. Stacking gel buffer (5 mL per gel): 6% (w/v) acrylamide–bis solution; 0.125 M Tris–HCl, pH 6.8; 0.1% (w/v) SDS; 0.3% APS; 0.1% (v/v) TEMED in ultrapure water.

4. Running Buffer 10×: 1% (w/v) SDS, 0.25 M Tris, 1.92 M glycine in ultrapure water, the 1× solution was always prepared freshly and supplemented with 10 mM EDTA pH 8.0 prior to run.

2.2 Cell Growth and Harvesting

1. Test tubes and 100-mL conical flasks for growth of bacterial cultures.
2. LB-broth: 10 g tryptone, 5 g yeast extract, 10 g sodium chloride filled up to 1 L with water, autoclaved and stored at room temperature (*see* **Note 1**).
3. 50% (w/v) glucose in water: Autoclaved and stored at room temperature.
4. 37 °C incubator with shaking function.
5. Photometer and plastic cuvettes.
6. 50 mL plastic tubes for harvesting.
7. PBSG buffer: 1× phosphate buffered saline (PBS): 137 mM NaCl, 2.7 mM KCl, 10 mM Na_2HPO_4, 1.8 mM KH_2PO_4, pH adjusted to 7.4, supplemented with 0.1% (w/v) glycerol, stored at 4 °C, kept on ice for 2–3 h prior to cell harvesting.

2.3 Cross-Linking and Preparation of Input Samples and Agarose Plugs

1. Bis-maleimidoethane (BMOE): 20 mM stock solution in DMSO, solution can be stored at −20 °C but should not undergo multiple freeze–thaw cycles, final concentration is 1 mM (*see* **Note 2**).
2. 2-mercaptoethanol: 572 mM stock solution in PBSG, solution is prepared freshly, final concentration for quenching the cross-linking reaction is 28.6 mM (*see* **Note 3**).
3. Low melting agarose (Bio-Rad): 2% (w/v) agarose in ultrapure water, prepared freshly and equilibrated at 70 °C for max. 1 h before the casting of agarose plugs.
4. Enzyme master mix 1 (per input sample): 400 units Ready-Lyse Lysozyme, 12.5 units benzonase, 1 μM HaloTag Oregon Green substrate (*see* **Note 4**), 1× Protease inhibitor cocktail in a total volume of 5.9 μL PBSG. Always prepare freshly and keep on ice protected from light.
5. Enzyme master mix 2 (per agarose plug): Same as enzyme master mix 1 but *without* benzonase.
6. Agarose plug molds (Bio-Rad): Before use seal the bottom with tape.
7. 2× Sample Loading Buffer: 200 mM DTT and 2× LDS Sample Buffer (NuPage) in PBSG.
8. Thermo mixer at 70 °C.
9. Parafilm

2.4 Plug Washing

1. Wash Buffer: 0.01 mM EDTA, pH 8.0; 0.5 mM Tris, pH 7.5; 0.5 mM $MgCl_2$, 0.01% (w/v) SDS in ultrapure water.
2. Benzonase mix: 10 units/μL benzonase dilution in Wash Buffer.
3. Freezer at −80 °C (if not available, −20 °C can be used instead).

2.5 Recovery of Proteins from Agarose Plugs

1. 0.45 μm cellulose acetate spin columns (Costar, nonsterile).
2. SpeedVac Concentrator (Thermo Scientific): Run conditions: 2.5 h, no heating, vacuum setting 5 Torr/min.
3. 1× Sample Loading Buffer: 100 mM DTT in 1× LDS Sample Buffer (NuPage) diluted in ultrapure water.
4. SDS-PAGE gel: 3–8% Novex Tris acetate gel (Thermo Scientific) or 4–12% Novex Bis–Tris gel (*see* **Note 5**).
5. Laser scanner (Typhoon FLA9000, GE Healthcare; or similar) for the detection of fluorescently labeled proteins (HaloTag).

3 Methods

3.1 Cell Culture and Harvesting

1. Inoculate *B. subtilis* strains from a −80 °C glycerol stock or from a fresh agar plate into a test tube containing 5 mL of L-broth supplemented with 0.5% glucose. Incubate overnight at 37 °C with agitation (*see* **Note 6**).
2. Measure the OD_{600} of the overnight culture and dilute the culture in 20 mL of fresh L-broth (in a 100-mL conical flask) to an OD_{600} of 0.005. Incubate at 37 °C with agitation until the culture reaches an OD_{600} of 0.2–0.3 (mid-exponential phase).
3. Prepare PBSG buffer and keep on ice. Prepare 50-mL tubes for harvesting. Fill each tube with 6 g of ice and keep at −20 °C (*see* **Note 7**).
4. Once the culture has reached the desired optical density, pour the culture into the prepared 50-mL tube with ice. Centrifuge the cells at 14,000 × *g* for 2 min at room temperature. Discard the medium and wash the cell pellet with 10 mL ice-cold PBSG buffer by carefully pipetting the cells up and down. Centrifuge the suspension again, and finally resuspend the cells in 1 mL ice-cold PBSG buffer. Transfer the cell suspension into a 1.5-mL tube and place on ice until all samples have been collected. After harvesting, cells are always kept on ice and each step is carried out swiftly and gently to prevent premature cell lysis.
5. Thaw the BMOE stock solution and prepare the 2-mercaptoethanol dilution freshly. Keep both solutions at

room temperature. The 2% low-melt agarose should be prepared and equilibrated at 70 °C.

6. To ensure equal biomass in every sample, remeasure the OD_{600} of the cell suspension by dilution of 10 μL with 990 μL PBSG in a cuvette (*see* **Note 8**). Transfer an equivalent of 1 mL culture at $OD_{600} = 3.75$ to a fresh 1.5-mL tube. Centrifuge at 10,000 × *g* for 1 min at 4 °C. Discard the supernatant and resuspend the pellet in 120 μL PBSG (expected $OD_{600} = 31.25$).

3.2 Cross-Linking and Casting of Agarose Plugs

1. Add 6 μL of the 20 mM BMOE-stock solution to 120 μL of cell suspension (final concentration 1 mM) and briefly mix by pipetting up and down. Incubate for 10 min on ice.
2. Meanwhile prepare the enzyme master mix 1 and 2 and keep the solutions on ice, protected from light (*see* **Note 9**).
3. To quench the cross-linking reaction add 6.3 μL of the 560 mM 2-mercaptoethanol solution to the cross-linked cells (final concentration 28.6 mM) and incubate for 3 or more minutes on ice.
4. The cell suspension should now be aliquoted into input samples and samples subjected to the entrapment assay (referred to as "agarose plug" samples). To do so, 44.1 μL of cell suspension is transferred to a new 1.5-mL tube, labeled with "input." For two agarose plugs, take 88.2 μL of the cell suspension into a new 1.5-mL tube.
5. For plug preparation all steps should be carried out as quickly as possible—one sample at a time (*see* **Note 10**). Before starting, prepare a pipet tip (for 100 μL volumes) by cutting the tip with scissors, so that the opening has a diameter of 2–3 mm (one tip needed per plug).
6. Place the first cell suspension in a 37 °C heat block for 10 s before adding 11.8 μL of the enzyme master mix 2 to a total volume of 100 μL. Mix by pipetting up and down twice (*see* **Note 11**). Quickly add 100 μL of the 2% Low Melt Agarose equilibrated to 70 °C using a cut pipet tip. Pipet up and down twice carefully (to avoid air bubbles) and pipet the mixture into two wells of the agarose plug mold. Two agarose plugs are thus cast per sample (*see* **Note 12**).
7. Cover the agarose plug mold gently by Parafilm and place in a 37 °C incubator without agitation, protected from light. Incubate for 25 min, and then for 10 min at 4 °C to allow the plugs to solidify.
8. Meanwhile pipet 5.9 μL of enzyme master mix 1 to 44.1 μL of input samples, mix by pipetting up and down and incubate for 25 min at 37 °C without agitation, protected from light.

Finally, add 50 μL of the 2× Sample loading buffer, mix carefully by pipetting up and down and store input samples at −20 °C. Final volume of input samples is 100 μL (concentration of DTT is 100 mM).

3.3 Plug Washing

1. Prepare 1× Running Buffer containing 10 mM EDTA. Unpack a 6% polyacrylamide gel, remove the comb and rinse slots with PBSG (*see* **Note 13**).
2. The agarose plugs are nearly translucent due to complete cell lysis. Carefully release plugs from the plug mold onto a smooth surface (e.g., a glass plate). Next, carefully transfer the plug into a slot of the polyacrylamide gel. Make sure that the plug stays intact and that no air bubbles are trapped between gel and plug (*see* **Note 14**).
3. After loading, assemble the SDS-PAGE apparatus and fill with 1× Running Buffer. Let the gel run for 1 h at 25 mA (per gel), at room temperature, protected from light.
4. After electrophoresis, open the gel tank and carefully recover the plugs from the gel. Transfer each plug into a separate 1.5-mL tube. Add 1 mL Wash Buffer to each tube and incubate for 10 min at room temperature under gentle agitation, protected from light. Replace Wash Buffer once and incubate for another 10 min (*see* **Note 15**).
5. Remove Wash Buffer and add 100 μL fresh Wash Buffer per plug. Add 5 μL of the benzonase mix (50 units final) to each tube and incubate for 30 min at 37 °C, protected from light.
6. Place each tube containing one plug into a thermo mixer, pre heated to 70 °C. Let the tubes shake at 1400 rpm until the agarose plug is completely dissolved (*see* **Note 16**). Transfer tubes to ice and let cool down for 5 min. Store samples at −80 °C for at least 1 h (up to several days).

3.4 Protein Recovery and Concentration

1. Thaw the samples at room temperature. Centrifuge for 10 min at 10,000 × *g* at 4 °C to extract liquid from the agarose gel phase.
2. Pipet the mixture onto a cellulose acetate spin column, and centrifuge for 1 min at 10,000 × *g* at room temperature to remove any agarose clumps from the solution (*see* **Note 17**).
3. Discard the column and place the tubes with open lid into a SpeedVac concentrator. The concentration is performed in vacuum without heating. It takes about 2–3 h to remove all water.
4. Resuspend the pellet in 10 μL of 1× Sample loading buffer by carefully pipetting up and down (*see* **Note 18**).

5. The sample can now be stored at −20 °C or used for SDS PAGE analysis immediately as described below.
6. Load 10 μL input (=10% of agarose plug input) and 10 μL of the agarose plug sample (=100% of agarose plug) on a SDS-PAGE gel (*see* Fig. 2). Protect the gel from light during the run.
7. Wash the gel briefly in ultrapure water, place it on a Typhoon FLA-9000 imager and scan with Cy2-DIGE filter setup (for HaloTag Oregon Green labeled samples). If needed, the gel can be analyzed subsequently by Western blotting or Coomassie staining.
8. Exclusively protein species previously cross-linked into rings should be detectable in agarose plug samples. Some cross-link reversal might occur. In contrast, all protein species should be present in the input samples (*see* Fig. 2 and **Notes 19–21**).

4 Notes

1. This protocol is also applicable for *B. subtilis* cells grown in minimal salt medium (15 mM ammonium sulfate, 80 mM dipotassium hydrogen phosphate, 44 mM potassium dihydrogen phosphate, 3.4 mM trisodium citrate, 0.8 mM magnesium sulfate, 6 g/L potassium hydrogen phosphate supplemented with 5 g/L glucose, 20 mg/L tryptophan, 1 g/L glutamate).
2. We developed the assay using BMOE, but other sulfhydryl reactive cross-linkers such as dibromobimane (bBBr) or bis-methanethiosulfonate variants might work as well.
3. Prepare the 2-mercaptoethanol stock solution in a laminar air flow system to prevent inhalation.
4. Any other HaloTag substrate, e.g., HaloTag TMR Ligand (Promega) might be used instead.
5. For good resolution of cross-linked protein bands of high molecular weight we use commercially available 3–8% gradient precast gels on Tris/Acetate basis, run for 2.5 h at 4 °C and 35 mA per gel. For low molecular weight proteins we use 4–12% gradient precast gels on Bis/Tris basis and let them run for 1 h at room temperature at 200 V.
6. If more samples are needed the protocol can be scaled up. The culture volume should not exceed 20% of the total volume of the culture flask.
7. We usually fill a 50-mL tube with one third of crushed ice compared to the volume of the liquid cell culture (e.g., 6 g ice for a 20 mL culture) and store it at −20 °C for 2–3 h before cell harvesting. After transferring the cell culture into this tube, it will cool down quickly and stop all metabolism. Usually the

ice does not melt completely during centrifugation. This is no problem as only the cell pellet is needed in the next step and the remaining ice cubes and the supernatant will be discarded.

8. When measuring the OD_{600} of each harvested culture, mix the cells thoroughly in the cuvette and wait for 1–2 min before measuring.
9. If not using the HaloTag technique as read out, protection from light is not necessary.
10. The culture volumes in this protocol are sufficient for one input sample and two agarose plug samples. Additional samples may serve as a backup if problems during plug casting occur. Always select an agarose plug lacking air bubbles. All agarose plugs should have the same size.
11. Do not heat up the cells too long before adding the enzyme master mix 2 to prevent premature lysis of cells.
12. Casting agarose plugs without bubbles is critical for the assay. If not familiar with the agarose plug technique, perform test experiments to optimize the handling of agarose solutions and agarose plug molds.
13. The polyacrylamide gels for plug washing can be stored for several weeks at 4 °C, when kept in moist conditions.
14. When mounting the agarose plugs onto the polyacrylamide gel, make sure that the agarose plug is in tight contact with the gel. Prevent any air layer in-between. This is critical for efficient removal of proteins from the plug.
15. Avoid the loss of agarose pieces during the washing steps.
16. Try to heat and shake the samples as short as possible when melting the agarose. Normally complete melting of one agarose plug at 70 °C takes 1–2 min. Prolonged incubation at high temperature leads to BMOE cross-link reversal.
17. When transferring the sample onto a cellulose acetate column, make sure to transfer the entire content of the tube and do not leave any agarose clumps behind.
18. We observed that the samples are most sensitive to cross-link reversal during and after the concentrating step. Prevent prolonged heating steps and load or freeze samples immediately.
19. We use internal positive controls (e.g., DnaN) to make sure the co-entrapment assay worked in case of analysis of nonentrapping mutants.
20. As negative control, samples are treated with benzonase during plus casting to digest the chromosome. No proteins should remain in the plug after chromosome digestion.
21. The cross-linking reaction is never 100% efficient. Therefore, intermediate cross-linking products will appear in the input sample (*see* Fig. 2).

References

1. Bell SP, Kaguni JM (2013) Helicase loading at chromosomal origins of replication. Cold Spring Harb Perspect Biol 5:a010124
2. Costa A, Hood IV, Berger JM (2013) Mechanisms for initiating cellular DNA replication. Annu Rev Biochem 82:25–54
3. Sherratt DJ, Arciszewska LK, Crozat E et al (2010) The Escherichia coli DNA translocase FtsK. Biochem Soc Trans 38:395–398
4. Lenhart JS, Pillon MC, Guarné A et al (2016) Mismatch repair in Gram-positive bacteria. Res Microbiol 167:4–12
5. Hauk G, Berger JM (2016) The role of ATP-dependent machines in regulating genome topology. Curr Opin Struct Biol 36:85–96
6. Indiani C, O'Donnell M (2006) The replication clamp-loading machine at work in the three domains of life. Nat Rev Mol Cell Biol 7:751–761
7. Gligoris TG, Scheinost JC, Bürmann F et al (2014) Closing the cohesin ring: structure and function of its Smc3-kleisin interface. Science 346:963–967
8. Samel SA, Fernández-Cid A, Sun J et al (2014) A unique DNA entry gate serves for regulated loading of the eukaryotic replicative helicase MCM2-7 onto DNA. Genes Dev 28:1653–1666
9. Wilhelm L, Bürmann F, Minnen A et al (2015) SMC condensin entraps chromosomal DNA by an ATP hydrolysis dependent loading mechanism in *Bacillus subtilis*. Elife 4:1–18
10. Bürmann F, Shin H-C, Basquin J et al (2013) An asymmetric SMC-kleisin bridge in prokaryotic condensin. Nat Struct Mol Biol 20:371–379
11. Mawer JSP, Leach DRF (2013) Pulsed-field gel electrophoresis of bacterial chromosomes. In: Makovets S (ed) DNA electrophoresis: methods and protocols, vol 1054. Humana Press, Totowa, NJ, pp 187–194

Chapter 11

Tethered Particle Motion Analysis of the DNA Binding Properties of Architectural Proteins

Ramon A. van der Valk, Niels Laurens, and Remus T. Dame

Abstract

Architectural DNA binding proteins are key to the organization and compaction of genomic DNA inside cells. Tethered Particle Motion (TPM) permits analysis of DNA conformation and detection of changes in conformation induced by such proteins at the single molecule level in vitro. As many individual protein–DNA complexes can be investigated in parallel, these experiments have high throughput. TPM is therefore well suited for characterization of the effects of protein–DNA stoichiometry and changes in physicochemical conditions (pH, osmolarity, and temperature). Here, we describe in detail how to perform Tethered Particle Motion experiments on complexes between DNA and architectural proteins to determine their structural and biochemical characteristics.

Key words Tethered particle motion, DNA conformation, DNA bending, DNA stiffening

1 Introduction

1.1 Tethered Particle Motion

In Tethered Particle Motion assays a bead tethered to a surface via a DNA molecule exhibits restricted Brownian motion. The excursion of the bead reflects the properties of the DNA tether. The excursion of the bead changes in response to altered effective DNA stiffness or DNA contour length following the binding of proteins. The motion of the bead over time can be followed by video microscopy, tracked, and quantitated in terms of Root Mean Squared (RMS) displacement. A decrease or increase in effective DNA stiffness or DNA contour length is observed as reduction or increase in RMS respectively.

Tethered Particle Motion is a single-molecule technique, which provides information on the properties of individual tethered DNA molecules. Single-molecule techniques are a powerful addition to the conventional biochemistry toolkit as they permit analysis of populations of molecules without averaging, thus providing insight into the existence and characteristics of possible subpopulations. Different from more sophisticated single-molecule techniques

Olivier Espéli (ed.), *The Bacterial Nucleoid: Methods and Protocols*, Methods in Molecular Biology, vol. 1624,
DOI 10.1007/978-1-4939-7098-8_11,

(involving the application of force), TPM is easily implemented in any biochemistry lab; it only requires a microscope suitable for imaging of beads and bead tracking software. The associated costs are also low. Due to its simplicity it is also straightforward to integrate temperature control, a feature often overlooked in single-molecule instrumentation (Fig. 1). The absence of applied force, reducing the spatial resolution, is its main limitation compared to other single molecule manipulation techniques. It is therefore less suited for detailed quantitative analysis of DNA based motor enzyme activities. The activity of motor enzymes is usually investigated with single-molecule techniques capable of applying force on the tether such as optical or magnetic tweezers [1–4]. TPM is however a powerful tool to analyze active systems involving looped states of the DNA tether on the timescale of tenth of seconds to minutes and passive systems altering the properties of the DNA tether at equilibrium.

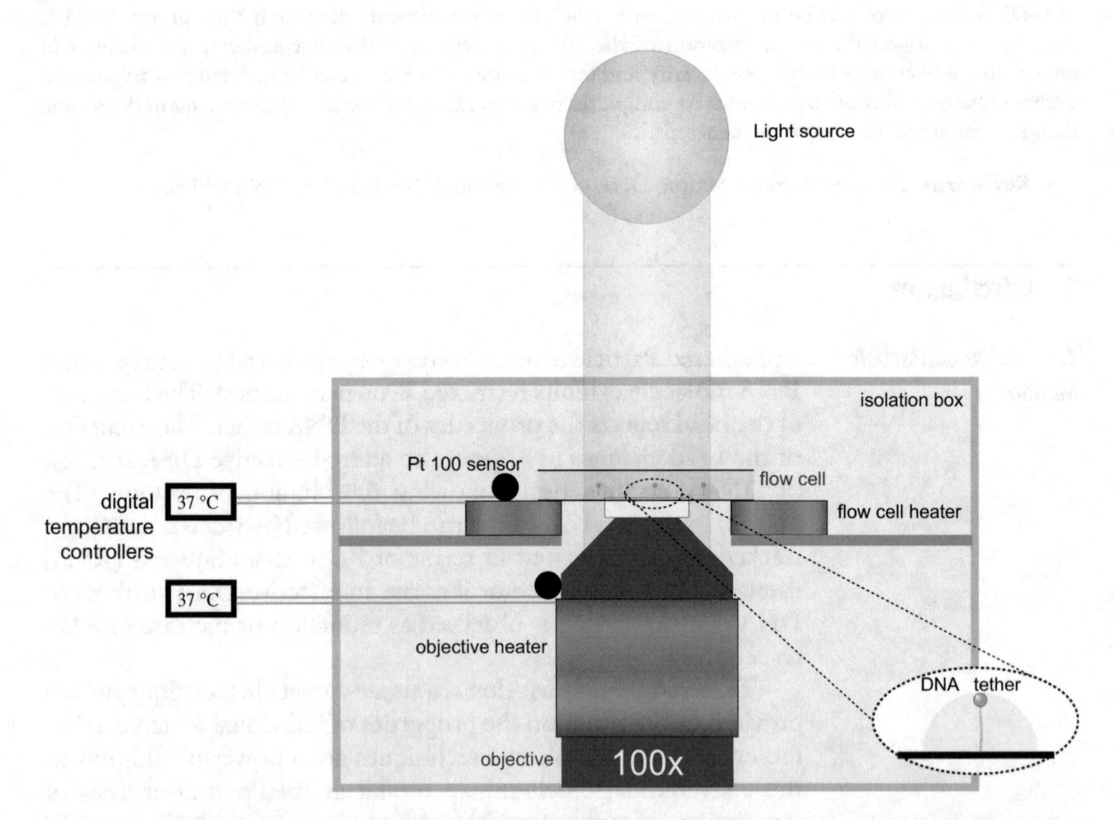

Fig. 1 Schematic overview of the Tethered Particle Motion setup instrument with temperature control. The temperature of the incubation chamber is maintained by heating the flow cell holder (by surrounding the location of the flow cell with a heating element consisting of 40 parallel resistors (~5.5 W)) and the objective (using an objective heating element (Bioscience tools TC-HLS-025)). The temperature is controlled by two SA200 PID digital temperature controllers, using pt100 sensors placed on the objective and on the edge of the flow cell heater. The system is insulated to ensure temperature stability [9]

1.2 Tethered Particle Motion Studies of Architectural Proteins

TPM has proven to be very useful in quantitatively analyzing the behavior of proteins capable of DNA looping. Using TPM, different looped states are observed as distinct RMS levels. Transition rates between these RMS levels yield quantitative insight in DNA looping kinetics [5–8]. The technique has also proven very useful in the characterization of the structural and biochemical properties of architectural proteins. TPM studies have shown that due to binding of proteins such as HU, Cren7, Sul7, Sso10a, NF-Y, and TFAM, DNA attains a compact conformation, reflected in reduced RMS [9–13]. The binding of some proteins—including HU and Sso10a (at high concentrations)—results in an increased RMS, interpreted as protein–DNA filament formation [10, 13, 14].

2 Materials

Prepare all solutions using ultrapure water (prepared by purifying deionized water, to attain a sensitivity of 18 MΩ cm at 25 °C) and analytical grade reagents. Prepare and store all stock solutions at −20 °C (unless indicated otherwise). You also need access to some routine biochemical techniques [15]

2.1 Stock Solutions

1. Coupling Buffer (CB): 10 mM Tris–HCl pH 7.5, 150 mM NaCl, 1 mM EDTA, 1 mM DTT, 3% (w/v) glycerol, 100 μg/mL acetylated BSA.
2. Passivation Solution (PS): 4 mg/mL Blotting Grade Blocker Non-fat Dry Milk (Bio-Rad) dissolved in CB.
3. Experimental Buffer (EB): This buffer corresponds with the chosen experimental conditions. Here we utilize a buffer consisting of 20 mM Tris–HCl pH 7.5 and 75 mM KCl.

2.2 Generation of DNA Substrates Using PCR

To generate a DNA substrate for TPM we use Polymerase Chain Reaction.

1. A DNA template containing the sequence of interest (in this example we generate a sequence of 685 bp in length using the pRD118 plasmid as a template; the plasmid is available upon request).
2. 5′ biotinylated primer (Table 1).
3. 5′ digoxygenin labeled primer (Table 1).
4. Dream-Taq DNA polymerase.
5. Deoxyribose nucleotide triphosphate (dNTP).
6. Dream Taq-polymerase reaction buffer.
7. GenElute PCR cleanup kit (Sigma-Aldrich).
8. Eppendorf® PCR tubes.
9. Bio-Rad T100 Thermocycler or any other available PCR machine.

Table 1
Primer sequences

Primer name	Sequence	Modification
32% GC 685 bp Fw	ATACATATGCAACTTGAACGGCGTAA AAGAGGAACAATGG	5′ biotin
32% GC 685 bp Rev	GGTGGATCCTTTTCATCCCTTTAGTT CTTCCAG	5′ Dig

10. 1% agarose gel in 1× TBE.
11. NanoDrop®.
12. GeneRuler (Thermo Fisher).

2.3 Preparation and Assembly of Incubation Chamber

1. Anti-Dig 200 μg polyclonal antibodies (Roche).
2. Circular cover glasses with a diameter of 28 and 35 mm (*see* **Note 1**) (If a dedicated holder as used in our lab is not available, regularly sized rectangular microscope slides and square coverslips can be used).
3. A method for cutting Parafilm as described below.
4. Whatman paper.
5. 0.46 μm streptavidin-coated polystyrene beads (G. Kisker GbR.).
6. Heating block capable of reaching 100 °C.
7. Fine-pointed tweezers.
8. 90–100% ethanol.
9. Acetone.
10. Sonication bath.
11. KIMTECH Precision wipes.
12. SHIELDskin™ ORANGE NITRILE™ 260.

2.4 Microscopy Equipment

1. TMC Vibracontrol clean top isolation table.
2. Inverted microscope Nikon Diaphot 300.
3. 100× oil-immersion objective (with a numerical aperture of at least 1.25).
4. Heat chamber.
5. Thorlabs CMOS camera DCC1545M.
6. Climatized room.
7. Nail polish/thermal glue.
8. Immersion oil.
9. Lens paper.

2.5 Particle Tracking and Analysis

1. Computer.
2. Particle tracking software. In our lab we employ the real-time particle tracking engine which is part of the software suite described by Sitters et al. and is freely available online [16]. We also have good experience tracking particles in recorded movies using PolyParticleTracker [17]. An objective comparison of particle tracking methods available is given by Chenouard et al. [18].
3. Custom MatLab routine (available upon request from the authors) for post-processing and analysis of raw xy-position data.

3 Methods

3.1 Generation of DNA Substrates Using PCR

1. Reagents are combined in an Eppendorf® PCR tube according to the scheme below.

Reagent	Quantity
DNA template	10 ng
Forward primer	10 pmol
Reverse primer	10 pmol
dNTP	5 μL
Taq polymerase buffer	5 μL
Taq polymerase	0.2 μL (1 enzyme unit)
H_2O	Add to total volume of 50 μL

2. This reaction mix is kept on ice as much as possible and the PCR is initiated using the following protocol in a Bio-Rad T100 Thermocycler.

Cycles	Temperature (°C)	Duration
1	95	1 min
25	95	30 s
	65	30 s
	72	4 min
1	72	10 min
1	15	∞

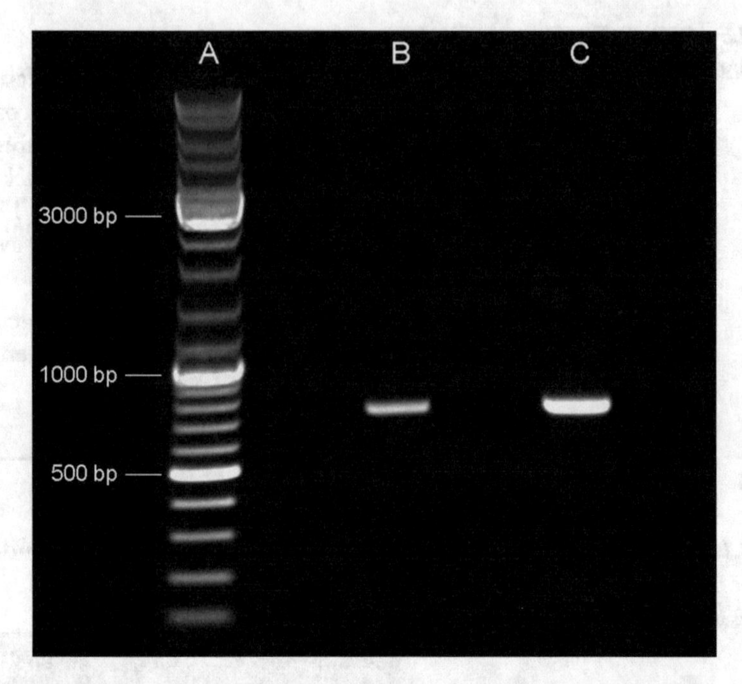

Fig. 2 1% agarose gel containing (**a**) 2 μL of the GeneRuler DNA molecular weight marker, (**b**) 2 μL of the PCR product, (**c**) 2 μL of the purified and DNA ready for use in Tethered Particle Motion experiments

3. The PCR product is purified using the GenElute PCR cleanup kit.
4. Two microliters of the purified PCR is loaded on a 1% agarose gel in TBE buffer alongside a DNA molecular weight marker for verification that a product of the expected length is formed.
5. An example of a successful PCR and purification of the obtained PCR product is shown in Fig. 2.
6. Finally, the concentration of purified PCR-generated DNA needs to be determined accurately. Use the NanoDrop® (or any other device capable of measuring UV absorbance) to estimate the purity (260/280) and concentration of the DNA. If no other method is available, the concentration of DNA can also be approximated using a DNA dilution series run on an agarose gel compared to a reference marker. Dilute the DNA in CB at a concentration of ~150 pM and store in aliquots at −20 °C.

3.2 Making Flow Cells

1. Before assembly of the flow cell the cover glasses must first be cleaned thoroughly by submerging them in a beaker with acetone and placing the beaker in a sonication bath for 10 min followed by the same sonication procedure in ethanol. After the last sonication step, while wearing unpowdered

Fig. 3 Metallic plates made to hold and equally distribute heat and pressure when melting the Parafilm and finalizing the flow cell

gloves, gently dry the cover glass using KIMTECH precision wipes. Leave exposed to air for 10 min to dry completely.

2. To make the flow cells as displayed in Fig. 1 Parafilm must be cut to the correct dimensions. The shape and size of the Parafilm must be identical to the 28 mm cover glass except for a single 6-mm wide channel running through the center. If possible, we advise using laser-cutting technology to obtain the exact dimensions. Otherwise it is also possible to cut by hand using a printed template and a scalpel.
3. Carefully align the Parafilm to the edges of the 28 mm cover glass.
4. Sandwich Parafilm between lower (large) and upper (small) round cover glass.
5. Compound the two halves by heating to 100 °C for 30 s, in between two massive metal blocks (Fig. 3) placed on top of a regular heating block. It is advised to apply a small amount of compression force to ensure that the halves are combined evenly. A complete flow cell can be seen in Fig. 4.

3.3 Preparing the Bead Solution

1. Dilute the beads to 0.01% w/v in CB.
2. Mix thoroughly using either a vortex or pipetting up and down.

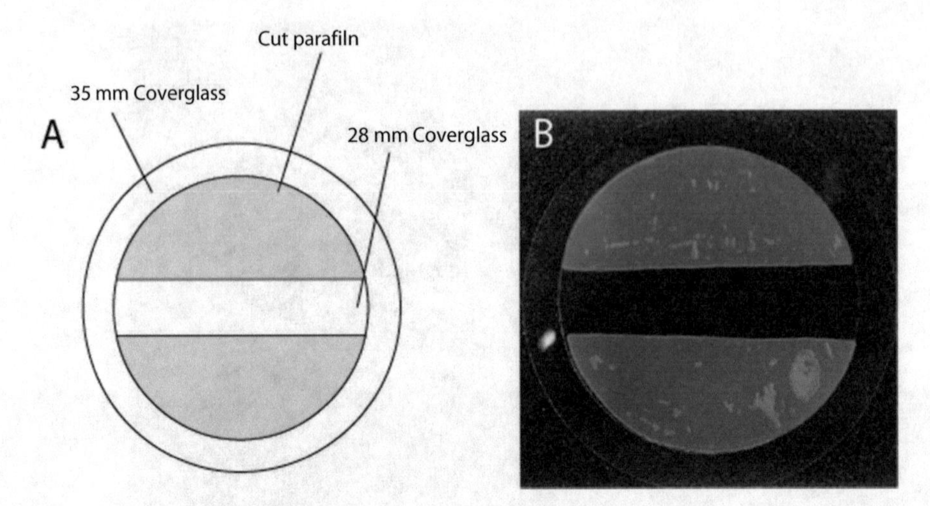

Fig. 4 TPM flow cell made by sandwiching Parafilm between two cover glasses and heating to "fix" Parafilm layer. (**a**) Schematic of a flow cell. (**b**) Picture of a "real" flow cell

3. Sonicate for 10–30 min to ensure proper suspension of the beads in sonication bath.
4. Check the bead suspension by flowing 50 μL into a prepared flow cell and observing under the microscope, while looking for bead clusters. (If very few bead clusters, <1%, are observed the Bead stock is considered good for use, store at 4 °C).

3.4 Preparing Flow Cells

1. Wash and incubate the flow cell with 100 μL of **CB** containing 20 μg/mL of anti-Dig for 10 min at room temperature.
2. Wash the flow cell with 100 μL of **PS** buffer and incubate for 10 min at room temperature.
3. Wash the flow cell with 100 μL of **CB**.
4. Wash the flow cell with 100 μL of DNA solution (100–250 nM) diluted in **CB**, incubate for 10 min at room temperature.
5. Wash the flow cell with 100 μL of **CB**.
6. Wash the flow cell with 100 μL of bead solution.
7. Wash the flow cell with 100 μL of **IB**.
8. Wash the flow cell with 100 μL of protein solution and incubate at room temperature for 10 min.
9. Wash the flow cell with 100 μL of protein solution and seal the flow cell using nail polish. (for relevant protein concentrations *see* Figs. 8, 9, and 10).

3.5 Microscopy (for the Instrument See Fig. 5)

1. Place the flow cell in the holder (seen in Fig. 6).
2. Turn on the lamp and heating stage.

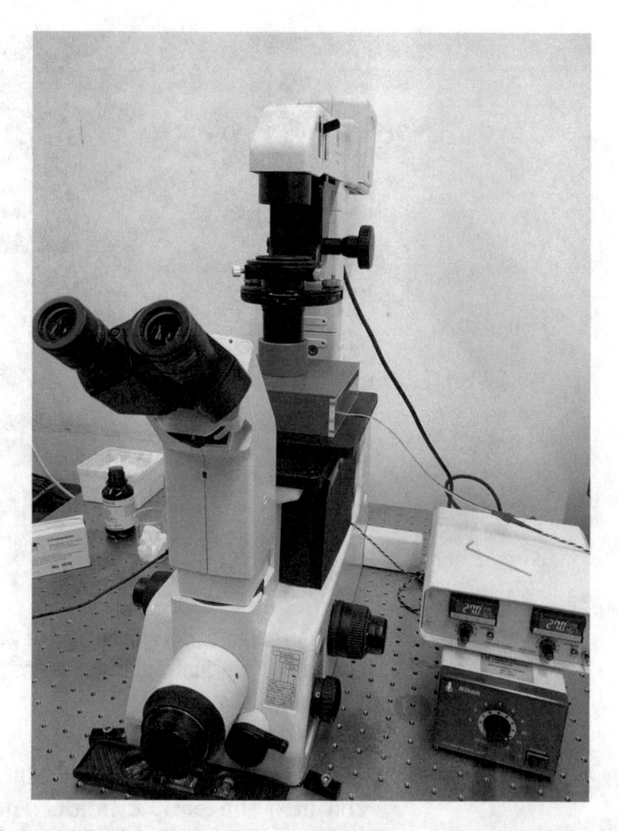

Fig. 5 Tethered Particle Motion instrument built on an inverted microscope. The flow cell is mounted on the microscope stage (not visible), which is embedded in an isolation chamber (*center*). The controller of the heating system (*right*) controls the temperature of the flow chamber as well as the objective used for imaging. Feedback ensures a temperature accuracy of 0.3 °C

Fig. 6 Top view of the heated flow cell holder. The central part of the holder has some degrees of freedom to allow for the acquisition of multiple fields of view. The attachment point for the heat regulation is also visible

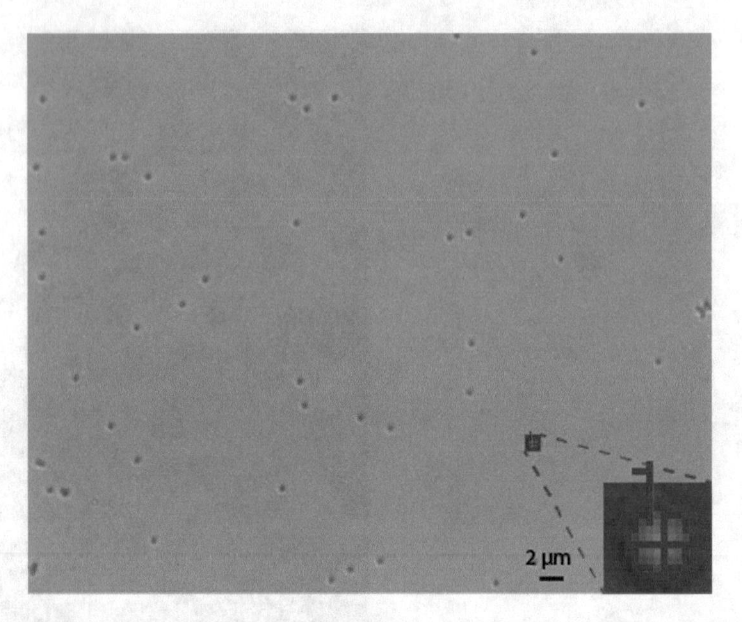

Fig. 7 Screen capture of a typical field of view with several surface-tethered and moving beads. Individual beads are selected and the position of the center of the beads is tracked over time

3. Place immersion oil on the objective (if needed), load the flow cell into the sample holder and raise the objective until the beads come into focus.
4. Close the isolation chamber and wait for the temperature to equilibrate. This may take 5–20 min depending on the desired temperature during the experiment; the system is equilibrated when no more focal drift is observed.
5. Verify sample quality by looking for moving beads (*see* **Note 2**).

3.6 Measurements

1. Select a region of interest in the field of view (as seen in Fig. 7), and track beads for a minimum of 1500 frames. Images are taken at 25 Hz, with a camera exposure time of 20 ms. Attempt to measure at least 300 regions of interest per experimental condition to ensure sufficient "good" bead–tether combinations for quantitation after quality check (see below).
2. Determine the *X* and *Y* coordinates of the beads in the field of view using particle-tracking software. This can be done in two ways: (1) Real-time particle tracking and (2) Post-measurement particle tracking. Real-time analysis has the advantage of being able to directly verify the quality of the tracking. The downside of real-time particle tracking is that tracking many beads in parallel is computationally demanding. In our lab we employ the real-time particle tracking system, which is part of the software suite described by Sitters et al. and freely available online [16]. For other options *see* Subheading 2.5.

3.7 Data Analysis

1. Calculate the anisotropic ratio—a measure of symmetry—of the scatter plots of the beads using Eq. 1. Exclude from further analysis beads with an anisotropic ratio greater than 1.2 as these may be attached to more than a single DNA tether (*see* **Note 3**); a high anisotropic ratio may also result from poor surface passivation of the flow cell (*see* **Note 4**).

$$\alpha = l_{\text{major}}/l_{\text{minor}} \tag{1}$$

Equation 1: Calculation of anisotropic ratio (α), where l_{major} and l_{minor} represent the major and minor axis of the *xy*-scatter plot respectively.

2. Calculate the RMS and the standard deviation of the RMS of the remaining beads.
3. Exclude all beads that have a standard deviation larger than 6% of their RMS. This is done to exclude beads that exhibit transient sticking to the surface of the flow cell. Note: If you believe that transient shortening events may be occurring in your sample please follow the instructions on "Highly dynamic DNA binding" described below.

$$\text{RMS} = \sqrt{\frac{1}{n}\sum_{i=1}^{n}\left[(x_i - \bar{x})^2 + (y_i - \bar{y})^2\right]} \tag{2}$$

Equation 2: Calculation of the root mean squared displacement (RMS) using the *x* and *y* coordinates of the beads to discern the displacement.

4. Make a histogram of all the RMS values for each experimental condition.
5. Fit the histogram using a Gaussian fitting curve to determine the average RMS and error (*see* **Note 5**).

3.8 Analysis of Protein–DNA Complexes Using TPM

Architectural proteins can interact with DNA in a plethora of ways, resulting in different, distinct changes in RMS. The DNA binding modes best studied by TPM are: DNA bending, DNA wrapping and multimerization along the DNA [19]. Some architectural proteins may exhibit sequence-specific interactions in addition to general nonspecific interactions. The structures formed upon binding at a specific—high affinity—sequence may be distinct from the structures formed upon general binding to DNA (this is further

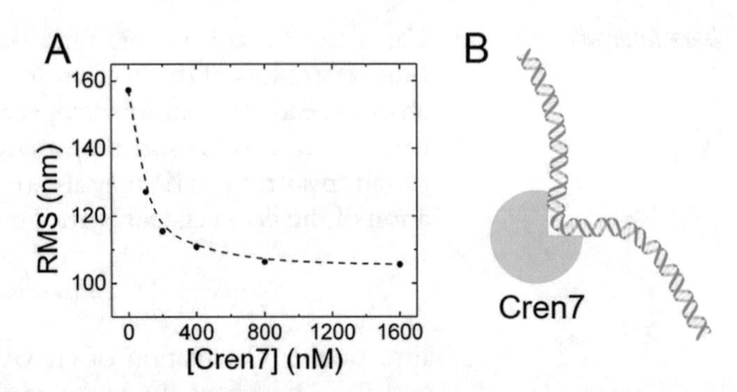

Fig. 8 Compaction of DNA by the DNA bending protein Cren7. (**a**) Decrease of the Root Mean Squared (RMS) displacement of the bead due to binding of increasing amounts Cren7. (**b**) Schematic model of DNA bending by Cren7

discussed in Subheading 3.7, **step 2**). DNA binding may also occur in a cooperative or noncooperative manner. The degree of cooperativity can be determined using analysis methods described in Subheading 3.7, **step 5**. Finally, the protein–DNA complexes assembled upon protein binding may be highly dynamic and depending on the relevant time scales, TPM is capable of sampling and quantifying dynamic changes. In Subheading 3.7, **step 4** we discuss strategies to not only measure but also quantify these effects. We also discuss (in Subheading 3.7, **step 5**) how to relate RMS to more robust standards such as the persistence length of DNA.

3.8.1 DNA Bending Proteins

DNA bending proteins induce bends in DNA upon binding; these bends may be formed toward or away from the protein (*see* **Notes 6** and 7). Here we illustrate the effect of the binding of such proteins on the RMS using the noncooperative sequence unspecific DNA bending protein Cren7 [20]. Titration of Cren7 yields a progressive concentration dependent reduction in RMS of the tether from 160 to 110 nm, indicating DNA compaction (Fig. 8).

3.8.2 DNA Wrapping Proteins

A more extreme form of DNA bending is DNA wrapping. Here we illustrate the effects of DNA wrapping on the RMS using HMfB. HMfB binds to DNA at specific high affinity sequences and form a wrapped structure (Fig. 5b). This results in a reduction of the RMS from 150 to 130 nm in the presence of 20 nM of HMfB (*see* Fig. 9a; unpublished).

3.8.3 DNA Binding Proteins Capable of Multimerizing Along the DNA

Some proteins multimerize along DNA, using DNA as a catalyst for multimer formation. An example of such a protein is H-NS. This protein exists as a dimer in solution, but assembles into large multimers along DNA. The binding of H-NS along a DNA molecule results in an increase in RMS from 145 nm up to 165 nm (Fig. 10; [14]).

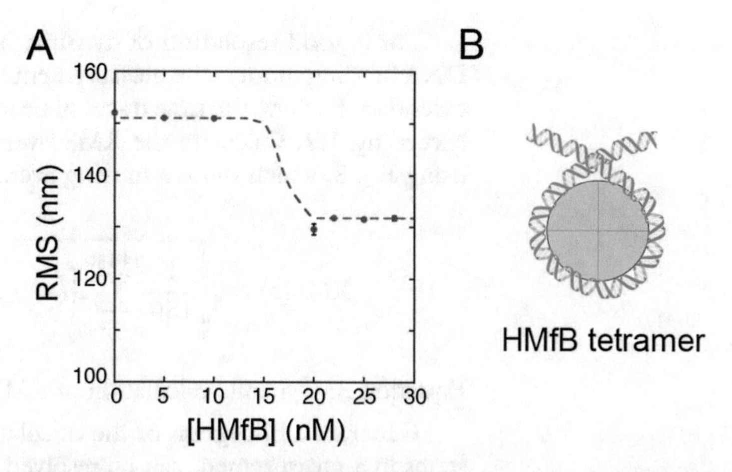

Fig. 9 Compaction of DNA by the DNA wrapping protein HMfB. (**a**) Reduction of the Root Mean Squared (RMS) displacement of the bead as a result of a titration with HMfB. (**b**) Model depicting the DNA wrapped structure formed by HMfB

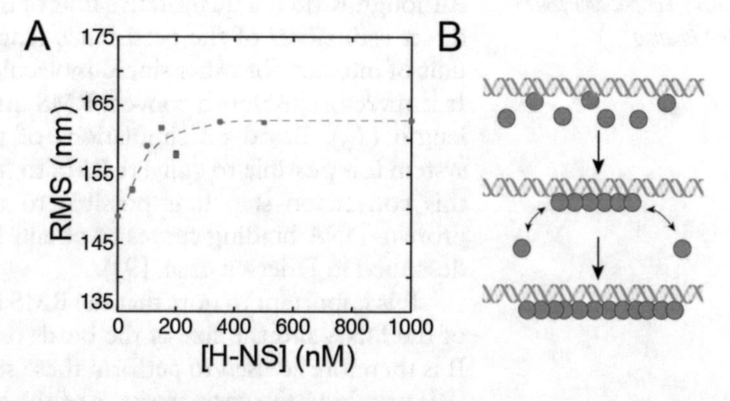

Fig. 10 DNA stiffening by H-NS. (**a**) Increase in Root Mean Squared (RMS) displacement of the bead as a result of a titration with H-NS (**b**) Model depicting the binding and multimerization along DNA

3.8.4 Highly Dynamic DNA Binding

The structure of protein–DNA complexes can be highly dynamic. In that case it may be difficult to resolve the RMS of your bead using the standard analysis of your data. A clear indication of a dynamic protein–DNA complex is that a large fraction of the tracked beads is discarded at the quality check stage. In such cases it is essential to confirm that particle disqualification is not a consequence of transient sticking of the bead to the glass surface (see above). Note that strong transient effects by the protein may be difficult to differentiate from random interactions of the bead with the surface. In such cases it is advised to confirm the observations using another technique.

For a good resolution of dynamic binding events or different DNA binding modes, the measurement time frame may need to be extended. Exclude the time traces of beads with an anisotropic ratio exceeding 1.2. Calculate the RMS over time of these time traces using Eq. 3, which yields a moving average.

$$\mathrm{RMS}(t) = \sqrt{\frac{1}{150}\sum_{i=t}^{t+150}\left[(x_i - \bar{x})^2 + (y_i - \bar{y})^2\right]} \tag{3}$$

Equation 3: Dynamic calculation of RMS over time

Generate a histogram of the calculated RMS values. Different states in a measurement can be resolved if they represent a significant subpopulation of all states. Analysis of the data can be refined using a hidden Markov model [21] to determine the different RMS levels and their durations.

3.8.5 Further Analysis of DNA Binding

Although RMS is a quantitative unit of measure, it does not permit direct estimation of the persistence length of DNA, the standard unit of measure for other single-molecule techniques (*see* **Note 8**). It is therefore useful to convert RMS into an *apparent* persistence length (L_P). Based on simulations of the surface-tethered bead system it is possible to convert RMS to *apparent* L_P [9]. Following this conversion step it is possible to quantitatively analyze the protein–DNA binding curves to obtain DNA binding affinities (as described in Driessen et al. [9]).

It is important to note that the RMS is dependent on the length of the DNA and the size of the beads used for TPM (*see* **Note 9**). It is therefore advised to perform these simulations (as described in [9]) matching the characteristics of the system investigated.

4 Notes

1. Although other shapes and types of cover glass can be used, in our hands these have shown the greatest ease of use.
2. Tethered beads can be absent. In that case the quality and integrity of the DNA substrate can be verified by electrophoretic mobility shift analysis [15]. Incubate 2 μL of purified PCR product with either 2 μL of 20 μg/μL anti-digoxygenin, or 2 μL of your bead stock. Load these samples on a 1% w/v agarose gel and visualize by staining (e.g., using ethidium bromide). The fraction of DNA shifted in the presence of either anti-digoxygenin or beads provides a quantitative measure of label presence and integrity.

3. Beads can be tethered to more than a single DNA molecule. This is visible as asymmetry in the xy-position plot of the bead. If a low fraction of the tethers exhibit this behavior it is not a problem as these are discarded at quality control. If it is a large fraction, the amount of DNA used in the assay needs to be reduced.
4. Beads can stick permanently or transiently to the glass surface. This indicates poor passivation of the glass surface and is undesired. Once a working protocol has been established, this may occur occasionally. If this occurs, the best solution is to discard the flow cell.
5. Beads can be tethered aspecifically. This is visible as a different RMS than expected for the length of the used tether. This would result in a wide distribution of RMS values in flow cells without DNA binding proteins. If this occurs it is advised to remake the DNA preparation from scratch.
6. As in any quantitative essay it is essential to accurately determine protein concentration and to perform precise and reproducible protein dilution.
7. Poor reproducibility of data at a given protein concentration. This indicates surface binding of your protein. To counteract loss of protein of interest due to surface binding, BSA can be included in the experimental buffer.
8. Interpretation: caution is needed as observed effects on RMS may be due to different types of binding, which cannot be directly discerned using TPM. It is therefore useful to verify structural models using direct visualization techniques such as AFM [22, 23], (single molecule) FRET [24, 25], or conventional biochemical techniques.
9. The motion of a surface-tethered bead is strongly dependent on the length of the DNA and the size of the bead. This implies that for each combination of DNA and beads a separate simulation series is needed to establish the relationship between RMS and Lp.

Acknowledgments

Alson van der Meulen and Rosalie Driessen are acknowledged for optimizing experimental and analysis procedures used for TPM experiments.

This work was supported through a VIDI [864.08.001] and VICI [016.160.613] grant from The Netherlands Organization for Scientific Research (NWO), NanonextNL a micro and nanotechnology program of the Dutch Government and 130 partners.

References

1. Neuman KC, Nagy A (2008) Single-molecule force spectroscopy: optical tweezers, magnetic tweezers and atomic force microscopy. Nat Methods 5(6):491–505. doi:10.1038/nmeth.1218
2. Herbert KM, Greenleaf WJ, Block SM (2008) Single-molecule studies of RNA polymerase: motoring along. Annu Rev Biochem 77:149–176. doi:10.1146/annurev.biochem.77.073106.100741
3. Bai L, Santangelo TJ, Wang MD (2006) Single-molecule analysis of RNA polymerase transcription. Annu Rev Biophys Biomol Struct 35:343–360. doi:10.1146/annurev.biophys.35.010406.150153
4. Bustamante C, Macosko JC, Wuite GJ (2000) Grabbing the cat by the tail: manipulating molecules one by one. Nat Rev Mol Cell Biol 1(2):130–136. doi:10.1038/35040072
5. Beausang JF, Zurla C, Manzo C, Dunlap D, Finzi L, Nelson PC (2007) DNA looping kinetics analyzed using diffusive hidden Markov model. Biophys J 92(8):L64–L66. doi:10.1529/biophysj.107.104828
6. Laurens N, Bellamy SR, Harms AF, Kovacheva YS, Halford SE, Wuite GJ (2009) Dissecting protein-induced DNA looping dynamics in real time. Nucleic Acids Res 37(16):5454–5464. doi:10.1093/nar/gkp570
7. Laurens N, Rusling DA, Pernstich C, Brouwer I, Halford SE, Wuite GJ (2012) DNA looping by FokI: the impact of twisting and bending rigidity on protein-induced looping dynamics. Nucleic Acids Res 40(11):4988–4997. doi:10.1093/nar/gks184
8. Zurla C, Samuely T, Bertoni G, Valle F, Dietler G, Finzi L, Dunlap DD (2007) Integration host factor alters LacI-induced DNA looping. Biophys Chem 128(2-3):245–252. doi:10.1016/j.bpc.2007.04.012
9. Driessen RP, Sitters G, Laurens N, Moolenaar GF, Wuite GJ, Goosen N, Dame RT (2014) Effect of temperature on the intrinsic flexibility of DNA and its interaction with architectural proteins. Biochemistry 53(41):6430–6438. doi: 10.1021/bi500344j
10. Driessen RP, Lin SN, Waterreus W-J, van der Meulen AL, van der Valk RA, Laurens N, Moolenaar GF, Pannu NS, Wuite GJ, Goosen N, Dame RT (2016) Diverse architectural properties of Sso10a proteins: evidence for a role in chromatin compaction and organization. Sci Rep 6:29422. doi: 10.1038/srep29422
11. Farge G, Laurens N, Broekmans OD, van den Wildenberg SM, Dekker LC, Gaspari M, Gustafsson CM, Peterman EJ, Falkenberg M, Wuite GJ (2012) Protein sliding and DNA denaturation are essential for DNA organization by human mitochondrial transcription factor A. Nat Commun 3:1013. doi:10.1038/ncomms2001
12. Guerra RF, Imperadori L, Mantovani R, Dunlap DD, Finzi L (2007) DNA compaction by the nuclear factor-Y. Biophys J 93(1):176–182. doi:10.1529/biophysj.106.099929
13. Nir G, Lindner M, Dietrich HR, Girshevitz O, Vorgias CE, Garini Y (2011) HU protein induces incoherent DNA persistence length. Biophys J 100(3):784–790. doi:10.1016/j.bpj.2010.12.3687
14. van der Valk RA, Vreede J, Moolenaar GF, Hofmann A, Goosen N, Dame RT (2016) Environment driven conformational changes modulate H-NS DNA bridging activity. bioRXiv, doi:10.1101/097436.
15. Sambrook J, Fritsch EF, Maniatis T (1998) Molecular cloning: a laboratory manual, 2nd edn. Cold Spring Harbor Laboratory Press, Cold Spring Harbor, NY
16. Sitters G, Kamsma D, Thalhammer G, Ritsch-Marte M, Peterman EJ, Wuite GJ (2015) Acoustic force spectroscopy. Nat Methods 12(1):47–50. doi:10.1038/nmeth.3183
17. Rogers SS, Waigh TA, Zhao X, JR L (2007) Precise particle tracking against a complicated background: polynomial fitting with Gaussian weight. Phys Biol 4(3):220–227. doi:10.1088/1478-3975/4/3/008
18. Chenouard N, Smal I, de Chaumont F, Maska M, Sbalzarini IF, Gong Y, Cardinale J, Carthel C, Coraluppi S, Winter M, Cohen AR, Godinez WJ, Rohr K, Kalaidzidis Y, Liang L, Duncan J, Shen H, Xu Y, Magnusson KE, Jalden J, Blau HM, Paul-Gilloteaux P, Roudot P, Kervrann C, Waharte F, Tinevez JY, Shorte SL, Willemse J, Celler K, van Wezel GP, Dan HW, Tsai YS, Ortiz de Solorzano C, Olivo-Marin JC, Meijering E (2014) Objective comparison of particle tracking methods. Nat Methods 11(3):281–289. doi:10.1038/nmeth.2808
19. Dame RT (2008) Single-molecule micromanipulation studies of DNA and architectural proteins. Biochem Soc Trans 36(Pt 4):732–737. doi:10.1042/BST0360732
20. Driessen RP, Meng H, Suresh G, Shahapure R, Lanzani G, Priyakumar UD, White MF, Schiessel H, van Noort J, Dame RT (2013) Crenarchaeal chromatin proteins Cren7 and Sul7 compact DNA by inducing rigid bends. Nucleic Acids Res 41(1):196–205. doi:10.1093/nar/gks1053

21. Markov AA (1907) Investigation of a specific case of dependent observations. Izv Imper Akad Nauk (St-Petersburg) 3:61–80
22. Dame RT, Wyman C, Goosen N (2003) Insights into the regulation of transcription by scanning force microscopy. J Microsc 212(Pt 3):244–253
23. Dame RT, van Mameren J, Luijsterburg MS, Mysiak ME, Janicijevic A, Pazdzior G, van der Vliet PC, Wyman C, Wuite GJ (2005) Analysis of scanning force microscopy images of protein-induced DNA bending using simulations. Nucleic Acids Res 33(7):e68. doi:10.1093/nar/gni073
24. Buning R, van Noort J (2010) Single-pair FRET experiments on nucleosome conformational dynamics. Biochimie 92(12):1729–1740. doi:10.1016/j.biochi.2010.08.010
25. Dragan AI, Privalov PL (2008) Use of fluorescence resonance energy transfer (FRET) in studying protein-induced DNA bending. Methods Enzymol 450:185–199. doi:10.1016/S0076-6879(08)03409-5

Chapter 12

Biochemical Analysis of Bacterial Condensins

Zoya M. Petrushenko and Valentin V. Rybenkov

Abstract

Condensins help establish compactness of bacterial chromosomes and assist in their segregation during cell growth and division. They act as elaborate macromolecular machines that organize the chromosome on a global scale and link it to the pan-cell dynamics. The mechanism of condensins in its entirety is yet to be elucidated. However, many aspects of condensin activity have been recuperated in vitro. This report described purification of the *Escherichia coli* condensin MukBEF, its reassembly from purified components, and reconstitution of DNA supercoiling and DNA bridging activities of the complex.

Key words MukB, MksB, SMC, DNA bridging, Chromatin structure, DNA topology, Condensins

1 Introduction

Condensins are essential for global folding of the chromosome (reviewed in [1–3]). These proteins were found in organisms from all kingdoms of life [4] and have been implicated virtually in every aspect of higher order chromatin dynamics. Condensins act as multisubunit complexes. They contain at the core a pair of structural maintenance of chromosome (SMC) proteins, which dynamically associate with the regulatory non-SMC subunits and are prone to oligomerization.

In bacteria, three families of condensins have been identified. The first discovered condensin, MukBEF, is found in a subset of gamma-proteobacteria, including *Escherichia coli* [5–7]. Many other bacteria, including archaea carry the SMC-ScpAB complex, whose SMC subunit is highly homologous to its eukaryotic counterparts [8, 9]. The third family, MksBEF, occurs sporadically in a wide range of eubacteria and is characterized by low sequence conservation [10]. It bears distant resemblance to the *E. coli* MukBEF, suggesting that they might be viewed as members of the same superfamily. However, their functions and activities are sufficiently diverged, which calls for caution when comparing the two families.

Olivier Espéli (ed.), *The Bacterial Nucleoid: Methods and Protocols*, Methods in Molecular Biology, vol. 1624,
DOI 10.1007/978-1-4939-7098-8_12, © Springer Science+Business Media LLC 2017

The mechanism of condensins is complicated and includes activities required for its intracellular recruitment and DNA organization [3]. In MukBEF, the DNA reshaping activities reside in its SMC subunit, MukB, whereas MukF and MukE are responsible for regulation of MukB and its correct subcellular localization [7, 11]. Whereas the mechanism of intracellular recruitment is yet to be determined, much became clear about DNA reshaping activities of condensins. We will focus on the latter in this review.

We begin with a protocol for purification of active condensins. Somewhat atypically, MukBEF does not act as a holoenzyme but rather forms a complex with a dynamic architecture. Attempts to purify the "holoenzyme" using affinity tags placed on either MukB or MukE invariably yield a broad distribution of complexes with varied composition [7]. Thus, we recommend purifying MukB and MukEF as two separate complexes and then reconstituting the holoenzyme as needed. So far, we were able to detect only inhibition of MukB-DNA interaction by MukEF. Thus, the highest DNA reshaping activity was observed with MukB alone. At least several other SMCs, including the yeast SMC2/4 complex [12], the *Bacillus subtilis* SMC [13], and the *Pseudomonas aeruginosa* MksB [10] were active in the absence of their cognate non-SMC subunits. However, ATPase activity of MukB is stimulated by MukEF [11, 14]. Purification of MukEF is rather uneventful and yields large quantities of protein. In contrast, MukB elutes as a mixture of two metastable conformations that can be converted into each other using chromatography.

Reconstitution of MukBEF (the second protocol) is fairly straightforward, but it too bears surprises. Depending on conditions, the complex emerges in one of two possible conformations with the stoichiometry $B_2E_4F_2$ or B_2E_2F, which differ in DNA binding [7]. It is unclear how common such diversity is for other condensins. Reconstitution of the *P. aeruginosa* MksBEF produced only one of the complexes, $B_2E_4F_2$ [10].

We next describe two biochemical assays that we routinely use in condensin studies: the DNA supercoiling assay (and its modification, protection from relaxation) [15] and the magnetic bead pull-down assay (MBPA). The latter one is particularly well suited for studies of condensin-mediated DNA bridging [16]. Here we skip the DNA knotting assay, which has been instrumental to understanding the proteins during early days of condensin studies [15, 17]. Nowadays, its utility has been reduced due to the emergence of alternative approaches, including the aforementioned MBPA and single molecule techniques.

2 Materials

2.1 Proteins Purification

2.1.1 Plasmids

Plasmids pBB10 [15] and pBB08 [7] are pBAD/Myc-HisB (Invitrogen)-based plasmids which encode, respectively, *E. coli mukB-His10* gene and the *mukFE-His9* fragment of the *smtA-mukF-mukE-mukB* operon under the control of arabinose-inducible promoter.

pBR322 (NEB) is a double-stranded 4361-bp plasmid.

pUC40 is a 4.0-kb pUC18-based plasmid [18].

pBIO is a linear DNA composed of the 8.34-kb BamH I–Nco I fragment of pBB10 ligated at the BamH I end to a 400-bp PCR-generated DNA with multiple biotins [16].

2.1.2 Reagents and Materials

1. LB broth containing 0.5% NaCl (Difco).
2. 100 mg/ml ampicillin, filter sterilized.
3. Phenylmethylsulfonyl fluoride (PMSF): freshly prepared 200 mM stock solution in ethanol.
4. 20% L-arabinose, filter sterilized, store at −20 °C.
5. Isopropyl β-D-1-thiogalactopyranoside (IPTG).
6. Dithiothreitol (DTT): 1 M stock solution in ddH_2O, stored at −20 °C.
7. 0.5 M ethylenediamine tetraacetic acid (EDTA), pH 8.0.
8. Polyethylene glycol (PEG), MW 20,000.
9. IGEPAL CA-630.
10. 15 ml column packed with the Novagen His-Bind resin (EMD Chemicals Inc, NJ).
11. 13 ml column packed with heparin agarose (Type 1, H6508, Sigma).
12. 1 ml HiTrap Hheparin column (GE Healthcare).
13. 5 ml HiTrap His-Bind column (GE Healthcare).
14. Complete His-bind resin (Roche Applied Sciences).
15. Sephacryl S300 (GE Healthcare).
16. Buffer A: 20 mM HEPES–KOH, pH 7.7, 300 mM NaCl, 1 mM PMSF.
17. Buffer B: 20 mM HEPES–KOH, pH 7.7, 5% glycerol, 2 mM EDTA, 1 mM DTT, 1 mM PMSF.
18. Storage Buffer 1: 20 mM HEPES–KOH, pH 7.7, 200 mM NaCl, 50% glycerol, 2 mM EDTA, 1 mM DTT, 1 mM PMSF.
19. Storage Buffer 2: 20 mM HEPES, pH 7.7, 200 mM NaCl, 20% glycerol, 1 mM EDTA, 1 mM DTT, 1 mM PMSF.

2.2 MukBEF Reconstitution

1. 4× Reconstitution Buffer 1: 60 mM HEPES–KOH, pH 7.7, 9 mM $MgCl_2$, 3 mM DTT.
2. 4× Reconstitution Buffer 2: 60 mM HEPES–KOH, pH 7.7, 600 mM NaCl, 3 mM EDTA, 3 mM DTT.

2.3 Supercoiling Assay

1. Wheat germ topoisomerase I, purified as in [19].
2. pBR322 plasmid, relaxed with wheat germ topoisomerase I.
3. SYBR Gold (Molecular probes).
4. Proteinase K (Sigma-Aldrich).
5. Ultrapure phenol–chloroform–isoamyl alcohol (25:24:1, v/v) (Invitrogen).
6. Reaction Buffer 1: 20 mM HEPES–KOH, pH 7.7, 50 mM NaCl, 2 mM $MgCl_2$, 5% glycerol, 1 mM DTT.
7. 4× Master Buffer: 60 mM HEPES–KOH, pH 7.7, 9 mM $MgCl_2$, 3 mM DTT.
8. Stop Buffer: 50 mM Tris–HCl, pH 7.9, 0.5% SDS, 20 mM EDTA, 200 mM NaCl, 0.5 mg/ml proteinase K.
9. 4× Stop buffer for topo-protection assay: 200 mM Tris-HCl, pH 7.9, 0.4% SDS, 80 mM EDTA, 800 mM NaCl, 2 mg/ml proteinase K.
10. 50× TAE buffer: 2 M Tris–acetate, pH 8.3, 50 mM EDTA.
11. Speed-vacuum evaporator (Savant).

2.4 MukB DNA Bridging

1. C1 streptavidin-coated magnetic beads; Dynabeads C1 (Invitrogen).
2. pBIO DNA (*see* Subheading 2.1.1).
3. Tween 20.
4. "Foreign" DNA: pUC40 of various topologies or linear PCR products of various lengths.
5. 1× PBS: 137 mM NaCl, 2.7 mM KCl, 4.3 mM Na_2HPO_4, 1.4 mM KH_2PO_4.
6. TE Buffer: 10 mM Tris–HCl, pH 7.9, 1 mM EDTA.
7. TEN buffer: 10 mM Tris–HCl, pH 7.9, 1 mM EDTA, 1 M NaCl, 2 μM poly-L-glutamic acid, 2 μM oligonucleotide 41 bases, 1 mg/ml BSA, and 0.02% Tween 20.
8. Bridging Reaction Buffer 1: 20 mM HEPES–KOH, pH 7.7, 50 mM NaCl, 2 mM $MgCl_2$, 5% glycerol, 1 mM DTT, and 0.02% Tween 20.
9. Bridging Reaction Buffer 2: 20 mM HEPES–KOH, pH 7.7, 4.5 mM $MgCl_2$, 1.5 mM DTT, 0.04% Tween 20.
10. BSA mixture: 20 mM HEPES–KOH, pH 7.7, 4 mg/ml BSA, 4 mM spermidine.

11. Elution buffer: 10 mM Tris–HCl, pH 7.9, 1 mM EDTA, 200 mM NaCl, 0.2%SDS, 0.5 mg/ml proteinase K.

3 Methods

3.1 Proteins Purification

3.1.1 Purification of E. coli MukB

MukB is purified using its C-terminal ten-histidine tag, which does not affect functionality of the protein [15]. When nickel-chelate chromatography alone is used, the protein is barely active in DNA binding and reshaping [15]. Further purification of MukB through heparin agarose increases the activity of the protein. Rather unusually, MukB elutes from a heparin agarose column as a mixture of two isoforms, only one of which is active in DNA binding [15]. The two isoforms can be readily separated using chromatography on a HiTrap heparin column (Fig. 1) and can then be kept on ice for several weeks for biochemical analysis. During such storage, the active fraction of MukB slowly reverts to its inactive form and must be reactivated by another round of heparin chromatography prior to further studies. This approach allowed the authors to obtain reproducible results over numerous preparations of MukB. The following purification protocol is optimized to increase the yield of active MukB.

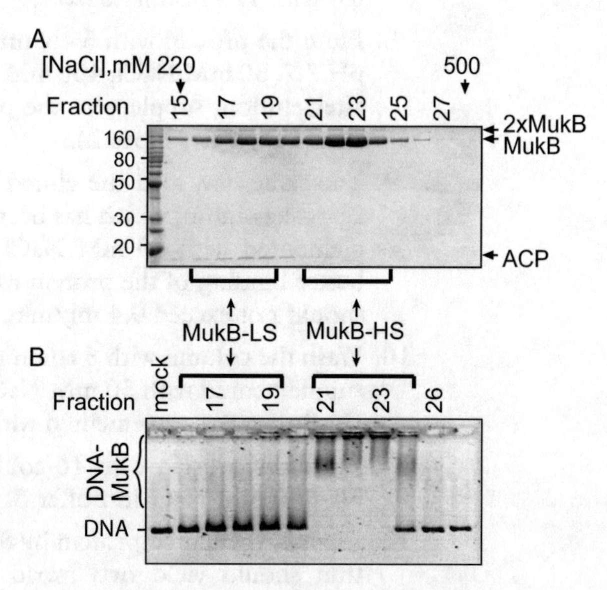

Fig. 1 Purification of MukB yields two distinct complexes that can be resolved by heparin chromatography. Shown are the silver-stained SDS-PAGE (**a**) and DNA gel-shift analysis (**b**) of the fractions eluted from the second heparin column. *Brackets* indicate the inactive low salt (MukB-LS) and the active high salt (MukB-HS) peaks of MukB. Also shown are the acyl carrier protein (ACP), which forms a 1:1 complex with MukB-LS, and presumably cross-linked MukB dimers (2xMukB)

1. Inoculate 20 ml of the overnight culture of *E. coli* DH5α cells, harboring pBB10 plasmid, into 2 lL of prewarmed (37 °C) LB medium supplemented with 100 μg/ml ampicillin. Grow cells at 37 °C up to OD_{600} of 0.6. Add 0.1% L-arabinose and continue to grow cells for 3 h.
2. Collect cells by centrifugation in JLA8.1 rotor for 25 min at 4000 rpm (4000 × *g*), 4 °C.
3. Wash cells with 100 ml of ice-cold 20 mM Tris–HCl, pH 7.5, 150 mM NaCl and pellet them by a 10-min centrifugation at 4000 rpm (3200 × *g*), 4 °C in Eppendorf 5810R centrifuge. Store the pellet at −80 °C.
4. Thaw cells on ice and then resuspend them in 50 ml of ice-cold buffer A supplemented with 20 mM imidazole.
5. Break cells open by passing them three times through French press at 12,000 psi.
6. Clarify cell extract by centrifugation at 17,000 rpm (37,000 × *g*) for 30 min (Avanti rotor JA-20).
7. Load clarified lysate onto a 15 ml nickel-charged His Bind column at 1 ml/min (*see* **Note 1**). Wash the column with 10-column volume of buffer A supplemented with 20 mM imidazole and then with 6-column volume of buffer A supplemented with 125 mM imidazole.
8. Elute the protein with 5 column volumes of 20 mM HEPES, pH 7.7, 50 mM NaCl, 400 mM imidazole, 1 mM PMSF. Right after elution, supplement the protein with 2 mM EDTA and 1 mM DTT (*see* **Note 2**).
9. The same day, load the eluted MukB onto a 13 ml heparin-agarose column, which has been equilibrated in buffer B, supplemented with 50 mM NaCl and 400 mM imidazole. For better binding of the protein to heparin agarose, the flow rate should not exceed 0.4 ml/min.
10. Wash the column with 5 column volumes each of (1) Buffer B, supplemented with 50 mM NaCl and 400 mM imidazole, and (2) Buffer B, supplemented with 50 mM NaCl.
11. Elute the protein with 16-column volume linear gradient of 50–800 mM NaCl in Buffer B. Collect 4 ml fractions.
12. Analyze the eluted protein by SDS-PAGE. Successful purification should yield two broad peaks. Pool fractions, eluted between 270 and 470 mM NaCl (*see* **Note 3**).
13. Concentrate the protein by dialysis against 2.5 lL of 20% PEG 20,000 in Buffer B, supplemented with 200 mM NaCl (*see* **Note 4**).
14. Dialyze the protein against two changes of Storage buffer 1 and store it at −20 °C (*see* **Note 5**).

3.1.2 Reactivation of MukB

1. Load 1 mg of MukB, which has been purified using nickel chelate and heparin chromatography, onto a 1 ml HiTrap heparin column at the flow rate 0.2 ml/min. Before loading, dilute the protein in Buffer B to a final NaCl concentration of 50 mM.
2. Wash the column with 6-column volume of Buffer B supplemented with 50 mM NaCl.
3. Elute the protein with the 15-column volume linear gradient of 50–800 mM NaCl in Buffer B. Collect 250 μl fractions.
4. Measure protein concentration and run SDS-PAGE for every fraction. The inactive, "low-salt" MukB is usually eluted between 220 and 320 mM NaCl; the active, "high-salt" MukB, is usually eluted between 350 and 450 mM NaCl (Fig. 1).
5. Pool the low- and high-salt fractions of MukB separately and dialyze the protein against Storage Buffer 2.
6. Store the protein on ice in a cold room (*see* **Notes 6** and **7**).

3.1.3 Purification of E. coli MukEF

1. Overproduce MukEF in 1 lL of DH5α cells harboring the pBB08 plasmid.
2. Purify MukEF using nickel chelate chromatography as described above for MukB (**steps 1–8**, Subheading 3.1.1).
3. After elution from the His-bind column, concentrate the protein by dialysis against 20% PEG 20,000 in Buffer B supplemented with 200 mM NaCl (*see* **Note 8**).
4. Dialyze MukEF against two changes of Buffer B supplemented with 200 mM NaCl.
5. Purify further MukEF by gel-filtration chromatography on a Sephacryl S300 column in Buffer B supplemented with 200 mM NaCl.
6. Measure the protein concentration and analyze its purity by SDS-PAGE.
7. Combine the peak fractions, which are usually >95% pure, and concentrate MukEF by dialysis against PEG 20,000 and then 50% glycerol as described above for MukB (**steps 13** and **14**, Subheading 3.1.1). Store the protein at −20 °C.

Purified MukEF complex has the E_4F_2 composition and is stable at diverse conditions.

3.2 Reconstitution of MukBEF

1. Mix 6 μl of water and 3 μl of 4× Reconstitution Buffer. Use Reconstitution Buffer 1 for assembly of $B_2E_4F_2$ and Reconstitution Buffer 2 for B_2E_2F (*see* **Note 9** and Fig. 2).
2. Add MukEF in 1.5 μl of Storage Buffer 2, gently mix by pipetting up and down (*see* **Note 10**).

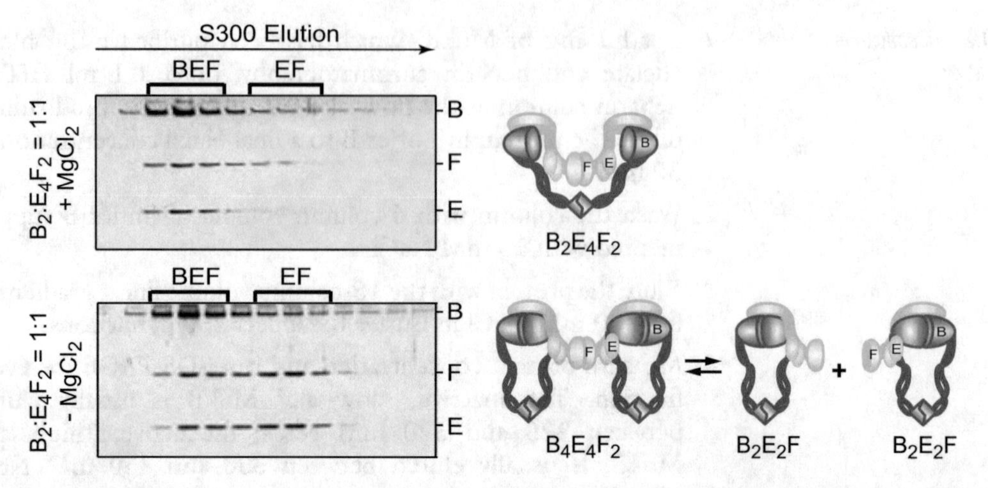

Fig. 2 Two MukBEF complexes. Depending on the buffer conditions, reconstitution of MukBEF from MukB and MukEF produces either $B_2E_4F_2$ (*top*) or B_2E_2F (*bottom*) complex. B_2E_2F can undergo further dimerization via the N-terminal winged helix domain of MukF [21] The two stoichiometries can be observed using gel filtration of reconstituted MukBEF on a Sephacryl S300 column or similar in a buffer that either contains (*top*) or lacks (*bottom*) magnesium chloride

3. Add 10 μg of "High-salt" MukB (the one that was passed through a HiTrap heparin column) in 1.5 μl of Storage Buffer 2; gently mix avoiding bubbles (*see* **Note 11**).
4. Incubate the protein mixture for 10 min at room temperature.

3.3 DNA Supercoiling Assay

Condensins introduce only transient changes into DNA structure, which disappear upon the removal of the protein for further analysis. These changes, however, can be captured in a coupled reaction using DNA topoisomerases and converted into alterations in DNA topology, which survives deproteinization. We describe here the DNA supercoiling assay, which helps detect supercoils introduced by the protein into DNA [15, 20]. When relaxed DNA is used as a substrate, formation of supercoils by condensins leads to extrusion of compensatory supercoils in the rest of the DNA (Fig. 3a). These compensatory supercoils are removed by a type-1 topoisomerase, leaving behind only the protected supercoils. The DNA supercoiling assay can be modified into a "nonrelaxation" (or topoprotection) assay, which reports the ability of the protein to protect supercoiled DNA from relaxation by topo-1 (Fig. 3c) (*see* **Note 12**). The net supercoiling varies depending on the used condensin. The frog condensin stabilizes positive supercoils in ATP-dependent manner [20], MukB supercoils are negative and ATP-independent [15], and the yeast SMC does not stabilize any supercoils [12].

1. Prepare 4× Master Buffer.
2. Prepare reaction mixture in 0.6 ml tube: 2.6 μl of water, 2.25 μl of 4× Master Buffer, 0.9 μl of BSA (10 mg/ml), 1 μl (10 ng) of relaxed pBR322 plasmid DNA (or any other relaxed DNA), mix carefully. Keep the reaction mixture at room temperature.

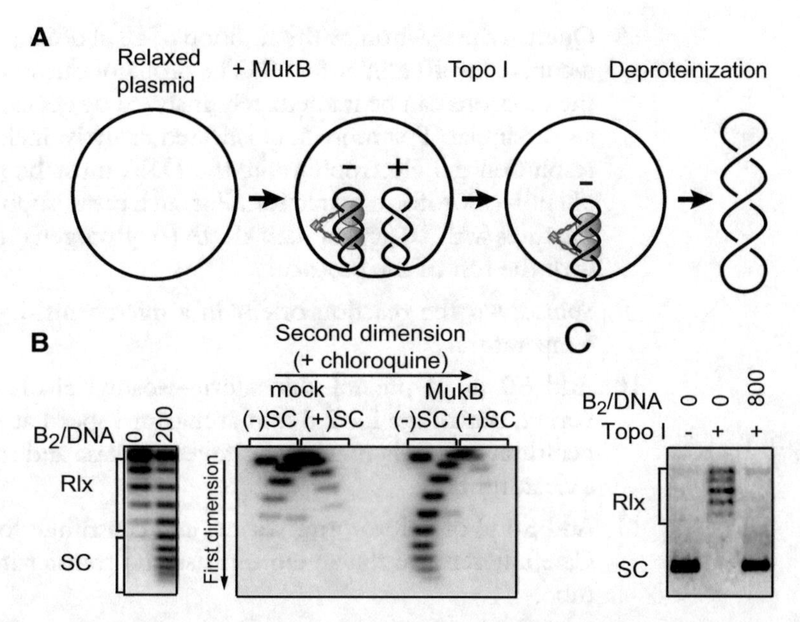

Fig. 3 DNA supercoiling assay. (**a**) A diagram of the assay. The binding of MukB to covalently closed DNA traps negative supercoils, which generates compensatory positive supercoils in the rest of the molecule. The subsequent treatment of the complex with a eukaryotic type-1 topoisomerase removes the compensatory supercoils, but not those protected by the bound protein. (**b**). Gel electrophoretic analysis of the DNA supercoiling. The sign of the supercoils can be determined using two-dimensional (2D) electrophoresis [15]. Following electrophoresis in the first dimension, supercoiled topoisomers migrate faster than the relaxed ones (*left panel*). The gel is then saturated with 0.8 μg/ml chloroquine, and electrophoresis continued in an orthogonal direction. The binding of chloroquine, an intercalating agent, to DNA unwinds it and introduces compensatory positive supercoils. This, in turn, increases the mobility of positive topoisomers but decreases it for the negative ones. As a result, two characteristic arcs can be found on the 2D gel (*right panel*), one for positive topoisomers and the other for negative topoisomers. As seen from the gel, MukB binding to DNA protects negative supercoils. (**c**). The topo-protection assay. At high protein levels, MukB protects the bound DNA from topoisomerases. Thus, treatment of supercoiled DNA with topo I leads to DNA relaxation in the absence but not presence of MukB

3. Dilute MukB into ice-cold Storage Buffer 2 to the desired concentration; keep on ice.
4. Dilute wheat germ topoisomerase I into 1× Reaction buffer supplemented with 1 mg/ml BSA. The topoisomerase–DNA molar ratio in the reaction should not exceed 10:1.
5. Add 2.25 μl of MukB (50–1000 ng) to the reaction mixture and gently mix. Final concentrations of the reagents in the reaction mixture should be as follows: 20 mM HEPES–KOH, pH 7.7, 50 mM NaCl, 2 mM $MgCl_2$, 5% glycerol, 1 mM DTT, 1 mg/ml BSA.
6. Add 1 μl of diluted topoisomerase I.
7. Incubate the reaction mixture for 30 min at 37 °C.

8. Quench the reaction by the addition of 40 μl of Stop Buffer and incubate for 40 min at 55 °C. The protocol can stop here, and the reactions can be immediately analyzed by gel electrophoresis or similar. For more sophisticated analysis, including high resolution gel electrophoresis, the DNA must be purified by phenol–chloroform extraction. For such cases, supplement the mixtures with 10 μg/ml yeast tRNA (or glycogen) and proceed with the rest of the protocol.
9. Spin down the reaction briefly in a microcentrifuge at room temperature.
10. Add 50 μl of phenol–chloroform–isoamyl alcohol mixture, vortex, centrifuge for 2.5 min at maximal speed at room temperature. Carefully remove the aqueous phase and transfer it to a clean tube.
11. Add 50 μl of chloroform, vortex, and centrifuge for 2.5 min. Carefully remove the aqueous phase and transfer it to a clean tube.
12. Add 130 μl of ice-cold 100% ethanol. Incubate at −20 °C for at least 2 h.
13. Spin down DNA for 10 min at maximal speed, 4 °C. Immediately aspirate the supernatant, being careful not to touch the pellet. Wash the pellet with 75% ice cold ethanol and leave at −20 °C for 1 h. Spin down and aspirate the liquid.
14. Dry the pellet in a SpeedVac evaporator.
15. Dissolve the pellet in 10 μl of TE buffer.

3.3.1 The Topo-Protection Assay

1. Prepare reaction mixture as described for the supercoiling assay (**step 2**), except that supercoiled DNA must be used as a substrate instead of the relaxed one (*see* **Note 12**).
2. Add MukB as desired and incubate the mixture for 10 min at 37 °C.
3. Add topoisomerase I and continue incubation for another 20 min.
4. Quench the reaction by the addition of 2.5 μl of 4× Stop buffer and further incubate for 40 min at 55 °C.
5. Resolve DNA using gel electrophoresis through a 0.8% agarose gel in 1× TAE buffer at 23 °C for 4 h at 8 V/cm followed by staining with SYBR Gold.

3.4 Magnetic Bead Pull-Down Assay

DNA bridging by condensins can be conveniently studied using the magnetic bead pull-down assay (MBPA). The assay allows one to sample diverse mixing protocols and DNA substrates, and has been essential in revealing the sequential manner of MukB-mediated DNA bridging [16]. The protocol below describes sequential addition of protein and DNA, but other mixing orders can be easily accommodated (Fig. 4).

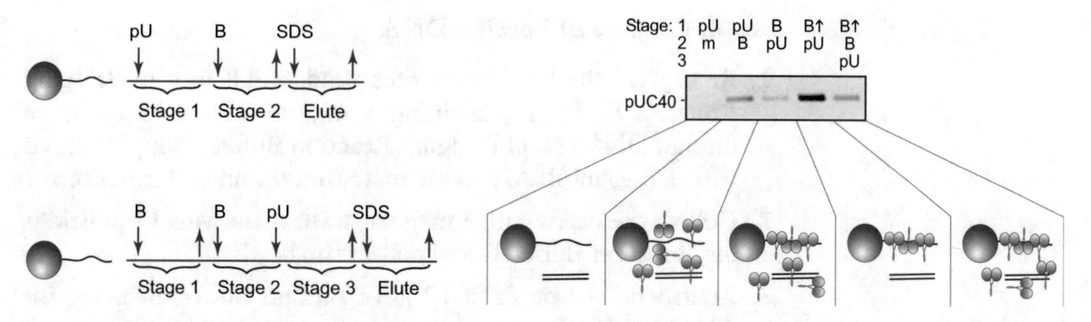

Fig. 4 DNA bridging assessed using the magnetic bead pull-down assay, MBPA. Free-floating DNA (pUC40) can be captured by MukB that is bound to bead-tethered double stranded pBIO DNA. Various mixing protocols can be evaluated using this format. Reactants can be added to (*down arrow*) and removed (*up arrow*) to the beads in any desired sequence. The captured DNA is then recovered from the beads by SDS treatment and visualized using gel electrophoresis. *pU* pUC40 DNA; *B* MukB; *m* mock reaction. Diagrams on the *left* illustrate reactions analyzed in lanes 2 and 5 of the gel. Note that the free-floating MukB interferes with bridging, and the highest bridging efficiency is observed when all unbound MukB is removed

Step 1: Binding of pBIO DNA to Streptavidin-Coated Beads

The following protocol is designed for five reactions.

1. Swirl carefully the bottle with Dynabeads C1 until the beads are uniformly suspended.
2. Transfer 1.5 μl of beads into a 0.6 ml Eppendorf tube and collect them using a magnetic separator.
3. Wash beads twice with 5 μl 1× PBS and once with 5 μl TEN. Resuspend the beads in 7.5 μl TEN.
4. Add 5 ng of pBIO DNA in 7.5 μl TEN buffer.
5. Incubate DNA for 1 h at room temperature on a rotator.
6. Collect beads using a magnetic separator and wash them with 15 μl of Bridging Reaction Buffer 1. Distribute the beads among reaction tubes 3 μl each. Collect the beads using a magnetic separator.

Step 2: Binding of MukB to pBIO DNA

1. Resuspend the beads with attached pBIO DNA in 6 μl of Bridging Reaction Buffer 2, add 3 μl of the BSA mixture, and gently mix by pipetting.
2. Add 3 μl of MukB in Storage buffer 2, mix carefully (*see* **Note 13**).
3. Incubate MukB with beads for 30 min at room temperature on a rotator.
4. Collect the beads using a magnetic separator and remove the supernatant.
5. Wash the beads with 12 μl of Bridging Reaction Buffer 1.

Step 3: Capture of Foreign DNA

1. Resuspend the beads with prebound MukB in 6 µl Bridging Reaction Buffer 1 containing 1 mg/ml BSA, add 10 ng of foreign DNA in 6 µl Bridging Reaction Buffer 1 supplemented with 1 mg/ml BSA, and incubate for 30 min with rotation.
2. Collect the beads with a magnet, wash them with 12 µl Bridging Reaction Buffer 1, and collect the beads.
3. Resuspend the beads in 12 µl of Elution Buffer; incubate for 40 min at 55 °C.
4. Briefly spin down the tubes in a microfuge, collect the beads with a magnet.
5. Transfer the supernatant to a clean tube and add DNA loading buffer.
6. Analyze eluted DNA alongside with a serially diluted foreign DNA, which will serve as a calibration standard, using gel electrophoresis through a 0.8% agarose gel in 1× TAE buffer for 2 h at 6 V/cm.
7. Stain the gel with SYBR Gold and visualize DNA on a transilluminator.
8. Quantify the recovered DNA using the serially diluted input DNA as a standard.

4 Notes

1. Prepare and charge the column in advance according to the manufacturer's instructions and then equilibrate it with 30 ml buffer A, supplemented with 20 mM imidazole shortly before use.
2. A significant fraction of MukB elutes in the flow-through fraction. This protein fails to bind DNA or the heparin column and is apparently misfolded.
3. MukB that is purified through nickel-chelate chromatography only is contaminated with highly active ATPases which can be removed by further purification of MukB through Heparin agarose and HiTrap heparin columns. Avoid collecting the shoulder fractions of protein peaks, which contain trace amounts of contaminants with powerful ATPase and nuclease activities. This problem is especially acute given the very low intrinsic ATPase activity of MukB.
4. Protein concentration occurs over several hours and must not be left overnight to avoid drying out the protein. When the volume of the solution is about 3 ml, rinse the bag with buffer B containing 200 mM NaCl and continue dialysis against

Storage buffer 1. Note that the final concentration of MukB should not exceed 3 mg/ml because the protein precipitates at higher concentrations

5. After such purification, MukB exists as a mixture of two populations, only one of which is active in DNA binding and reshaping. To separate these two populations and reactivate the protein, MukB should be further passed through a HiTrap heparin column.
6. The high salt MukB can be stored on ice for up to 1 month without visible loss of any of its activities. Additional purification can be achieved using gel filtration through Sephacryl S300. Note, however, that MukB switches to an inactive conformation during gel filtration and must be reactivated using Hi-Trap heparin chromatography [15].
7. Purification of the PA MksB is very similar to that for the *E. coli* MukB with one key exception. Unlike MukB, MksB elutes from the columns as a single sharp peak and does not alternate its activity during chromatography. Purification of the PA MksEF required the use of the pET21 expression system and a special protocol, which is described in detail elsewhere [10].
8. MukEF can be concentrated by diafiltration using Centricon or Microcon spin columns without substantial loss of protein. Microcon columns must not be used with MukB, MksB and MksEF because of protein precipitation.
9. The saturated MukBEF, $B_2E_4F_2$, is stable only in the presence of $MgCl_2$ and low salt concentration (50 mM NaCl). It can only bind DNA upon dissociation of MukEF. The unsaturated B_2E_2F complex is formed in the absence of $MgCl_2$ and is stable at moderate to high salt concentrations (200 mM NaCl and higher). The DNA reshaping properties of the unsaturated MukBEF are identical to those of MukB.
10. We found that MukB is prone for aggregation at low ionic strength; especially in the presence of magnesium. Therefore, it is important to add MukEF to the reconstitution mixture prior to MukB.
11. Final concentrations of the reagents during reconstitution of $B_2E_4F_2$ should be: 20 mM HEPES–KOH, pH 7.7, 50 mM NaCl, 2 mM $MgCl_2$, 5% glycerol, 1 mM DTT. For B_2E_2F, the concentrations are: 20 mM HEPES–KOH, pH 7.7, 200 mM NaCl, 1 mM EDTA, 5% glycerol, 1 mM DTT.
12. With MukB, treatment of supercoiled DNA with topo I leads to DNA relaxation in the absence but not presence of levels of the protein (Fig. 3c). This assay can serve as an alternative to the DNA gel shift analysis. A frequent concern in the gel shift studies is the possibility of artifacts due to the transfer of the

sample into the gel buffer and dissociation of the complex during gel electrophoresis. The nonrelaxation assay does not require any such transfers and can be used under a broad range of conditions.

13. MukB can stick to magnetic beads. Therefore, control experiments without biotinylated DNA should be done in parallel with the major experiment. The addition of spermidine dramatically reduces MukB stickiness.

Acknowledgments

This work was supported by the awards for the project number HR14-042 from Oklahoma Center for Advancement of Science and Technology to VVR, and HDTRA1-14-1-0019 from the Department of the Defense, Defense Threat Reduction Agency. The content of the chapter does not necessarily reflect the position or the policy of the federal government, and no official endorsement should be inferred.

References

1. Kleine Borgmann LA, Graumann PL (2014) Structural maintenance of chromosome complex in bacteria. J Mol Microbiol Biotechnol 24 (5-6):384–395
2. Reyes-Lamothe R, Nicolas E, Sherratt DJ (2012) Chromosome replication and segregation in bacteria. Annu Rev Genet 46:121–143
3. Rybenkov VV, Herrera V, Petrushenko ZM, Zhao H (2014) MukBEF, a chromosomal organizer. J Mol Microbiol Biotechnol 24 (5–6):371–383. PMC4377647
4. Cobbe N, Heck MM (2004) The evolution of SMC proteins: phylogenetic analysis and structural implications. Mol Biol Evol 21 (2):332–347
5. Danilova O, Reyes-Lamothe R, Pinskaya M, Sherratt D et al (2007) MukB colocalizes with the oriC region and is required for organization of the two Escherichia coli chromosome arms into separate cell halves. Mol Microbiol 65(6):1485–1492
6. Niki H, Jaffe A, Imamura R, Ogura T et al (1991) The new gene mukB codes for a 177 kd protein with coiled-coil domains involved in chromosome partitioning of E. coli. EMBO J 10(1):183–193
7. Petrushenko ZM, Lai CH, Rybenkov VV (2006) Antagonistic interactions of kleisins and DNA with bacterial condensin MukB. J Biol Chem 281(45):34208–34217
8. Mascarenhas J, Soppa J, Strunnikov AV, Graumann PL (2002) Cell cycle-dependent localization of two novel prokaryotic chromosome segregation and condensation proteins in Bacillus subtilis that interact with SMC protein. EMBO J 21(12):3108–3118
9. Soppa J, Kobayashi K, Noirot-Gros MF, Oesterhelt D et al (2002) Discovery of two novel families of proteins that are proposed to interact with prokaryotic SMC proteins, and characterization of the Bacillus subtilis family members ScpA and ScpB. Mol Microbiol 45 (1):59–71
10. Petrushenko ZM, She W, Rybenkov VV (2011) A new family of bacterial condensins. Mol Microbiol 81(4):881–896
11. She W, Mordukhova E, Zhao H, Petrushenko ZM et al (2013) Mutational analysis of MukE reveals its role in focal subcellular localization of MukBEF. Mol Microbiol 87 (3):539–552
12. Stray JE, Crisona NJ, Belotserkovskii BP, Lindsley JE et al (2005) The Saccharomyces cerevisiae SMC2/4 condensin compacts DNA into (+) chiral structures without net supercoiling. J Biol Chem 280(41):34723–34734
13. Hirano M, Hirano T (1998) ATP-dependent aggregation of single-stranded DNA by a bacterial SMC homodimer. EMBO J 17 (23):7139–7148

14. Shin HC, Lim JH, Woo JS, Oh BH (2009) Focal localization of MukBEF condensin on the chromosome requires the flexible linker region of MukF. FEBS J 276(18):5101–5110
15. Petrushenko ZM, Lai CH, Rai R, Rybenkov VV (2006) DNA reshaping by MukB. Right-handed knotting, left-handed supercoiling. J Biol Chem 281(8):4606–4615
16. Petrushenko ZM, Cui Y, She W, Rybenkov VV (2010) Mechanics of DNA bridging by bacterial condensin MukBEF in vitro and in singulo. EMBO J 29(6):1126–1135. PMC2845270
17. Kimura K, Rybenkov V, Crisona N, Hirano T et al (1999) 13S Condensin actively reconfigures DNA by introducing global positive writhe: implications for chromosome condensation. Cell 98:239–248
18. Plasterk RHA, Ilmer TAM, van de Putte P (1983) Site-specific recombination by Gin of bacteriophage Mu: inversions and deletions. Virology 127:24–36
19. Dynan WS, Jendrisak JJ, Hager DA, Burgess RR (1981) Purification and characterization of wheat germ DNA topoisomerase I (nicking-closing enzyme). J Biol Chem 256:5860–5865
20. Kimura K, Hirano T (1997) ATP-dependent positive supercoiling of DNA by 13S condensin: a biochemical implication for chromosome condensation. Cell 90:625–634
21. Woo JS, Lim JH, Shin HC, Suh MK et al (2009) Structural studies of a bacterial condensin complex reveal ATP-dependent disruption of intersubunit interactions. Cell 136:85–96

Chapter 13

Exploring Condensins with Magnetic Tweezers

Rupa Sarkar and Valentin V. Rybenkov

Abstract

The signature activity of condensins as DNA reshaping machines is their ability to impose the giant loop architecture onto the chromosome. At the heart of this activity lies the propensity of the proteins to assemble into macromolecular clusters that bring distant DNA segments together. This gives rise to a rich dynamic behavior when the proteins are presented with the DNA substrate. The protocols in this section describe how the interaction between *Escherichia coli* condensin MukB and DNA proceeds in real time as observed using magnetic tweezers.

Key words MukB, MksB, SMC, Magnetic tweezers, Chromatin structure, DNA topology, Condensins

1 Introduction

The ability of condensins to control the global folding of the chromosome results from a combination of several elementary DNA distorting activities [1–5]. These distortions are transient in nature and occur on a scale of several-to-many nanometers, which makes these proteins highly suitable for studies using various single DNA stretching approaches [4, 5]. In this chapter, we describe a protocol for magnetic tweezers manipulation of bacterial condensin MukB.

Two experimental designs proved to be highly informative in force spectroscopy studies of MukB. The first, conventional approach involves stretching a single DNA using magnetic tweezers and then exploring energetics and kinetics of protein-mediated DNA condensation and decondensation [4, 6]. This setup is particularly useful for investigation of cooperativity of condensin–DNA interactions and geometry and energetics of the protein-induced DNA distortions. The second approach focuses on DNA bridging and involves stretching of two DNAs between the surface of a capillary and a magnetic bead [5]. In both cases, a particular sequence of DNA stretching and twisting events needs to

Olivier Espéli (ed.), *The Bacterial Nucleoid: Methods and Protocols*, Methods in Molecular Biology, vol. 1624,
DOI 10.1007/978-1-4939-7098-8_13, © Springer Science+Business Media LLC 2017

be designed to address one or another aspect of condensin–DNA interaction.

2 Materials

1. Magnetic tweezers. This is still a custom-made instrument that required a significant effort for the assembly. The experiments described here used the tweezers that are based on a Leica DMIRE2 microscope equipped with a JAI CV M30 camera and controlled by a home-written program MagTweezer [4].
2. Reaction buffer: 20 mM Hepes–KOH, pH 7.7, 2 mM $MgCl_2$, 40 mM NaCl, 7% (v/v) glycerol, 1 mM DTT, 0.1% (w/v) Tween 20, 1 mg/ml BSA.
3. PBS: 137 mM NaCl, 2.7 mM KCl, 4.3 mM Na_2HPO_4, 1.4 mM KH_2PO_4.
4. Passivation buffer: 1× PBS, 10 mg/ml BSA, 10 mM NaN_3.
5. Capillary Storage buffer: 1× PBS, 0.1% Tween 20, and 10 mM NaN_3.
6. Hollow square glass capillaries of 1.00 **mm** square ID, 0.20 **mm** wall thickness, borosilicate glass, 50 **mm** length (VitroCom # ST-8100-050).
7. Biotin-16-dUTP (Roche Life Sciences #11093070910).
8. Digoxigenin-11-dUTP (DIG-dUTP) (Roche Life Sciences # 11573152910).
9. dATP, dGTP, dCTP, dTTP (Invitrogen).
10. Taq DNA polymerase (New England Biolabs).
11. 1× Taq reaction buffer (New England Biolabs).
12. pBB10 or pBB10BD DNA (*see* **Note 1**).
13. Anti-digoxigenin antibody, Fab fragments (Roche Applied Sciences #11214667001).
14. Streptavidin-coated magnetic beads; Dynabeads M-270 (Invitrogen).
15. Toluene.
16. Polystyrene, MW 44,000 (Aldrich # 330345).
17. A Labquake Rotator.
18. A PCR thermocycler.

3 Methods

The experiments described here are done using a simple setup [4, 7] that employs square glass capillaries coated inside with an

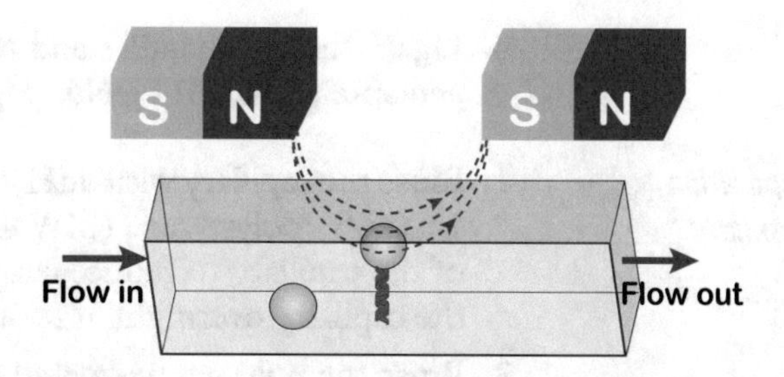

Fig. 1 The flow chamber setup for magnetic tweezers. DNA is stretched using magnetic field between a paramagnetic bead and the surface of a glass capillary

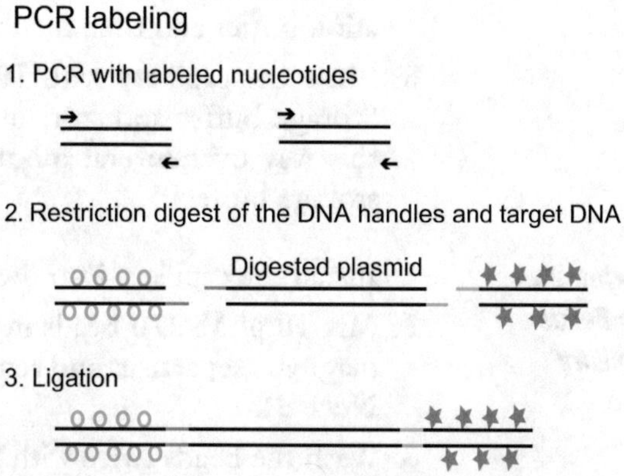

Fig. 2 A strategy for DNA end labeling. DNA handles with multiple incorporated biotin- or DIG-derivatized dUTP are prepared by PCR. A sticky ended nonderivatized linear DNA is made by a restriction digest of a plasmid. DNA is digested with restriction enzymes in order to generate sticky ends compatible to the unlabeled fragment and, finally, is ligated to it

anti-DIG antibody (Fig. 1). We use the 8.3-kb pBB10 DNA, which was ligated to two 0.2-kb DNA fragments, which are multiply labeled with either biotin or digoxigenin (DIG) using PCR [4]. Any DNA with multiply labeled DNA ends can be used in these experiments. The DNA is attached to streptavidin-coated magnetic beads, injected into the capillary, and then stretched using a pair of permanent magnets placed above the capillary [4].

3.1 Preparation of the DNA

1. Generate DIG- and biotin-labeled 400-bp DNA handles using PCR in the presence of DIG-dUTP and biotin-dUTP, as appropriate (*see* **Note 2** and Fig. 2).
2. Digest DNA handles with restriction enzymes to create sticky ends.
3. Digest pBB10 plasmid DNA with appropriate restriction endonucleases to generate a linear 8.3 kb DNA with sticky extremities compatible with the handles.

4. Ligate the two handles and the cut pBB10 DNA together to generate pBB10BD DNA (Fig. 2).

3.2 Preparation of the Capillary

1. Rinse the capillary with ddH_2O and toluene and then fill it with 1% (w/v) polystyrene (MW 45,000) in toluene. Seal the ends of the capillary to reduce evaporation of toluene and incubate the capillary overnight at room temperature.
2. Rinse the polystyrene-coated capillary with 10 ml PBS.
3. Fill the capillary with PBS supplemented with 0.2 mg/ml anti-DIG polyclonal antibody and incubate it overnight at 37 °C.
4. Rinse the capillary with 10 ml PBS, then fill it with the Passivation buffer and then incubate overnight at 37 °C.
5. Rinse the capillary with 10 ml PBS, fill it with the Capillary Storage buffer and store at 4 °C (The capillary can be stored this way over several months, with occasional change of the storage buffer).

3.3 Anchoring DNA to Magnetic Beads and the Capillary

1. Install the capillary into the magnetic tweezers.
2. Mix 10 μl M-270 beads in 15 μl PBS, collect the beads using magnetic separator, and remove the buffer above the beads (*see* **Note 3**).
3. Wash the beads twice with 25 μl PBS as described in **step 2**.
4. Wash the beads twice with 25 μl Reaction buffer.
5. Dissolve the beads into 50 μl Reaction buffer. Add 0.5 ng pBB10BD DNA, mix the beads and DNA thoroughly by pipetting up and down and further incubate on a Labquake rotator for 5 min at room temperature or in the cold room. The bead-tethered DNA can be stored on ice for several hours with no visible nicking.
6. Move the magnets away from the flow cell. Mix 3 μl of bead-tethered DNA with 200 μl of Reaction buffer (the buffer should be room temperature before use to reduce formation of air bubbles during the experiment) and mix them well using pipette. Introduce the beads into the capillary. Incubate for 15 min. Rinse the capillary with 5–10 ml Reaction buffer until no free floating beads are seen. Immediately start searching for a bead with attached DNA (*see* **Note 4**).

3.4 Finding Beads with One and Two Attached DNAs

1. Apply a high force (approximately 10 pN) and visually inspect beads in the capillary. Most of the beads will be stuck to the surface and will appear immobile. Focus the objective onto the immobile beads. Beads that are linked to the surface via DNA will be out of focus and can be recognized by their diffused image.

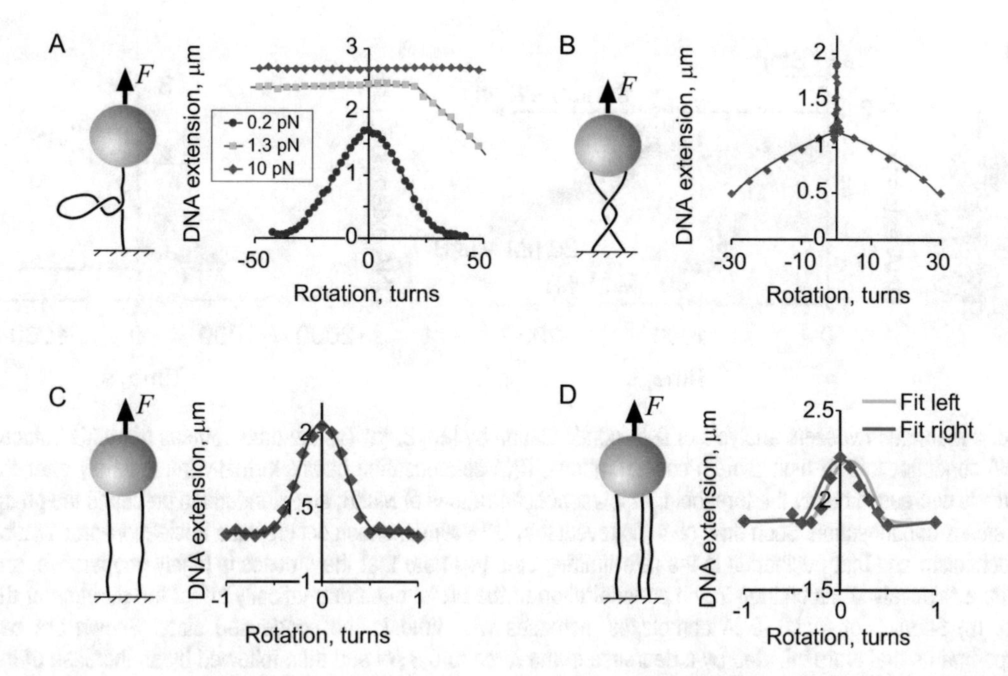

Fig. 3 Hat curves for beads with one (**a**), two (**b**, **c**), or three (**d**) DNAs. Beads with one DNA produce symmetric bell-shaped curves at low force. At higher forces, untwisting causes DNA melting whereas overtwisting induces formation of the P-DNA, both of which suppress formation of DNA supercoils and the resulting DNA contraction [12]. For beads with two DNAs, the hat curves appear symmetric even at high forces but contain a characteristic spike between −0.5 and +0.5 turns. In this range, DNA extension, z, can be computed using the equation: $z^2 = z_0{}^2 - 4E^2\sin^2(\pi n)$, where n is the rotation of the bead expressed in turns, z_0 is the highest DNA extension, which is observed when n equals zero, and $2E$ is the average distance between the anchor points of the two DNAs [11]. For beads with three or more DNAs, the spike is asymmetric and can be fit only poorly to this equation [5]

2. Depending on your needs, select a bead with one or two attached DNAs. To achieve this, rotate the magnets by +30 turns and −30 turns while monitoring the change in DNA extension. For beads with a single DNA, DNA extension is not expected to change. At lower forces, DNA extension should yield bell-shaped curves, which reflect formation of DNA supercoils (Fig. 3a). For beads with two DNAs, DNA extension will produce a characteristic bell-shaped curve with a sharp symmetric spike at ±0.5 turns (Fig. 3b, c). An asymmetric spike signals multiple attached DNAs (Fig. 3d).
3. Build the calibration image of the selected beads and measure the force at various DNA extensions as recommended by your bead tracking software.
4. Inject the protein and carry out single molecule analysis.

3.5 DNA Nanomanipulations

DNA nanomanipulations using magnetic tweezers is a true real time technique and, therefore, the sequence and duration of applied forces and bead rotations can be easily tailored to specific

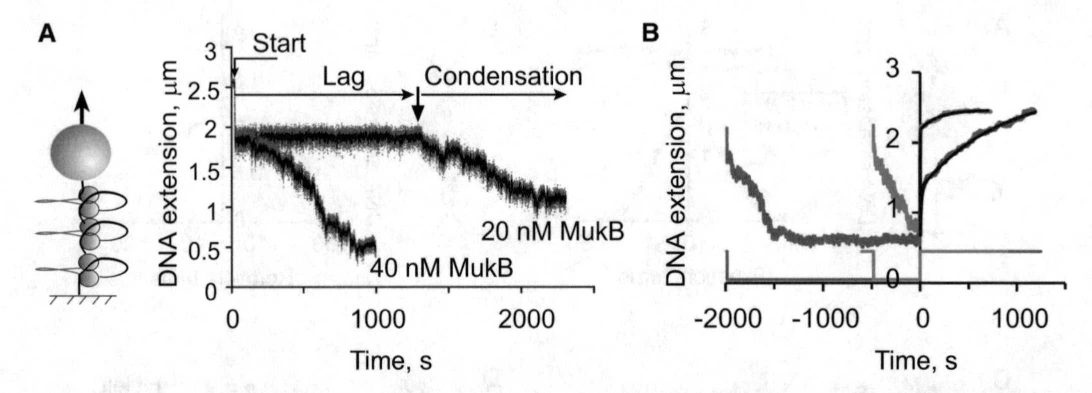

Fig. 4 Magnetic tweezers analysis of DNA condensation by MukB. (**a**) Typical time courses of MukB-induced DNA condensation. At high protein concentrations, DNA condensation begins virtually immediately after the force is decreased below the threshold. At lower concentrations of MukB, an extended lag precedes the phase of steady condensation. Such time courses reveal that DNA condensation occurs via a nucleation-propagation mechanism and that nucleation is the rate-limiting step [4]. Note that the process is highly cooperative, and even a relatively small change in the concentration of the protein can dramatically affect the duration of the lag. (**b**) Stability of MukB–DNA complexes increases with time in the condensed state. Shown are two experiments that were initiated by a decrease in the force to 0.3 pN and then followed by an increase of the force to 10 pN (time zero). *Solid lines* show profiles of the applied force. Note the biphasic nature of decondensation curves, which were fit to a double-exponential decay (*black lines*)

experimental questions. Our exploration of MukB revealed several intriguing aspects of its interaction with DNA and we focus below on some of these results. Other properties of the protein will likely emerge from altered DNA stretching protocols or reaction conditions.

We recommend starting the experiment at high stretching forces (~10 pN). Both MukB and the frog condensin I were only able to condense DNA at forces well below 1 pN [4, 6]. This protocol, therefore, minimizes DNA condensation during poorly controlled conditions of unsettled fluxes. Once the hydrodynamic flows in the chamber stabilize, the bead will rise back closer to the focal plane and away from the surface. DNA manipulations can then be started (Fig. 4).

Because of the high cooperativity of MukB–DNA interaction, the protein must be carefully titrated when starting the experiments. Using protein concentrations that are too high or too low might preclude detection of any activity even if the protein would be binding DNA. For the same reason, the protein must be completely removed from the flow chamber by extensive washing (5–10 ml) prior to the start of a new experiment, well past the moment when DNA appears completely decondensed. Failure to do so results in abnormally high condensation rates due to the residual protein.

The high cooperativity of condensin-induced DNA condensation is manifested as a characteristic lag followed by steady condensation, which is apparent at low but not high protein concentrations (Fig. 4a). Similar time courses could arise due to nonspecific effects such as protein degradation or its sorption onto the surface. To rule out such explanations, it is paramount to explore DNA condensation rates at various protein concentrations and fit the data to a meaningful kinetic model. An example of such quantitative analysis is presented in [4].

The studies on the effects of ATP on condensins must take into account that metals can induce conformational transitions in condensins [4] and that ATP is a powerful chelating agent. ATP must always be added into the reaction as a magnesium salt, MgATP. Otherwise, depletion of magnesium by ATP might mask the effects of nucleotide itself and lead to artifactual observations.

At high force, MukB–DNA complex unravels leading to DNA decondensation. In the first approximation, DNA decondensation can be modeled as a double-exponential decay (Fig. 4b), which suggests the existence of at least two distinct MukB–DNA complexes. Notably, the more stable complex accumulates during incubation at low force (Fig. 4b), and longer incubations yet (or those performed at higher protein concentrations) often yield complexes that are stable throughout the experiment. A similar phenomenon was observed during studies of MukB-induced DNA bridging using the double-DNA setup ([5] and Fig. 6). Such behavior is consistent with the notion that the observed increase in the complex stability is caused by an increase in the number of MukBs that hold DNA fragments together [4, 5]. It must be noted here that application of high forces is not a good way to accelerate disassembly of the stable MukB–DNA complexes. Indeed, high forces (greater than 20 pN) have been reported to unzip DNA without necessarily displacing MukB from DNA.

3.6 Step Sizes

Measurements of the DNA condensation step sizes are exacerbated by the high noise in the observed DNA extensions. This high noise does not appear from any instrumental shortcomings of the used apparatus but reflects the fact that condensins can condense DNA only against very small opposing force (~0.4 pN), when thermal fluctuations are prominent. Using smaller magnetic beads and shorter DNAs partially alleviates this problem [8]; however, the resulting time courses remain noisy, which often precludes unambiguous identification of all DNA shrinkage events. Several automatic algorithms have been designed that allow hands-free deconvolution of condensation time courses into series of distinct steps, and new algorithms are likely to appear in the future [4, 9, 10]. In all these cases, it is important not to over-solve the problem trying to identify each and every step. Rather, the researchers should be aware that the deconvoluted step sequence will inevitably contain false-positive and false-negative steps.

One way to address this problem is to include only statistically significant steps into the final step sequence [4, 10]. Statistical significance of the resulting steps can be determined, for example, using the Student's t-test, which evaluates the probability that the two sequences of time points (i.e., before and after the putative step) occur so by chance. Only steps with T-values higher than a preselected threshold must be included into the deconvolution step series.

Another point to keep in mind is that not all data points are statistically independent. This problem becomes acute at low forces, when thermal fluctuations slow down to typical data acquisition rates, about 30 Hz. In such cases, the Student's T-value must be computed using the generalized least squares approach:

$$T = s/\sqrt{\sigma^2 \cdot \mathrm{VIF} \cdot (1/N_1 + 1/N_2)}, \quad (1)$$

where s is the size of the putative step, σ^2 is the variance of the distribution of the measured DNA extensions, N_1 and N_2 are the numbers of points before and after the step, and VIF is the variance inflation factor, which corrects for the correlation between the acquired time points, x_i. VIF can be computed according to Eq. 2

$$\mathrm{VIF} = \frac{1}{1 - R^2}, \quad (2)$$

where $<x>$ is the time-averaged DNA extension, and R is a measure of auto-correlation within the time series:

$$R = \frac{\sum_i (x_i - <x>)(x_{i+1} - <x>)}{\sum_i (x_i - <x>)^2} \quad (3)$$

This approach is illustrated in Fig. 5.

3.7 DNA Bridging

The use of beads with two attached DNAs offers a powerful tool for exploration of the DNA bridging activity of condensins. These experiments are carried out at high stretching force (greater than ~1 pN) to preclude intramolecular DNA condensation (which would enormously complicate interpretation of the data). The synapse reaction is initiated by bringing the two DNAs into proximity by rotating the bead by half a turn or more. Because of the geometric constraints of the system, such rotation results in a significant DNA contraction, whereas untwisting the DNAs should immediately extend the molecules (Fig. 6 and [5, 11]). A delay in DNA extension signals the formation of the protein-mediated bridge, and the length of the delay reflects the stability of the protein–DNA complex.

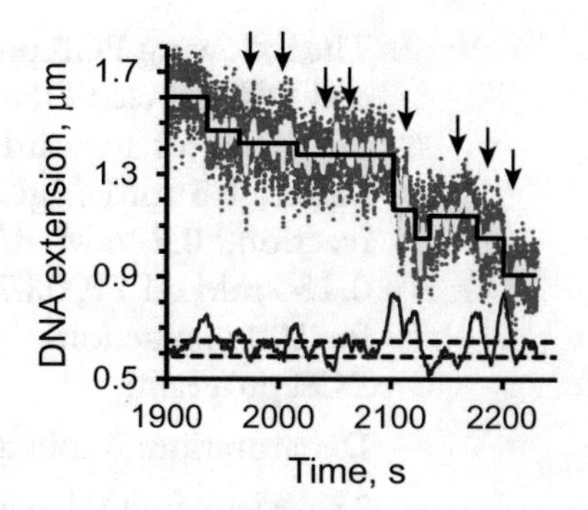

Fig. 5 Deconvolution of DNA condensation into steps. The recorded time points are shown in *dark grey*, the signal filtered at 1 Hz is *light grey*, and the step approximation in *black*. The approach is illustrated using the algorithm developed in [4]. Potential steps are first identified at the time points where the change in the denoised bead position occurs faster than a threshold (*line* on the *bottom*). The time course is then approximated as a series of trial steps, which is further filtered to generate the final step series. *Arrows* mark the trial steps that were discarded during filtration. The cutoff *T*-value for the acceptance of the trial and final steps was set at 1 and 3, respectively

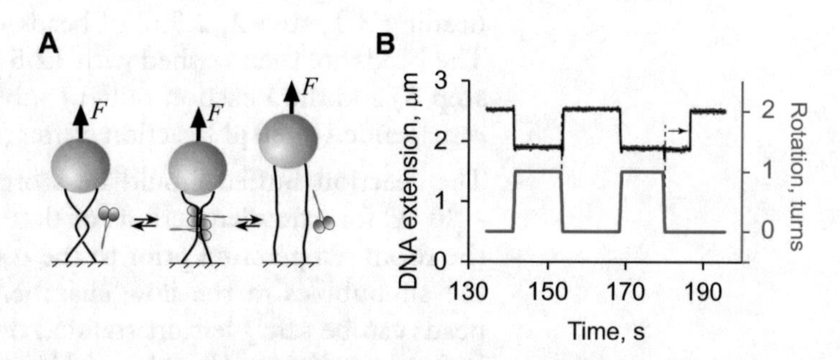

Fig. 6 A diagram (**a**) and an example (**b**) of magnetic tweezers analysis of DNA bridging. The reaction is initiated by rotating the magnets by one turn to bring the two DNAs into contact and then allowed to proceed for a desired time. The DNAs are then untwisted, by bringing the rotation back to zero, and the delay in the restoration on the DNA extension (marked with an *arrow*) serves as a measure of the life-time of the protein–DNA bridge. Note that the bridge is formed during the second crossing event only

4 Notes

1. The authors used the 8.3-kb pBB10 plasmid cut with NcoI and BamHI [3] in the described studies. Any DNA without large long AT-rich stretches can be used instead. The DNA was appended with DNA handles as described in the Subheading 3.1 to generate pBB10BD (Fig. 2).

2. The following PCR protocol can be used for generating biotin- and DIG-labeled DNA handles:

   1 μM of forward and reverse primers, 1× Taq reaction buffer, 1.5 mM $MgCl_2$, 30 ng of pBB10 DNA for a 100 μl reaction, 0.2 mM dATP, 0.2 mM dGTP, 0.2 mM dCTP, 0.130 mM dTTP, 0.070 mM DIG-dUTP, 4 U Taq Polymerase for 100 μl reaction.

   PCR program:

   Denaturation: 3 min at 95 °C

   23 cycles of: (1) denaturation for 40 s at 94 °C, (2) annealing for 40 s at 56 °C, (3) extension for 2 min at 72 °C

   Final Extension: 5 min at 72 °C

   The long extension time (2 min) in the PCR program is needed to achieve efficient amplification of DIG-labeled DNA fragment. For biotin, one can reduce this time to 1 min.

3. The shown protocol is optimized to ensure high yield of beads with a single attached DNA. If beads with two DNAs are needed, the protocol needs to be modified as follows: In Subheading 3.1, **step 2**, 2.5 μl of beads is mixed with 10 μl PBS. The beads are then washed with 12.5 μl PBS (Subheading 3.1, **step 3**) and the Reaction buffer (Subheading 3.1, **step 4**) and resuspended in 20 μl Reaction buffer (Subheading 3.1, **step 5**).

4. The reaction buffer should be stored frozen at −20 °C or −80 °C for extended periods of time but must be brought to the room temperature prior to the use to reduce formation of the air bubbles in the flow chamber. The capillary with the beads can be safely left unattended overnight if filled with any buffer containing 10 mM NaN_3, which prevents bacterial growth.

Acknowledgments

This work was supported by the awards for the project number HR14-042 from Oklahoma Center for Advancement of Science and Technology to VVR, and HDTRA1-14-1-0019 from the Department of the Defense, Defense Threat Reduction Agency. The content of the chapter does not necessarily reflect the position or the policy of the federal government, and no official endorsement should be inferred.

References

1. Hirano M, Hirano T (1998) ATP-dependent aggregation of single-stranded DNA by a bacterial SMC homodimer. EMBO J 17 (23):7139–7148
2. Petrushenko ZM, Lai CH, Rybenkov VV (2006) Antagonistic interactions of kleisins and DNA with bacterial condensin MukB. J Biol Chem 281(45):34208–34217
3. Petrushenko ZM, Lai CH, Rai R, Rybenkov VV (2006) DNA reshaping by MukB. Right-handed knotting, left-handed supercoiling. J Biol Chem 281(8):4606–4615
4. Cui Y, Petrushenko ZM, Rybenkov VV (2008) MukB acts as a macromolecular clamp in DNA condensation. Nat Struct Mol Biol 15 (4):411–418
5. Petrushenko ZM, Cui Y, She W, Rybenkov VV (2010) Mechanics of DNA bridging by bacterial condensin MukBEF in vitro and in singulo. EMBO J 29(6):1126–1135. PMC2845270
6. Strick TR, Kawaguchi T, Hirano T (2004) Real-time detection of single-molecule DNA compaction by condensin I. Curr Biol 14 (10):874–880
7. Strick TR, Allemand J-F, Bensimon D, Croquette V (1996) The elasticity of a single supercoiled DNA molecule. Science 271:1835–1837
8. Revyakin A, Ebright RH, Strick TR (2005) Single-molecule DNA nanomanipulation: improved resolution through use of shorter DNA fragments. Nat Methods 2(2):127–138
9. Kerssemakers JW, Munteanu EL, Laan L, Noetzel TL et al (2006) Assembly dynamics of microtubules at molecular resolution. Nature 442(7103):709–712
10. Sun Z, Tikhonova EB, Zgurskaya HI, Rybenkov VV (2012) Parallel lipoplex folding pathways revealed using magnetic tweezers. Biomacromolecules 13(10):3395–3400
11. Charvin G, Vologodskii A, Bensimon D, Croquette V (2005) Braiding DNA: experiments, simulations, and models. Biophys J 88 (6):4124–4136
12. Allemand JF, Bensimon D, Lavery R, Croquette V (1998) Stretched and overwound DNA forms a Pauling-like structure with exposed bases. Proc Natl Acad Sci U S A 95 (24):14152–14157

Chapter 14

Applications of Magnetic Tweezers to Studies of NAPs

Ricksen S. Winardhi and Jie Yan

Abstract

Nucleoid-associated proteins (NAPs) are important factors in shaping bacterial nucleoid and regulating global gene expression. A great deal of insights into NAPs can be obtained through studies using single DNA molecule, which has been made possible owing to recent rapid development of single-DNA manipulation techniques. These studies provide information on modes of binding to DNA, which shed light on the mechanism underlying the regulatory function of NAPs. In addition, how NAPs organize DNA and thus their contribution to chromosomal DNA packaging can be determined. In this chapter, we introduce transverse magnetic tweezers that allows for convenient manipulation of long DNA molecules, and its applications in studies of NAPs as exemplified by the *E. coli* H-NS protein. We describe how transverse magnetic tweezers is a powerful tool that can be used to characterize the DNA binding and organization modes of NAPs and how such information leads to better understanding of its roles in DNA packaging of bacterial nucleoid and transcription regulation.

Key words Protein–DNA interactions, Magnetic tweezers, Gene regulation, Chromosomal packaging, Nucleoid-associated proteins (NAPs), H-NS

1 Introduction

Bacterial nucleoid is a highly compact yet dynamic structure that contains essential genetic information [1, 2]. There are few factors that play a major role in nucleoid organization, namely DNA supercoiling [3], macromolecular crowding [4, 5], and a set of DNA-binding proteins collectively referred to as nucleoid-associated proteins (NAPs) [6–8]. In addition "Applications of Magnetic Tweezers to Studies of NAPs" to their DNA architectural role, NAPs also play an important role in regulating bacterial transcription, by gene silencing or antisilencing [1, 8]. This dual role of NAPs is achieved mainly via their DNA binding activities that alter the DNA conformation and topology. As such, studies of DNA–NAPs interaction will provide valuable insights into our understanding of bacterial nucleoid.

The interactions between DNA and NAPs have been extensively studied in biochemical bulk assays, usually related to their

Olivier Espéli (ed.), *The Bacterial Nucleoid: Methods and Protocols*, Methods in Molecular Biology, vol. 1624,
DOI 10.1007/978-1-4939-7098-8_14, © Springer Science+Business Media LLC 2017

function in regulating gene expression. Most of the commonly used methods include electrophoretic mobility shift assay (EMSA), DNase I footprinting assay, chromatin immunoprecipitation (ChIP), X-ray crystallography, and many others [9]. These techniques provide important information regarding regulatory site, function, and network, but they carry less information on the physical aspects of the regulation, such as how the conformation and topology of DNA–protein complex give rise to the regulatory functions. Increasing amount of evidences has also revealed that the physical aspect of protein binding is a crucial determinant of the protein's regulatory functions [10–12]. Another limitation of the traditional biochemical bulk assays is that most of them do not provide information on the roles of NAPs on nucleoid organization. As such, there is a gap between the knowledge derived from the biochemical bulk assays and the physical mechanism underlying their function. To fill this gap, technologies that allow probing of protein–DNA interactions at single DNA level are needed.

There are several methods that allow single DNA manipulation, namely magnetic tweezers, optical tweezers, and atomic force microscopy (AFM) [13]. These methods apply tensile force to the two ends of a single DNA to measure its mechanical response. In this chapter, we focus on magnetic tweezers, which is a simple and versatile tool that offers a few advantages over other techniques, including the capability of applying torque to DNA, force constraint operation, as well as low mechanical and thermal drift for long-time measurements [13–16].

1.1 Development of Transverse Magnetic Tweezers

Most magnetic tweezers are typically designed with direction of stretching force perpendicular to the focal plane, often referred to as vertical magnetic tweezers. In this design, the diffraction patterns of the bead are used to localize the bead along the force direction with a nanometer resolution [17, 18]. The sensitivity of this bead localization depends on several factors, including the distance from the focal plane. Another configuration is transverse magnetic tweezers [19, 20], in which the direction of stretching force is in the focal plane (Fig. 1a). The molecule of interest can be tethered on a fixed surface (e.g., a coverslip edge) on one end and a superparamagnetic bead on the other end. The accuracy of bead localization by determining its centroid is typically around a few nanometers depending on the objective lens magnification and the pixel size of camera. The standard error of extension determination depends on the level of fluctuation of the bead and the time window used to do the average (*see* Subheading 1.3). Thus, the extension of the molecule can be directly obtained from the centroid position of bead relative to a reference point.

Transverse magnetic tweezers were first reported in 2004, where the DNA was tethered in between a polystyrene bead fixed on a micropipette and a superparamagnetic bead [19, 20]. The

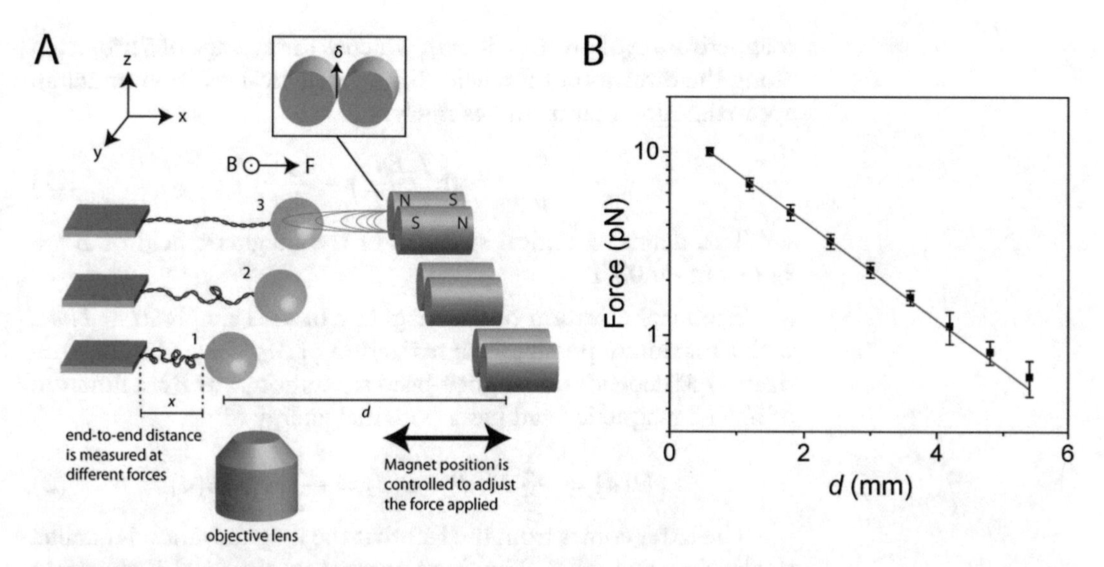

Fig. 1 (**a**) In transverse magnetic tweezers, the molecule of interest is typically tethered to a coverslip on one end and a superparamagnetic bead on the other end. Force is produced by the magnetic field generated from a pair of oppositely oriented permanent magnet, and controlled by adjusting bead–magnet distance *d*. In this configuration of magnetic tweezers, DNA molecules are stretched in the focal plane. End-to-end distance of single double-stranded DNA molecule varies depending on the stretching forces. At lower forces, the DNA adopts a more coiled conformation (scenario 1), while extended conformation is favored at higher forces (scenario 2 and scenario 3). (**b**) Force–distance profile in magnetic tweezers. Force decays exponentially with respect to distance away from the magnetic bead. This feature allows forces at other bead–magnet distance values to be obtained via interpolation.

extension was determined by the center-to-center distance between the two beads. Later on, the design was evolved to replace the bead micropipette with the edge of a coverslip [21–25] or the wall of a small square capillary tube [26, 27], which increased the throughput of tether formation. The DNA can be tethered to the surface and to the superparamagnetic bead in many ways, including streptavidin–biotin, digoxigenin–anti-digoxigenin, sulfo-SMCC conjugation to thiol, and click chemistry labeling. This setup is contained within a chamber that allows convenient buffer exchange (*see* Subheading 3.2.2), and a pair of magnets is used to apply force on the superparamagnetic beads.

1.2 Magnetization Properties of Superparamagnetic Beads

The superparamagnetic beads used in magnetic tweezers experiments contain many small single-domain magnetic nanoparticles of around 8 nm in size, which is made of maghemite [28]. Each particle can be considered as a magnetic dipole with a magnetic moment of $\mu = \text{volume} \times 340\ \text{kA/m} \approx 1.6 \times 10^{-19}\ \text{A m}^2$ (or J/T). Under a magnetic field $\vec{B} = B\hat{y}$, the dipoles interact with the field with a potential energy of $-\vec{\mu} \bullet \vec{B}$, which tends to align the dipoles along the direction of the field. Competition with thermal fluctuation results in a Boltzmann distribution of the alignment. The magnetization of one dipole, m_1, which is the average of $\vec{\mu}$ projected along the direction of the field, has an analytical solution under an approximation that it rotates freely:

$$\frac{m_1}{\mu} = \coth\left(\frac{B\mu}{k_B T}\right) - \frac{k_B T}{B\mu} \tag{1}$$

This defines a critical strength of the magnetic field of $B^* = k_B T/\mu$ of ~0.03 T.

The magnetization of the magnetic bead is simply $M = Nm_1$, with a maximum possible magnetization of $M_0 = N\mu$. The magnetization M depends on magnet–bead separation d, as B is a function of d. The magnetic bead has a potential energy of:

$$U(d) = -\frac{1}{2}\vec{M}(d) \bullet \vec{B}(d) = -\frac{1}{2}M(d)B(d) \tag{2}$$

The latter comes from the fact that the magnetization is parallel to the direction of B. The force exerted to the bead is therefore $F = -U'(d)$, where prime mark denotes a derivative. At small bead–magnet distance of d where $B \gg B^*$, $M \sim M_0$ can be approximated as a constant and force decays in proportion to B. At large bead–magnet distance of d where $B \ll B^*$, $M \propto B$, $U \propto B^2$, and therefore $F \propto BB'$.

1.3 Force Calibration

A bead tethered to a DNA under a force F along the x-direction undergoes thermal fluctuation. Along a direction perpendicular to the force (y-direction), the motion of the bead can be considered as a bead linked to a spring with the origin located at the intersection of x- and y-directions. The effective spring constant is $k_\perp = F/x_F$, where x_F is the average extension at force F. Along the force direction, the bead fluctuates around the equilibrium extension under force, which can also be approximated as a bead linked to a spring with the equilibrium extension as the origin. The spring constant is the derivative of the force–extension curve of the DNA, $k_{||} = f'(x_F)$.

The correlation time of bead motions along the two directions is $2\pi\xi/k_i$, where $\xi = 6\pi\eta r$ is the drag coefficient and $i = \perp$ or $||$ denotes the direction. η is the fluid viscosity and r is the Stokes radius of the particle. For example, using 2.8-μm magnetic bead to apply a force of 1 pN to a DNA with an extension of 10 μm, the correlation time in the direction perpendicular to the force direction is ~1.7 s, and the correlation time along the force direction is ~3.5 s. In order to accurately measure force based on fluctuation of the bead, the sampling rate should be much faster than the reciprocal of the correlation times, and the measurement time window should be much longer than the correlation times. As these conditions are met, the measurements of force–extension curves are done in thermal equilibrium, and the results and interpretations are independent from the size of the beads used. For example, the 2.8-μm M270 or M280 bead is often used by many magnetic tweezers groups because it can provide a large force range up to 100 pN [16], while 1-μm bead is often used for applying smaller force [29].

A pair of permanent magnet produces an external magnetic field, resulting in the bead experiencing a force directed toward the magnet (Fig. 1a). The DNA is thus stretched along the focal plane. The pair of magnets can be mounted on a micromanipulator, such that the bead–magnet distance *d* and hence the force *F* exerted on the magnetic bead can be adjusted. To measure the elastic properties of DNA, we need both force and extension data. The extension data at a given force is obtained from the projected end-to-end distance of DNA along the force direction (from glass surface to bead centroid, with an arbitrary offset). The force at a certain magnet position can be obtained by the following equation which is independent from the mechanical property of DNA (*see* appendix of ref. 30):

$$F = \frac{k_B T x}{{\delta_y}^2} \tag{3}$$

where k_B is the Boltzmann constant, T is the temperature, x is the end-to-end distance of DNA, and δ_y is the fluctuation of the bead in the y-direction.

In experiments, the force–distance profile for M270 Dynabeads in the magnetic field produced by neodymium rod magnets was shown to follow a single exponential decay function in the force range below 20 pN that is relevant to most of single-DNA manipulation studies of protein binding [16, 31, 32]. Utilizing this feature, one does not need to calibrate force based on bead fluctuation at each bead–magnet distance *d*. One simply needs to measure the force by Eq. 3 at a few bead–magnet distances (*see* Fig. 1b), and forces at other values of *d* for the same bead can be obtained by interpolation [16]. This procedure provides a quick and accurate way of determining force in magnetic tweezers. The force is transmitted to the tethered molecule, thus establishing a tension in the molecule. The extension can be measured at different forces by localization of the bead in real time. The dependence of extension on the applied force, the so-called force–extension curve, carries important information on the mechanical response of the molecule.

1.4 Force–Extension Curve

The force and extension data obtained in single DNA stretching experiments determine the elastic properties of DNA, which are well described by the worm-like chain model (WLC) with its bending stiffness characterized by the bending persistence length [33]. The force–extension data are thus fitted with Marko–Siggia formula [34], which is an interpolation formula that approximates the force–extension behavior for ideal worm-like chain:

$$\frac{FA}{k_B T} = \frac{1}{4}\left(1 - \frac{x}{L_0}\right)^{-2} - \frac{1}{4} + \frac{x}{L_0} \tag{4}$$

where A is the persistence length that describes the bending rigidity of DNA chain and L_0 is the contour length of the DNA chain. The persistence length of DNA is about 50 nm at physiological solution condition [35–37]. The DNA bending persistence length A also defines a characteristic force of $k_B T/A$, which is ~0.08 pN for dsDNA. At forces much greater than $k_B T/A$, the exact analytical solution was $\frac{FA}{k_B T} = \frac{1}{4\left(1 - x/L_0\right)^2}$. At forces much lower than the characteristic force, the exact solution was also known: $\frac{FA}{k_B T} = \frac{3}{2}\frac{x}{L_0}$. The Marko–Siggia interpolation formula converges and reproduces exact solution in these two force regimes. At forces comparable to $k_B T/A$, deviation occurs with maximum difference up to 10% [36, 38]. Due to its simplicity and usefulness, Marko–Siggia formula is widely used to fit single DNA stretching data.

As protein binding to DNA often results in local structural deformations, the elastic properties of DNA can be used to study protein–DNA interactions as reflected in the force–extension curves [39]. As sketched in Fig. 2, if a protein causes stiffening of the DNA backbone, the equilibrium extension of the DNA will be longer than that of the naked DNA, with more prominent effect at lower forces [10, 21, 40, 41]. In contrast, a DNA-bending protein will have an opposite effect [42–45]. A DNA-intercalating protein that elongates DNA backbone will cause a longer extension especially at larger forces as compared to naked DNA [46]. A protein that causes condensation of DNA will be manifested by nonequilibrium progressive extension reduction at low forces, which often

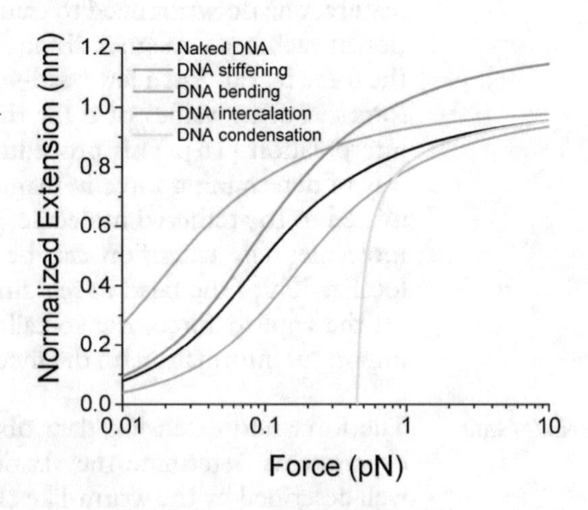

Fig. 2 The effect of DNA distortion upon protein binding to the force–extension curves. Protein binding to DNA causes physical deformation and the resulting nucleoprotein complexes have a different elastic behavior compared to naked DNA. The binding and organization mode of NAPs can be inferred from the changes in elastic properties as reflected in the force–extension curves

results in hysteresis between force–extension curves recorded in a force-decrease scan and a following force-increase scan [47–49]. These various changes to force–extension curves are used to decipher the mechanism of protein binding to DNA.

1.5 Application of Transverse Magnetic Tweezers

Transverse magnetic tweezers technology has contributed significantly to deciphering the binding and organization mode of many NAPs as inferred from their effect on DNA elasticity [21, 26, 40, 41, 44, 47, 48, 50, 51]. These studies also determined various crucial aspects of NAPs' activities, such as the nature of DNA distortion at the binding sites, the global organization of DNA at large length scales, binding affinity, binding cooperativity, and kinetics of binding [10, 36, 39, 47, 48, 52].

In this chapter, interaction between H-NS and DNA is taken as an example of how single DNA manipulation can be used to study protein–DNA interactions. H-NS is an abundant NAP, which plays important roles in global gene silencing and organization of bacterial nucleoid [53]. Previous single-molecule manipulation studies have revealed that H-NS predominantly forms nucleoprotein filaments on DNA at 50 mM KCl and <2 mM $MgCl_2$, while bridging starts to predominate at higher $MgCl_2$ concentrations [10, 21, 54]. These DNA binding and organization modes have provided important clues in understanding how H-NS functions in gene regulation and chromosomal DNA packaging [12].

In single DNA manipulation, the stiffening and bridging actions introduce different physical deformations that are reflected in the force–extension curves. In the case of H-NS, DNA binding and subsequent formation of a rigid nucleoprotein filament at low $MgCl_2$ concentration is expected to cause extension increase compared to naked DNA at the same force (*see* Fig. 3a, scenario 1 and 2). At higher $MgCl_2$ concentration, H-NS forms DNA bridges and nonequilibrium DNA condensation is expected. As a result, the extension of H-NS–DNA complex should progressively fall below that of the naked DNA at the same force (*see* Fig. 3a, scenario 3). Indeed, these scenarios were observed in experiments, as shown in Fig. 3b for naked DNA (black data), H-NS–DNA complex in 50 mM KCl (*blue* data), and H-NS–DNA complex in 50 mM KCl, 10 mM $MgCl_2$ (*purple* data). Dynamics of protein–DNA interaction can also be directly observed, e.g., DNA folding caused by H-NS bridges [21].

1.6 Interpretations of Data Obtained from Magnetic Tweezers

The data obtained from single DNA manipulation can reveal wealth of information regarding the mechanism of protein binding to DNA, especially on the structural distortions caused by protein binding (*see* Subheading 1.4). Single-molecule experiments also allow interaction between a single DNA with controlled concentration of protein. Therefore, it can quantify binding affinity and cooperativity of protein binding. For DNA stiffening proteins

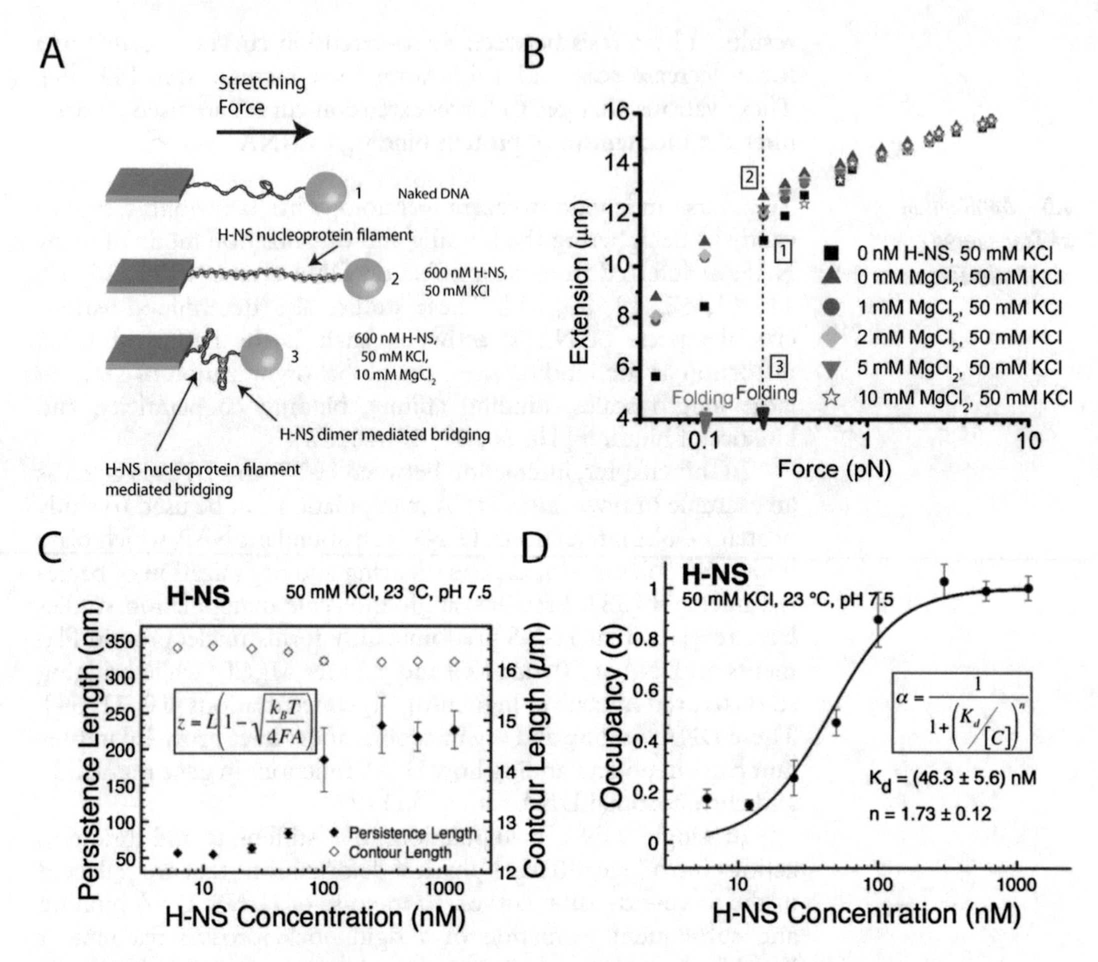

Fig. 3 (**a**) Schematic diagram depicting the length of naked DNA (scenario 1), H-NS–DNA complexes in 50 mM KCl (scenario 2), and H-NS–DNA complexes in 50 mM KCl, 10 mM $MgCl_2$ (scenario 3) at the same stretching force. The DNA length varies as a result of protein binding. The experimental data that corresponds to these illustrated scenarios is shown in panel (**b**) (see scenario 1–3). (**b**) Force–extension data of DNA and H-NS–DNA complexes in several different buffer conditions. H-NS predominantly forms rigid nucleoprotein filaments in 50 mM KCl, while DNA bridging formation is favored at sufficiently high $MgCl_2$ concentration (reproduced from ref. 21 with permission from Cold Spring Harbor Laboratory Press). (**c**) Bending persistence length and contour length of H-NS–DNA complexes at various concentrations were obtained by fitting with the Marko–Siggia formula. (**d**) DNA occupancy (computed based on persistence length) was fitted with the Hill equation to obtain the Hill coefficient and dissociation constant of H-NS binding to DNA (panel **c** and **d** are reproduced from ref. 50 with permission from American Society for Biochemistry and Molecular Biology)

such as H-NS, the persistence length of H-NS–DNA complexes can be measured at various concentrations (Fig. 3c). The protein occupancy on DNA can be approximated from the experimentally measured persistence length and fitted to Hills equation to obtain dissociation constant and Hill coefficient, which describe the cooperativity of protein binding (Fig. 3d). In principle, the affinity and cooperativity of DNA bending proteins can be determined in a similar manner. In the case of DNA intercalating proteins, the affinity and cooperativity can be determined based on the amount of contour length elongation ΔL_0, which is proportional to the number of bound molecules at different protein concentration.

The binding affinity obtained in such single-DNA stretching experiments depends on the force applied to DNA. Obviously, force will increase the affinity for DNA stiffening proteins, while the affinity for DNA bending proteins is decreased. In order to compare with the affinity measured in bulk assays, the results must be extrapolated to zero force. In general, the dissociation constant of a DNA binding protein under force, $K_D(F)$, is related to that measured in the absence of force in bulk assay, $K_D(0)$, by $K_D(F) = K_D(0)e^{\beta\Delta\phi_G(F)}$. Here $\Delta\phi_G(F) = -\int_0^F (x_p(f) - x_d(f))df$ is the force-dependent entropic conformational free energy difference between the bound and unbound states of DNA [55]. Since the force–extension data of the bound and unbound states of DNA can be measured, the extrapolation of K_D to zero force can be done easily.

1.7 Features and Limitations of Transverse Magnetic Tweezers

Transverse magnetic tweezers is capable of stretching very long DNA tethers, and is limited only by the camera field of view. Using 50× objective lens with a camera resolution of 640 × 480 pixels used in our experiments as an example, the maximum DNA extension that can be detected is ~100 μm. This setup also allows longer DNA (>100 μm) to be stretched by mounting the sample on a high-precision motorized X-Y stage, which allows tracking of the magnetic bead position beyond the camera field of view via sample stage movement. With such motorized X-Y stage, the size of the molecule that can be manipulated is virtually unlimited.

On the other hand, the minimum length of DNA molecule that can be reliably stretched depends on how the DNA is attached. On one end, the DNA is attached to a magnetic bead, while at the other end the DNA can be attached to a coverslip surface, a coverslip edge [21] (detailed in this chapter), or to another bead that is either held by a micropipette or immobilized on the surface [19]. If it is directly attached to the bottom surface, force is generated at an angle above the surface and the minimum extension of DNA that can be stretched is when the bead begins to interact with the surface, which is roughly the size of the bead. If the other end is attached to the edge of a coverslip, force is generated on the focal plane and the shortest extension that can be stretched is roughly 3 μm due to the coverslip edge shadow that prevents accurate determination of bead position. If the other end of DNA is attached to surface-immobilized bead or bead held by a micropipette [19], the shortest extension that can be detected is <1 μm. For its simplicity and capability of stretching DNA over a wide range of lengths, transverse magnetic tweezers is an ideal tool for investigation of physical organization of long DNA molecules by NAPs.

Although many insights regarding protein binding can be obtained, single DNA manipulation experiments alone do not provide information regarding the nucleoprotein structures formed. As such, single DNA manipulation is often complemented with single molecule imaging method such as atomic force microscopy (AFM). Single-molecule fluorescence imaging can also be performed to obtain more insights on the dynamics of protein binding to DNA, such as observation of nucleation and growth process of protein binding to DNA [56]. Combined together, the combination of single DNA manipulation and imaging can provide a holistic understanding of NAPs–DNA interaction. In summary, we have demonstrated the manipulation of single DNA using transverse magnetic tweezers in the study of bacterial NAPs as exemplified by H-NS. This method allows one to understand the underlying mechanisms of NAPs' action through their characteristic DNA binding and organization modes.

2 Materials

2.1 Instrumentation

The transverse magnetic tweezers setup consists of the following:

1. Inverted light microscope with a long working distance 50× objective lens and a light source (IX-71, Olympus).
2. Glass channel (homemade, *see* Subheading 3.2 for details).
3. A pair of permanent magnets mounted on motorized micromanipulator (MP-285, Sutter Instrument).
4. CCD camera for image acquisition (Pike, Allied Vision).
5. An in-house written program for hardware control and data acquisition (LabVIEW, National Instruments).

The use of long working distance 50× objective lens enables the objective lens to be detached from the sample stage, thereby temperature control via sample stage heater will not affect the optics. In addition, translational motion of the magnets close to the glass channel will not be blocked by objective lens. Thus, the magnets can approach closer to the glass channel to achieve higher forces.

2.2 Chemicals and Buffers

1. Biotin-labeled λ-DNA (*see* Subheading 3.1 for preparation).
2. M280 superparamagnetic bead (Dynabeads, Thermo Fisher Scientific).
3. Blocking and storage buffer (BSB) solution: 1% BSA in PBS buffer.

3 Methods

3.1 Preparation of Biotin-Labeled DNA

1. Prepare reaction tube containing the following reagents to label λ-DNA (48,502 bp) with biotin on both ends:
   λ-DNA 500 μg/mL (50 μL)
   1 mM biotin-dUTP (1 μL)
   1 mM dATP (1 μL)
   1 mM dCTP (1 μL)
   1 mM dGTP (1 μL)
   ThermoPol® Reaction Buffer 10× buffer (6 μL)
   100 mM $MgSO_4$ (2 μL)
   Vent (-exo) DNA polymerase (1 μL).
2. Incubate the reaction tube at 72 °C for 30 min.
3. Use PCR purification kit (Qiagen) to remove nucleotides, enzymes, salts, and other impurities from the DNA samples.

3.2 Transverse Magnetic Tweezers

3.2.1 Coverslip Polishing

The coverslip edge is used as the anchoring point for DNA (not the typical coverslip surface), and has to be polished prior to use in order to improve smoothness. Multiple coverslips can be glued together (using paper glue, for example), and polished with fine sandpaper. After polishing, the coverslips are washed with acetone and sonicated to remove the glue. Alternatively, the polishing job can be conveniently outsourced to companies.

3.2.2 Coverslip Preparation and Treatment

1. Wash the coverslip in piranha solution with 1:1:7 ratio of the following: 95% H_2SO_4, 30% H_2O_2, and deionized H_2O. Heat the solution at 150 °C for 2 h.
2. Rinse glass coverslips thoroughly with deionized H_2O and sonicate for 5 min in MilliQ H_2O. Do this three times.
3. Put coverslips in acetone and sonicate for 15 min.
4. Put coverslips in deionized H_2O and sonicate for 15 min.
5. Incubate coverslip edges in 1% solution of (3-aminopropyl) triethoxysilane (APTES) in methanol for 1 h (*see* **Note 1**).
6. Rinse and sonicate coverslips for 5 min in deionized H_2O.
7. Incubate coverslip edges in 2.5% solution of glutaraldeyde in PBS buffer for 4 h
8. Rinse and sonicate for 15 min in deionized H_2O.
9. Incubate coverslip edges in 1 mg/mL streptavidin in PBS buffer for 4 h
10. Rinse the coverslip edges with deionized H_2O
11. Incubate coverslip edges in 0.5 M ethanolamine for 3 h
12. Rinse the coverslip edges with deionized H_2O
13. Incubate coverslip edges in BSB buffer for blocking and prolonged storage (*see* **Note 2**).

3.2.3 Glass Channel Construction

Glass channel is constructed in such a way that it can contain a stack of coverslips (*see* Fig. 4a, b). The glass channel can be made using glass slides and coverslips, cut by diamond tip glasscutter to the desired size. These components are then glued together with UV-curing optical adhesives. At one end, a short capillary tube is connected to a small tube for buffer reservoir, and this serves as the inlet. On the other end, another short capillary tube is added for the buffer outlet. During buffer exchange, buffer is added to the inlet, and the solution can be drawn to the outlet by using tissue paper. Alternatively, the outlet capillary tubes can be connected to silicone tubing and the buffer is sucked out by syringe pump. This allows an automated and more consistent buffer exchange. The flow speed can also be reduced to prevent large drag force that can potentially break the DNA tether during buffer exchange. This glass channel can be reused from one experiment to another. One simply needs to take out and replace the stack of coverslips in between experiments.

Before each experiment, follow these steps:

1. Rinse the glass channel with deionized H_2O and sonicate for 15 min.
2. A functionalized #0 coverslip is then sandwiched in between #1.5 coverslips and Parafilm (*see* Fig. 4b), with the #0 functionalized coverslip protruding out slightly (~1 mm).
3. Heat the stack of coverslips on a hot plate (~150 °C) to melt the Parafilm and stick the coverslips together.
4. Insert the stack of coverslips into the glass channel (*see* **Note 3**) and seal the gap with wax (melt at ~150 °C before use).

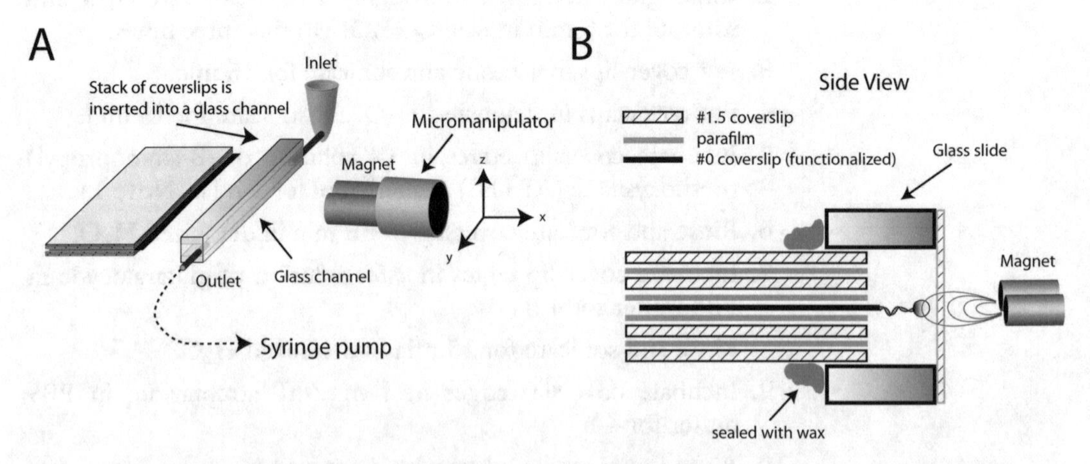

Fig. 4 (**a**) Schematic diagram of the glass channel used in our experiments. Stack of coverslips is inserted into the glass channel and sealed with wax to form a closed chamber. Solution inside the channel can be exchanged via the capillary tubes that serve as the inlet and outlet. (**b**) Side view of the construct showing the glass channel and the stack of coverslips. The DNA is tethered to the functionalized edge of the #0 coverslip on one end and to streptavidin-coated magnetic bead on the other end

5. The construct (stack of coverslips in glass channel) is then mounted on a platform (not shown) that can sit well on a microscope stage. Silicone glue or wax can be used to fix the construct on the platform.
6. Flow 1 mL of PBS buffer into the channel.
7. Flow 1 mL of BSB buffer into the channel and incubate for 30 min or longer for effective blocking.
8. Flow 1 mL of PBS buffer into the channel.

3.2.4 Tethering DNA Molecules to Coverslip Edges

1. Flow in 0.5 ng/μL of biotin-labeled λ-DNA into the channel and incubate for 5 min.
2. Rinse thoroughly with PBS buffer (~1 mL) to remove excess biotin-labeled λ-DNA.
3. Flow in 200× diluted M280 superparamagnetic bead in PBS buffer into the channel.
4. Turn the channel such that the magnetic bead will sink toward coverslip edge. This step helps to increase the number of DNA tethers attached to magnetic bead. Incubate for 2 min.
5. Rinse with 1 mL of PBS buffer to remove excess magnetic bead.

3.2.5 Image Acquisition and Analysis, Data Collection

To visualize tether formation in the channel, the coverslip edge is brought to the microscope viewing area (*see* Fig. 5 for an example). A homemade LabView program can be written to couple image acquisition and analysis with data collection. During experiments, images containing the bead and the coverslip edge are continuously

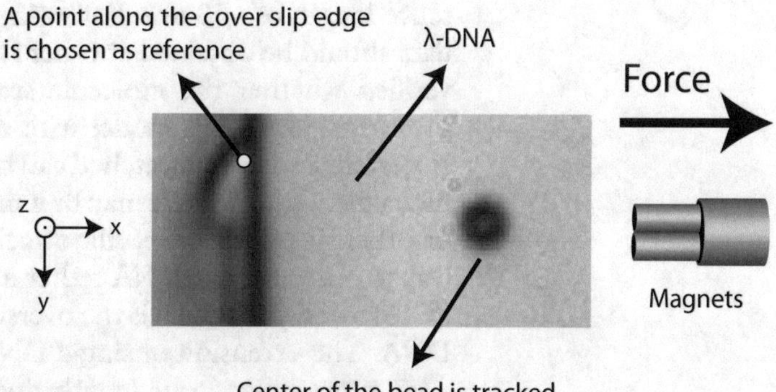

Fig. 5 A screenshot of DNA stretching experiments using transverse magnetic tweezers. λ-DNA is tethered in between coverslip edge and a superparamagnetic bead (the DNA cannot be seen under a light microscope). A point along the coverslip edge is chosen as a reference point, and the center of the bead is continuously tracked to give real-time data of end-to-end distance. Any events that affect DNA elasticity can be directly observed (reproduced from ref. 47 with permission from Oxford University Press)

acquired for real-time processing of the force and extension data. The workflow for the experiments is as follows:

1. To change the magnitude of force, the permanent magnet is translated along the *x*-direction (*see* Fig. 5). Magnet position has to be aligned well along the *y*- and *z*-direction, such that changing the magnet position along the *x*-direction will not change the force direction (*see* **Note 4**).
2. The movement of the bead is tracked by locating the center of the bead using centroid algorithm after applying a threshold to the image to increase accuracy and resolution. An arbitrary point on the coverslip edge is also tracked as a reference point to calculate the end-to-end distance of the DNA. This point is also used to minimize drift that may occur during experiments, using image correlation analysis.
3. The variance of the bead motion perpendicular to the force direction (*y*-direction) is obtained, and stretching force applied to the bead can be calculated (*see* Subheading 1.3). To accurately measure force based on bead fluctuation, the sampling rate should be much faster than the reciprocal of the correlation times of bead motions and the measurement time window should be longer than the correlation times of bead motions.
4. End-to-end distance of the DNA is obtained from the difference between the bead centroid to the arbitrary point on the edge of coverslip, minus an offset. The offset is to account for the bead radius (1.4-μm for M280 magnetic bead) and portion of the DNA that possibly attaches on the coverslip surface. The offset is arbitrarily chosen such that the extension at a given force obeys Marko–Siggia formula for a polymer with a persistence length of ~50 nm. Following this, a few other extension data should be obtained at other force values, and it should be verified whether the molecule stretched consistently follows the worm-like chain model with a similar persistence length. If it is, the molecule stretched can be identified as a single DNA molecule. Otherwise, it may be a multitethered DNA molecule or other possible nonspecific attachment. Another convenient way to ensure single DNA tether in the setup is to increase the force to ~65 pN, which is the overstretching transition force for DNA. The extension of single DNA tether will increase up to ~1.7 times its contour length due to the transition from B-DNA into S-DNA [25].
5. A set of force–extension data at a desired force range is then obtained after verifying that the molecule stretched is a single DNA molecule (Fig. 6). Force at a set of magnet position can be calculated using Eq. 3 (*see* **Note 5**). Alternatively, a quick

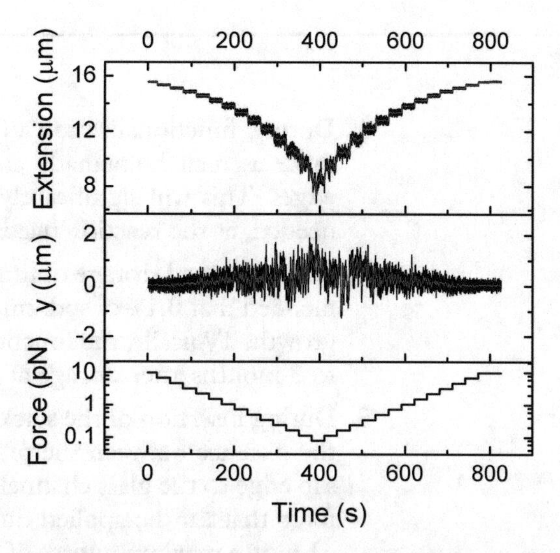

Fig. 6 DNA extension and bead transverse position at various stretching forces. The raw data was recorded at ~260 Hz (*black lines*) with 5 s smoothing window (*red lines*). The DNA extension decreases as the force is decreased, and increases as the force is increased. The position of the bead perpendicular to the force direction (*y*-direction) remains largely constant regardless of the force, indicating that the magnet is well aligned along the *y*-direction. Note that the transverse fluctuation decreases as force is increased due to the fluctuation being restrained by force

force calibration can be done by measurement at several magnet positions (e.g., between 1 and 10 pN), while forces at other magnet positions can be obtained via interpolation (*see* Fig. 1b). Note that due to the weight of the bead, the extension values at lower forces are generally lower than expected.

6. Once force–extension data of single naked DNA is obtained (*see* **Note 6**), binding partner such as proteins can be flowed in. The DNA is stretched at ~10 pN prior to introducing buffer to minimize bead–surface interaction that may occur during buffer exchange. Solutions containing protein sample is then transferred to the inlet tube, and the solution in the glass channel is exchanged by slowly withdrawing the solution via the outlet either by either tissue paper or syringe pump.
7. Force–extension data in the presence of protein is obtained and compared to that obtained in the absence of proteins to obtain insights on the DNA binding and organization mode of the protein of interest. Parameters such as bending persistence length, contour length, dissociation constant, and binding cooperativity can be extracted for more information on the DNA–protein interaction.

4 Notes

1. During functionalization of coverslips edges, it is useful to make a small homemade container that can fit the coverslip edges. This will significantly reduce the amount of chemicals needed, as the reaction needs to take place only at the edges.
2. For prolonged storage of functionalized coverslips, it is recommended that 0.1% of sodium azide is added to prevent bacterial growth. Typically, the functionalized coverslips can be used up to 3 months after storage at 4 °C.
3. During insertion of the stack of coverslips to the glass channel, the distance between the protruding #0 functionalized coverslip edge to the glass channel end will determine the maximum force that can be applied during experiment. At a distance of ~1 mm, a maximum force of 10–20 pN can be achieved using a 2.8-μm M280 superparamagnetic bead.
4. As the force applied to the magnetic bead can have a component along the *y*- and *z*-axis, the magnet position has to be well aligned to minimize these components. To align the magnet position along the *y*-direction, the magnet can be brought to the microscope viewing area using 10× or 20× objective lens, where the center between the two magnets can be found. Alternatively, one can occasionally find DNA tethered to two magnetic beads and align the magnet position such that the two beads are aligned in *y*-direction. To align the magnet position along the *z*-direction, a DNA molecule can be stretched, and the magnet position is adjusted such that maximum extension is obtained at ~10 pN. Note that it is crucial to do this *y*- and *z*-direction alignment step properly, as the end-to-end distance measured in experiments is the projected DNA extension to the *x*-direction. Any misalignment will cause the measured end-to-end distance to be shorter than expected. Figure 6 shows an example of DNA extension and transverse position *y* when force is varied. A minimum variation of average *y* at the measured force range indicates a good alignment along the *y*-direction.
5. For measurement of force using the Eq. 3, care should be taken to ensure that the system has accumulated sufficient data points to calculate δ_y, the bead fluctuation in the direction perpendicular to the stretching force. At lower forces, more time is required for force calibration because of the larger magnitude of bead fluctuation caused by thermal motion. *See* Subheading 1.3 for calculation of bead correlation time.
6. Care should be taken to ensure that no folding event occurs during naked DNA stretching. This can be easily detected by first measuring the extension at higher forces, then reduce the

force subsequently to lower forces, before increasing the force again through the same set of forces. Any hysteresis or discrepancies between the force-increase and force-decrease data indicate DNA folding events, and such DNA tether should not be used.

References

1. Dillon SC, Dorman CJ (2010) Bacterial nucleoid-associated proteins, nucleoid structure and gene expression. Nat Rev Microbiol 8(3):185–195. doi:10.1038/nrmicro2261
2. Tolstorukov MY, Virnik K, Zhurkin VB, Adhya S (2016) Organization of DNA in a bacterial nucleoid. BMC Microbiol 16(1):22. doi:10.1186/s12866-016-0637-3
3. Stuger R, Woldringh CL, van der Weijden CC, Vischer NO, Bakker BM, van Spanning RJ, Snoep JL, Westerhoff HV (2002) DNA supercoiling by gyrase is linked to nucleoid compaction. Mol Biol Rep 29(1-2):79–82
4. Murphy LD, Zimmerman SB (1994) Macromolecular crowding effects on the interaction of DNA with Escherichia coli DNA-binding proteins: a model for bacterial nucleoid stabilization. Biochim Biophys Acta 1219 (2):277–284
5. de Vries R (2010) DNA condensation in bacteria: interplay between macromolecular crowding and nucleoid proteins. Biochimie 92(12):1715–1721. doi:10.1016/j.biochi.2010.06.024
6. Dame RT (2005) The role of nucleoid-associated proteins in the organization and compaction of bacterial chromatin. Mol Microbiol 56(4):858–870. doi:10.1111/j.1365-2958.2005.04598.x
7. Luijsterburg MS, Noom MC, Wuite GJ, Dame RT (2006) The architectural role of nucleoid-associated proteins in the organization of bacterial chromatin: a molecular perspective. J Struct Biol 156(2):262–272. doi:10.1016/j.jsb.2006.05.006
8. Browning DF, Grainger DC, Busby SJ (2010) Effects of nucleoid-associated proteins on bacterial chromosome structure and gene expression. Curr Opin Microbiol 13(6):773–780. doi:10.1016/j.mib.2010.09.013
9. Cai YH, Huang H (2012) Advances in the study of protein-DNA interaction. Amino Acids 43(3):1141–1146. doi:10.1007/s00726-012-1377-9
10. Amit R, Oppenheim AB, Stavans J (2003) Increased bending rigidity of single DNA molecules by H-NS, a temperature and osmolarity sensor. Biophys J 84 (4):2467–2473. doi:10.1016/S0006-3495(03)75051-6
11. Dame RT, Wyman C, Wurm R, Wagner R, Goosen N (2002) Structural basis for H-NS-mediated trapping of RNA polymerase in the open initiation complex at the rrnB P1. J Biol Chem 277(3):2146–2150. doi:10.1074/jbc.C100603200
12. Winardhi RS, Yan J, Kenney LJ (2015) H-NS regulates gene expression and compacts the nucleoid: insights from single-molecule experiments. Biophys J 109(7):1321–1329. doi:10.1016/j.bpj.2015.08.016
13. Neuman KC, Nagy A (2008) Single-molecule force spectroscopy: optical tweezers, magnetic tweezers and atomic force microscopy. Nat Methods 5(6):491–505. doi:10.1038/nmeth.1218
14. Strick TR, Allemand JF, Bensimon D, Bensimon A, Croquette V (1996) The elasticity of a single supercoiled DNA molecule. Science 271 (5257):1835–1837
15. Strick TR, Allemand JF, Bensimon D, Croquette V (1998) Behavior of supercoiled DNA. Biophys J 74(4):2016–2028. doi:10.1016/S0006-3495(98)77908-1
16. Chen H, Fu H, Zhu X, Cong P, Nakamura F, Yan J (2011) Improved high-force magnetic tweezers for stretching and refolding of proteins and short DNA. Biophys J 100 (2):517–523. doi:10.1016/j.bpj.2010.12.3700
17. Gosse C, Croquette V (2002) Magnetic tweezers: micromanipulation and force measurement at the molecular level. Biophys J 82 (6):3314–3329. doi:10.1016/S0006-3495(02)75672-5
18. Lipfert J, Lee M, Ordu O, Kerssemakers JW, Dekker NH (2014) Magnetic tweezers for the measurement of twist and torque. J Vis Exp (87). doi:10.3791/51503
19. Yan J, Skoko D, Marko JF (2004) Near-field-magnetic-tweezer manipulation of single DNA molecules. Phys Rev E Stat Nonlin Soft Matter Phys 70(1 Pt 1):011905. doi:10.1103/PhysRevE.70.011905

20. Yan J, Maresca TJ, Skoko D, Adams CD, Xiao B, Christensen MO, Heald R, Marko JF (2007) Micromanipulation studies of chromatin fibers in Xenopus egg extracts reveal ATP-dependent chromatin assembly dynamics. Mol Biol Cell 18(2):464–474. doi:10.1091/mbc.E06-09-0800
21. Liu Y, Chen H, Kenney LJ, Yan J (2010) A divalent switch drives H-NS/DNA-binding conformations between stiffening and bridging modes. Genes Dev 24(4):339–344. doi:10.1101/gad.1883510
22. Fu H, Chen H, Marko JF, Yan J (2010) Two distinct overstretched DNA states. Nucleic Acids Res 38(16):5594–5600. doi:10.1093/nar/gkq309
23. Fu H, Chen H, Zhang X, Qu Y, Marko JF, Yan J (2011) Transition dynamics and selection of the distinct S-DNA and strand unpeeling modes of double helix overstretching. Nucleic Acids Res 39(8):3473–3481. doi:10.1093/nar/gkq1278
24. Zhang X, Chen H, Fu H, Doyle PS, Yan J (2012) Two distinct overstretched DNA structures revealed by single-molecule thermodynamics measurements. Proc Natl Acad Sci U S A 109(21):8103–8108. doi:10.1073/pnas.1109824109
25. Zhang X, Chen H, Le S, Rouzina I, Doyle PS, Yan J (2013) Revealing the competition between peeled ssDNA, melting bubbles, and S-DNA during DNA overstretching by single-molecule calorimetry. Proc Natl Acad Sci U S A 110(10):3865–3870. doi:10.1073/pnas.1213740110
26. Graham JS, Johnson RC, Marko JF (2011) Concentration-dependent exchange accelerates turnover of proteins bound to double-stranded DNA. Nucleic Acids Res 39 (6):2249–2259. doi:10.1093/nar/gkq1140
27. Graham JS, Johnson RC, Marko JF (2011) Counting proteins bound to a single DNA molecule. Biochem Biophys Res Commun 415(1):131–134. doi:10.1016/j.bbrc.2011.10.029
28. Fonnum G, Johansson C, Molteberg A, Mørup S, Aksnes E (2005) Characterisation of Dynabeads® by magnetization measurements and Mössbauer spectroscopy. J Magn Magn Mater 293(1):41–47
29. Lionnet T, Allemand JF, Revyakin A, Strick TR, Saleh OA, Bensimon D, Croquette V (2012) Single-molecule studies using magnetic traps. Cold Spring Harb Protoc 2012 (1):34–49. doi:10.1101/pdb.top067488
30. Yan J, Kawamura R, Marko JF (2005) Statistics of loop formation along double helix DNAs. Phys Rev E Stat Nonlin Soft Matter Phys 71(6 Pt 1):061905. doi:10.1103/PhysRevE.71.061905
31. Kruithof M, Chien F, de Jager M, van Noort J (2008) Subpiconewton dynamic force spectroscopy using magnetic tweezers. Biophys J 94(6):2343–2348. doi:10.1529/biophysj.107.121673
32. Lipfert J, Hao X, Dekker NH (2009) Quantitative modeling and optimization of magnetic tweezers. Biophys J 96(12):5040–5049. doi:10.1016/j.bpj.2009.03.055
33. Doi M, Edwards SF (1986) The theory of polymer dynamics. Clarendon, Oxford
34. Marko JF, Siggia ED (1995) Stretching DNA. Macromolecules 28(26):8759–8770. doi:10.1021/ma00130a008
35. Smith SB, Finzi L, Bustamante C (1992) Direct mechanical measurements of the elasticity of single DNA molecules by using magnetic beads. Science 258(5085):1122–1126
36. Bustamante C, Marko JF, Siggia ED, Smith S (1994) Entropic elasticity of lambda-phage DNA. Science 265(5178):1599–1600
37. Baumann CG, Smith SB, Bloomfield VA, Bustamante C (1997) Ionic effects on the elasticity of single DNA molecules. Proc Natl Acad Sci U S A 94(12):6185–6190
38. Bouchiat C, Wang MD, Allemand J, Strick T, Block SM, Croquette V (1999) Estimating the persistence length of a worm-like chain molecule from force-extension measurements. Biophys J 76(1 Pt 1):409–413
39. Yan J, Marko JF (2003) Effects of DNA-distorting proteins on DNA elastic response. Phys Rev E Stat Nonlin Soft Matter Phys 68 (1 Pt 1):011905. doi:10.1103/PhysRevE.68.011905
40. Lim CJ, Whang YR, Kenney LJ, Yan J (2012) Gene silencing H-NS paralogue StpA forms a rigid protein filament along DNA that blocks DNA accessibility. Nucleic Acids Res 40 (8):3316–3328. doi:10.1093/nar/gkr1247
41. Efremov AK, Qu Y, Maruyama H, Lim CJ, Takeyasu K, Yan J (2015) Transcriptional repressor TrmBL2 from thermococcus kodakarensis forms filamentous nucleoprotein structures and competes with histones for DNA binding in a salt- and DNA supercoiling-dependent manner. J Biol Chem 290(25):15770–15784. doi:10.1074/jbc.M114.626705
42. Ali BM, Amit R, Braslavsky I, Oppenheim AB, Gileadi O, Stavans J (2001) Compaction of single DNA molecules induced by binding of integration host factor (IHF). Proc Natl Acad

Sci U S A 98(19):10658–10663. doi:10.1073/pnas.181029198

43. van Noort J, Verbrugge S, Goosen N, Dekker C, Dame RT (2004) Dual architectural roles of HU: formation of flexible hinges and rigid filaments. Proc Natl Acad Sci U S A 101(18):6969–6974. doi:10.1073/pnas.0308230101
44. Lin J, Chen H, Droge P, Yan J (2012) Physical organization of DNA by multiple non-specific DNA-binding modes of integration host factor (IHF). PLoS One 7(11):e49885. doi:10.1371/journal.pone.0049885
45. Xiao B, Johnson RC, Marko JF (2010) Modulation of HU-DNA interactions by salt concentration and applied force. Nucleic Acids Res 38 (18):6176–6185. doi:10.1093/nar/gkq435
46. McCauley M, Hardwidge PR, Maher LJ 3rd, Williams MC (2005) Dual binding modes for an HMG domain from human HMGB2 on DNA. Biophys J 89(1):353–364. doi:10.1529/biophysj.104.052068
47. Winardhi RS, Fu W, Castang S, Li Y, Dove SL, Yan J (2012) Higher order oligomerization is required for H-NS family member MvaT to form gene-silencing nucleoprotein filament. Nucleic Acids Res 40(18):8942–8952. doi:10.1093/nar/gks669
48. Winardhi RS, Castang S, Dove SL, Yan J (2014) Single-molecule study on histone-like nucleoid-structuring protein (H-NS) paralogue in Pseudomonas aeruginosa: MvaU bears DNA organization mode similarities to MvaT. PLoS One 9(11):e112246. doi:10.1371/journal.pone.0112246
49. Lee SY, Lim CJ, Droge P, Yan J (2015) Regulation of Bacterial DNA Packaging in Early Stationary Phase by Competitive DNA Binding of Dps and IHF. Sci Rep 5:18146. doi:10.1038/srep18146
50. Winardhi RS, Gulvady R, Mellies JL, Yan J (2014) Locus of enterocyte effacement-encoded regulator (Ler) of pathogenic Escherichia coli competes off histone-like nucleoid-structuring protein (H-NS) through noncooperative DNA binding. J Biol Chem 289 (20):13739–13750. doi:10.1074/jbc.M113.545954
51. Qu Y, Lim CJ, Whang YR, Liu J, Yan J (2013) Mechanism of DNA organization by Mycobacterium tuberculosis protein Lsr2. Nucleic Acids Res 41(10):5263–5272. doi:10.1093/nar/gkt249
52. Lim CJ, Lee SY, Teramoto J, Ishihama A, Yan J (2013) The nucleoid-associated protein Dan organizes chromosomal DNA through rigid nucleoprotein filament formation in E. coli during anoxia. Nucleic Acids Res 41 (2):746–753. doi:10.1093/nar/gks1126
53. Dorman CJ (2004) H-NS: a universal regulator for a dynamic genome. Nat Rev Microbiol 2(5):391–400. doi:10.1038/nrmicro883
54. Dame RT, Wyman C, Goosen N (2000) H-NS mediated compaction of DNA visualised by atomic force microscopy. Nucleic Acids Res 28(18):3504–3510
55. Marko JF (2015) Biophysics of protein-DNA interactions and chromosome organization. Physica A 418:126–153. doi:10.1016/j.physa.2014.07.045
56. Bell JC, Plank JL, Dombrowski CC, Kowalczykowski SC (2012) Direct imaging of RecA nucleation and growth on single molecules of SSB-coated ssDNA. Nature 491 (7423):274–278. doi:10.1038/nature11598

Chapter 15

A User-Friendly DNA Modeling Software for the Interpretation of Cryo-Electron Microscopy Data

Damien Larivière, Rodrigo Galindo-Murillo, Eric Fourmentin, Samuel Hornus, Bruno Lévy, Julie Papillon, Jean-François Ménétret, and Valérie Lamour

Abstract

The structural modeling of a macromolecular machine is like a "Lego" approach that is challenged when blocks, like proteins imported from the Protein Data Bank, are to be assembled with an element adopting a serpentine shape, such as DNA templates. DNA must then be built ex nihilo, but modeling approaches are either not user-friendly or very long and fastidious. In this method chapter we show how to use GraphiteLifeExplorer, a software with a simple graphical user interface that enables the sketching of free forms of DNA, of any length, at the atomic scale, as fast as drawing a line on a sheet of paper. We took as an example the nucleoprotein complex of DNA gyrase, a bacterial topoisomerase whose structure has been determined using cryo-electron microscopy (Cryo-EM). Using GraphiteLifeExplorer, we could model in one go a 155 bp long and twisted DNA duplex that wraps around DNA gyrase in the cryo-EM map, improving the quality and interpretation of the final model compared to the initially published data.

Key words DNA modeling software, Cryo-electron microscopy, Molecular dynamics, Protein Data Bank, Macromolecular complex

1 Introduction

Cryo-electron microscopy allows to capture either in vitro or in vivo the shape of a macromolecular complex with a mass higher than 300 kDa. While the resolution is currently between 8 and 20 Å depending on the observed specimen, it is now reaching between 4 and 2 Å for particles with a mass as low as 64 kDa making true "the 20-year old dream of being able to use cryo-EM to determine

Electronic supplementary material: The online version of this chapter (doi: 10.1007/978-1-4939-7098-8_15) contains supplementary material, which is available to authorized users.

Olivier Espéli (ed.), *The Bacterial Nucleoid: Methods and Protocols*, Methods in Molecular Biology, vol. 1624, DOI 10.1007/978-1-4939-7098-8_15,

atomic-resolution structures" [1–3]. Once the cryo-envelope has been obtained, crystal structures of the individual components forming the complex are fitted into the envelope thanks to computational techniques. A "pseudo-atomic resolution" model of the complex is obtained with an effective resolution higher than the nominal resolution of the cryo-EM density map [4]. However, when one of the components is DNA, no ad hoc crystal structures are available, making necessary a 3D modeling approach to create the serpentine shape of the DNA, position it in the cryo-EM map with respect to the proteic elements, and complete the model.

Several open access tools are available to model DNA [5]. They fall into two categories: some try to predict the conformation DNA can adopt from its base pair sequence, while others consider that the shape is known and build DNA onto the anticipated path. 3DNA and its web-based counterpart w3DNA [6] belong to the first category as they are able to produce atomic structures of nucleic acids based on a sequence given by the user. Nucleic Acid Builder (NAB) [7] can do both (building of sequence-dependent atomic structures and of arbitrary conformations). However, NAB lacks a user interface so that direct interaction with the shape is not possible and instructions are given via command lines instead (*see* **Note 1**). GraphiteLifeExplorer [8], via a genuine user interface, enables the modeling of DNA of arbitrary form, of any length, within a few minutes. The tool incorporates a method in which the helical axis is modeled as a quadratic or cubic Bézier smooth curve in space [5, 8] (*see* **Note 2**).

As a proof of concept, we chose to work on the complex of the full-length *Thermus thermophilus* DNA gyrase with a 155 bp DNA whose pseudo-atomic structure was obtained for the first time using cryo-EM data by Papillon et al. [9]. DNA gyrase is a type 2 DNA topoisomerase that solves topological problems occurring in DNA during replication, chromosome segregation, or transcription. It is the only type 2 DNA topoisomerase that is able to introduce negative supercoils in DNA by transporting one DNA double helix (T-segment) through a double-strand break in another (G-segment) (Fig. 1). This essential role makes DNA gyrase a target of choice for antibiotics.

Papillon et al. [9] used a DNA molecule of 155 base pairs in their experimental procedure. The choice of this length was motivated by previous footprinting studies arguing that this length would be sufficient to completely wrap DNA around the protein and potentially generate a DNA crossover captured within the gyrase [10]. In addition to the possibility to generate a DNA crossover, the DNA length of 155 bp was also thought to be able to bind completely to the protein, preventing DNA ends to move freely and generate heterogeneous species detrimental to the quality of cryo-EM maps.

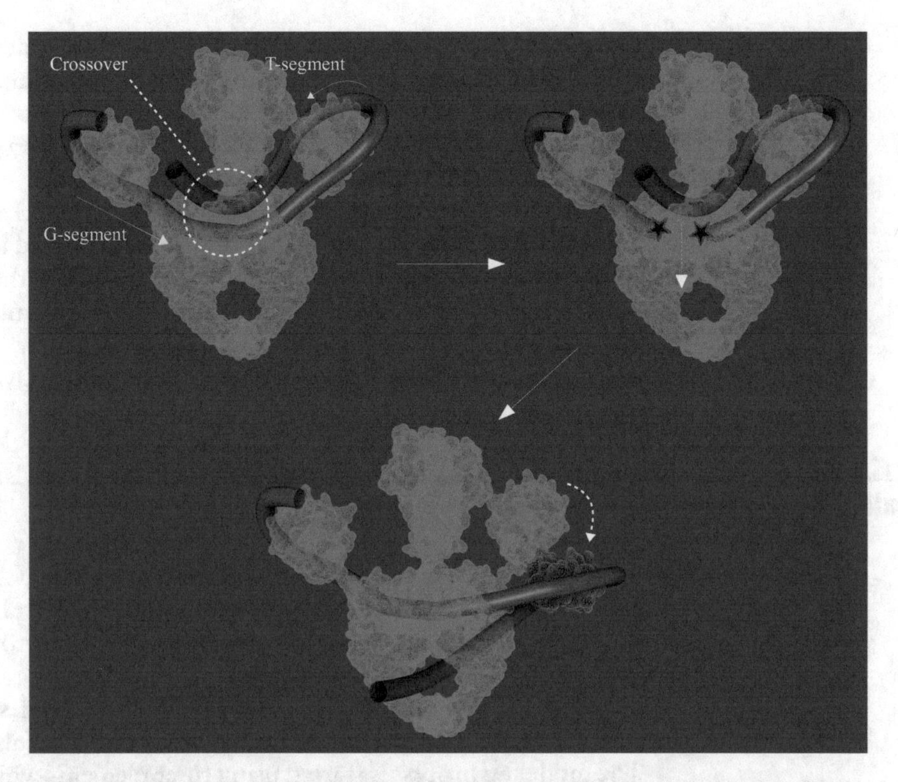

Fig. 1 Simplified description of the supercoiling DNA gyrase cycle: the topoisomerase first binds a DNA fragment (G-segment) and then captures a contiguous second duplex (T-segment) forming a positive crossover. As the G-segment is cleaved, the T-segment is transported through the double strand break while bound to the disk-shaped pinwheel. After transport, the double-strand break is resealed. Experimentally, the DNA duplex must be long enough to generate the crossover but short enough to avoid floating ends detrimental the quality of cryo-EM maps

The experimental study by Papillon et al. shed light on many aspects of the way DNA gyrase wraps and sequesters DNA. In particular, the C-terminal β-pinwheel domains (CTD) were resolved for the first time along with all the other domains of gyrase (Fig. 3). The DNA gyrase model obtained initially by the authors combining electron microscopy and DNA modeling was the starting point to address important and pending questions, for instance: what is the sequence of steps necessary to introduce negative supercoils in DNA; which domains of DNA gyrase are involved in this operation and how are they coordinated; how does gyrase recognize positive supercoils in DNA; where and how is the DNA crossover maintained within gyrase; and ultimately, how to disrupt DNA wrapping using inhibitors?

The visualization for the first time of the quaternary structure of DNA gyrase has answered some of these questions. However, only 130 bp of duplex DNA could be modeled out of the 155 bp sequence used in the experimental procedure since no similarly long DNA structure was available in the databanks. The authors faced major limitations to generate long DNA and introduce sharp curvatures with the available modeling tools at the time. As a consequence, the EM map could not be completely exploited and the DNA crossover could not be modeled, leaving many structural aspects of DNA supercoiling unanswered.

In this method chapter, we show how GraphiteLifeExplorer [8] can be used to generate a full DNA duplex (155 bp and longer) wrapping around DNA gyrase pseudo-atomic model that was obtained from the cryo-EM data. We show how different DNA lengths can be modeled effortlessly to help design new experiments, formulate hypothesis, or feed Molecular Dynamics simulations.

2 Materials

The tools necessary for this modeling protocol are UCSF Chimera [11] (https://www.cgl.ucsf.edu/chimera/) and GraphiteLifeExplorer (http://www.lifeexplorer.info/download/). Both tools are available for OS X, Windows, and Linux.

Throughout this chapter, the user is invited to read specific tutorials accessible at http://www.lifeexplorer.info/tutorials/.

The modeling process is started using three elements which can be downloaded as Supplementary Data from MMB online:

1. A file *gyrase.gsg* readable by GraphiteLifeExplorer and containing the whole gyrase as a point cloud (*see* **Note 3**). The coordinates of the gyrase, not freely available as a PDB file, were provided by Papillon and coworkers and converted into a point cloud. These coordinates were generated after docking and connecting the X-ray structures of the gyrase individual domains in the cryo-EM map EMDB-2361 [9]. It is assumed that the reader has already performed a fitting of their protein (s) of interest within a cryo-envelope to follow the method presented in this chapter. The methods to fit protein(s) within an envelope obtained from cryo-electron microscopy are not shown here (for an overview, *see* ref. 12).
2. A file *2xct-DNA.pdb* corresponding to a short 34 bp DNA duplex extracted from the structure of the gyrase DNA-binding domain (PDB code 2xct). This structure guides DNA modeling: it is centered inside the DNA-binding domain and has to be extended from both extremities to generate a 155 bp duplex as used for the experimental sequence.
3. The electronic density map EMDB 2361, fetched directly from the software UCSF Chimera but accessible also on the EMDB website (http://www.emdatabank.org/index.html).

All the necessary files are available through as Supplementary Data along with one of the DNA models created and discussed in this protocol chapter and a video showing how to create DNA with GraphiteLifeExplorer.

3 Methods

3.1 EM Map Conversion and Import in GraphiteLifeExplorer

Since we want to model a DNA duplex within the limits of a cryo-EM density map, the first step is to convert the map (initially in the CCP4 format recognized by file suffix *.map) into the Wavefront OBJ geometrical format (recognized by file suffix *.obj) readable by GraphiteLifeExplorer. This section describes the process.

1. Open UCSF Chimera, go to the File menu, Fetch by ID, check EMDB, and type 2361. The corresponding EM map displays along with the Volume Viewer popup window. In the Volume Viewer popup window, type 0.55 in the Level box as proposed by the authors of the map [9]. Go to the File menu, Export Scene... choose a location for the file, give the file a name, for instance *2361*, choose OBJ as the File type and Save. The EM map is now available with the name and extension *2361.obj*, in the geometrical 3D format Wavefront OBJ readable by GraphiteLifeExplorer (*see* **Note 4**).
2. Open GraphiteLifeExplorer, go to the File menu, select Load, and locate the element *2361.obj* and open it. Again, go to the File menu, select Load, and locate the file *gyrase.gsg* and open it. In the same way, open *2xct-DNA.pdb* (*see* **Note 5**). The 3D scene looks like Fig. 2.
3. In the outliner (Panel 2, *see* Fig. 2), make sure that the map "2361" is selected. In the light tab (Panel 4, *see* Fig. 2), uncheck "surface style" and check "mesh style." It is now possible to see through the map. The color of the mesh can be adjusted: the grey color for the mesh associated with the default grey color of the background is convenient to model DNA while the map is displayed. Select "gyrase" in the outliner, click on the Plain tab (right panel), and choose Molecule and set "points size" to 2. Select "2xct-DNA" in the outliner, click on the Plain tab, and choose Molecule and set "points size" to 2. The 3D scene looks like Fig. 3.

The 3D scene is now ready for adding DNA in the cryo-EM map. The goal consists of filling the spaces left empty by the gyrase in the map and supposed to be where DNA stands.

3.2 DNA Sketching with GraphiteLifeExplorer

The objective of this step is to draw a DNA duplex of fixed length that wraps around a specific domain of the DNA gyrase (called the pinwheel C-terminal domain), runs in the middle of the complex

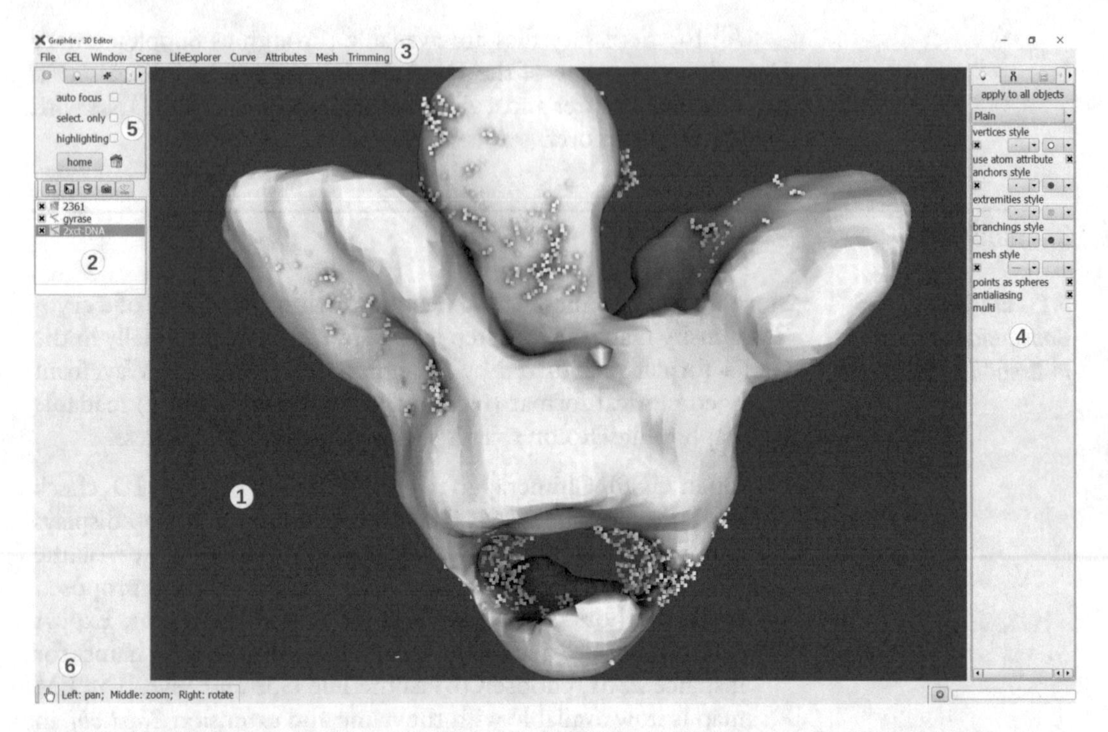

Fig. 2 GraphiteLifeExplorer User Interface. Zone 1 is the interactive 3D scene or 3D view where the user can explore a content made of objects of various geometrical nature: here, the cryo-EM map is a white surface (imported as a Wavefront OBJ file format) and the gyrase protein a point cloud (imported as a gsg file created from atomic coordinates). Zone 2 ("the outliner") is a list of all objects present in the scene. Zone 3 is the classical menu bar from which the user mainly imports objects in the 3D scene. Zone 4 gives access to more tabs: the one with the light is the shader tab that lets choose how an object selected in the outliner is displayed (a surface, a mesh, a point cloud, spheres, etc.); the one represented by a pair of pliers is the toolbox tab giving access to interactive tools available for the object selected in the outliner (particularly the tools dedicated to DNA modeling); the screen effect tab (not visible, represented by the acronym FX) allows to choose among various visual effects like ambient occlusion and cartoon representation. Zone 5 has several tabs allowing to change the background color, view the scene content from various camera positions, etc. Zone 6 shows in which mode the user stands: the "hand" icon here indicates that the user can navigate within the 3D scene without provoking unwanted modifications to the content. There is a floating panel (not shown here) from which the user can transform a point cloud into a surface or rewrite the atomic coordinates (write PDB button) of a component that has been moved in the scene

based on the position of the short DNA in the crystallographic structure (PDB ID: 2xct), and wraps around the second pinwheel (Fig. 3). The length of the DNA molecule is to be set to 155 bp as in the experimental setup.

To do so, GraphiteLifeExplorer is equipped with two drawing tools. One of these allows to edit cubic Bézier curves and is well suited to draw a complicated path like the one seen in a nucleosomal particle (*see* **Note 6**). Therefore, the cubic Bézier editor is used here to bend DNA and make a continuous path between and around the pinwheels. Figure 4 shows the path as a succession of

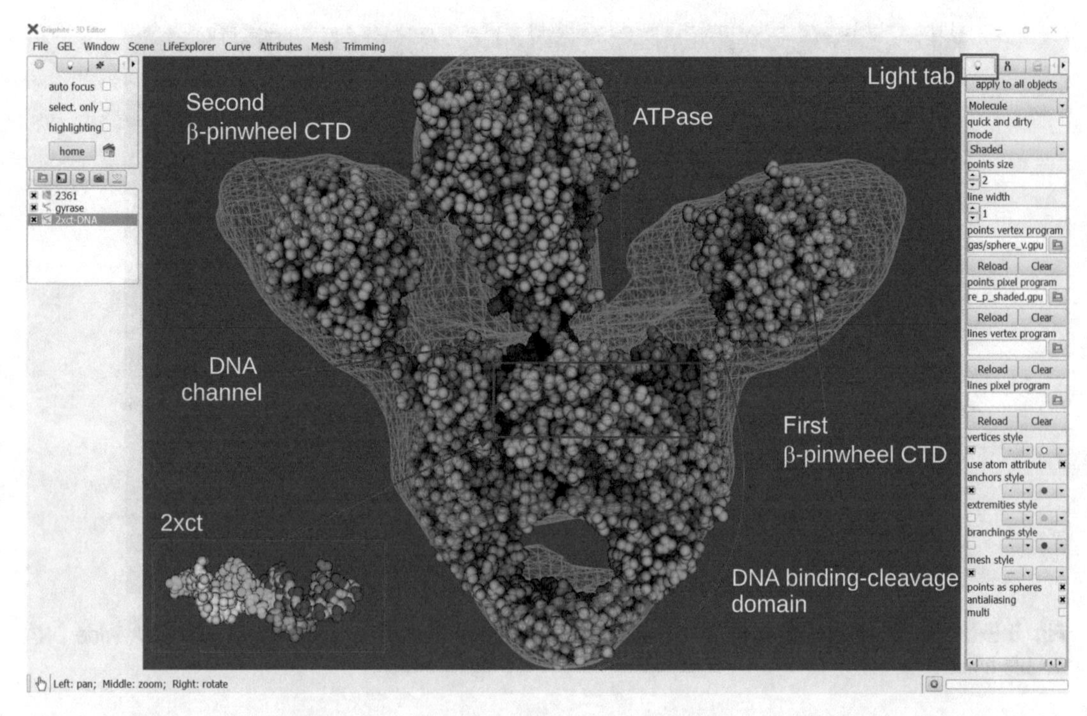

Fig. 3 DNA gyrase is a 320 kDa dimeric enzyme comprising an ATPase domain, a DNA-binding cleavage domain and a disk-shaped domain (β-pinwheel) attached to the DNA-binding cleavage domain. The short DNA duplex extracted from a crystal structure (PDB ID: 2xct) is used as a starting point to draw the path of the 155 bp modeled DNA

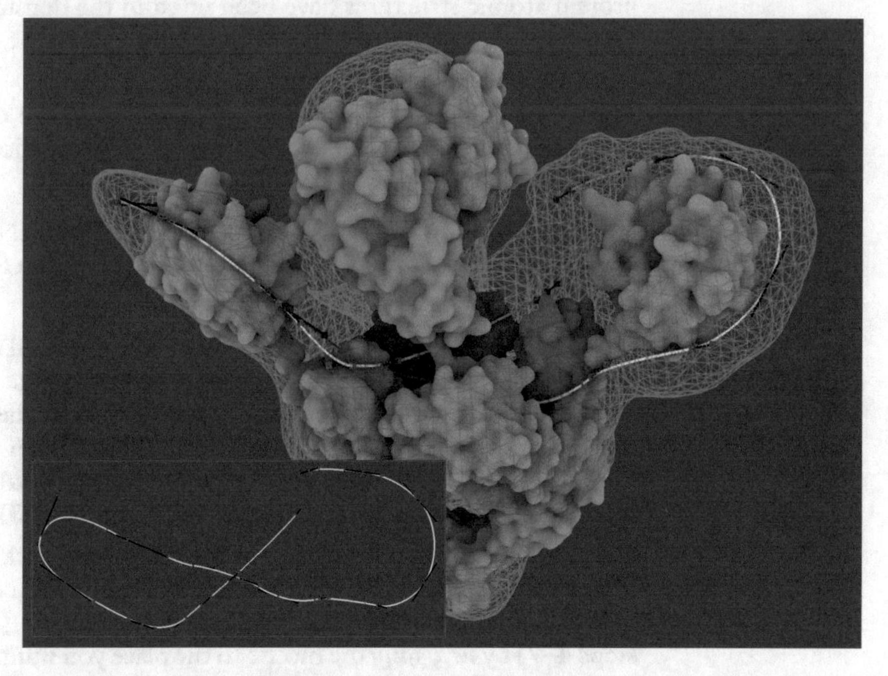

Fig. 4 The final DNA path made of control points that can be adjusted. In medallion, the path shown from the top and with the protein omitted. The protein is represented as an atomic surface representation (surface representation is not explained here, *see* Subheading 2 for relevant tutorials and *see* **Note 8**)

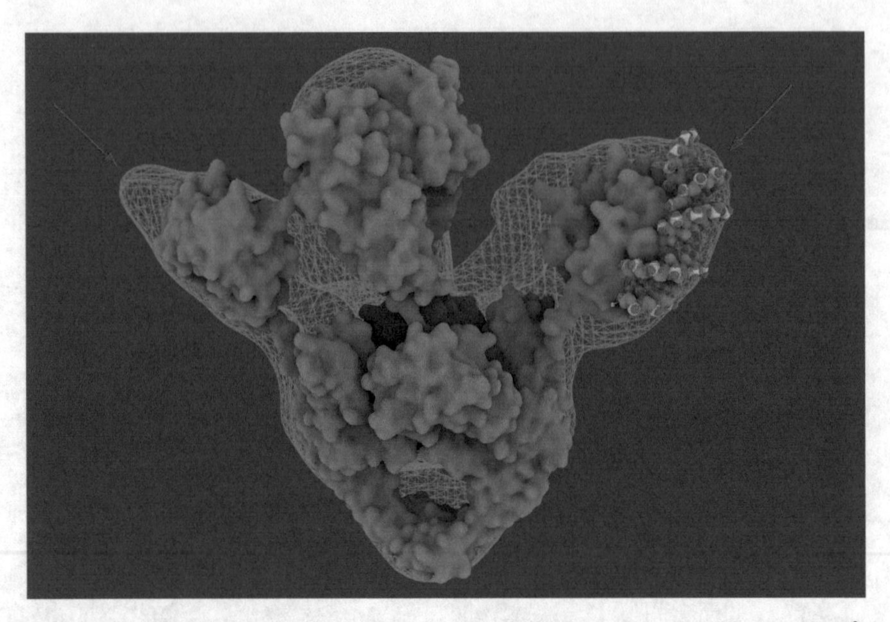

Fig. 5 Extra-densities (*red arrows*) around the pinwheels large enough to accommodate a 20 Å-wide DNA double helix

control points around the gyrase model within the cryo-EM map. The sketching is guided by the knowledge that:

1. DNA is in contact with the two pinwheels. An extra-density, clearly seen in Fig. 5, remains around the pinwheels once the protein atomic structures have been fitted in the density map; the size of these extra-densities corresponds well to the diameter of the dsDNA.
2. The DNA-binding cleavage domain of gyrase at the dimeric interface has been crystallized with a short DNA portion indicating a binding region in addition to the pinwheels.

Hereafter we show in details how to create a simple DNA path with the cubic Bézier editor (a video is also available from Supplementary Data):

1. From Subheading 3.1, **step 3** above, the current 3D scene contains the pseudo-atomic resolution model of the gyrase embedded in its cryo-EM envelope (Fig. 3). Go to the Scene menu (Panel 3, *see* Fig. 2), choose Create Object, then choose Line (HexGrid by default) and name the object DNA, press OK. "DNA" appears in the outliner (Fig. 6, **steps 1–3**).
2. To create the DNA molecule, go to the "tool" tab, click on the "Create cubic Bézier curves" button, and add two (control) points in the 3D scene with the Left Mouse Button (Fig. 6, **steps 4–7**) by *dragging* the mouse to the place you want to add a point (exactly as if you were drawing a line instead of a point).

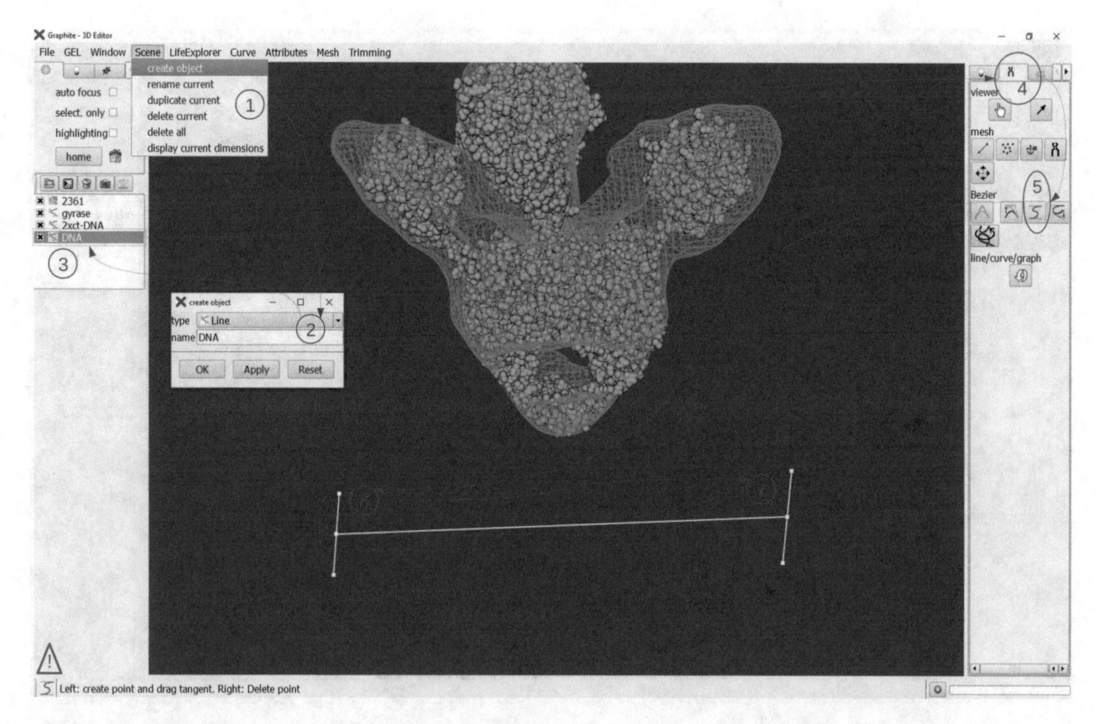

Fig. 6 DNA modeling: steps 1–3 corresponds to the addition of a new element, called DNA, in the outliner. This element is physically created in the 3D scene thanks to the steps 4–7. In this example, from the tool tab (step 4), the user selects the cubic Bézier editor (step 5), draws a first vertical line with the mouse (step 6) and a second vertical line (step 7). Watch also the corresponding video in the Supplementary Data

Go to the "light" tab and select "curve" instead of "plain." The DNA helical axis appears (Fig. 7) (*see* **Note 7**).

3. Add a third line to make the concept of a cubic Bézier construction clearer (Fig. 8). Go to the "light" tab, select "Atoms" instead of "None," an atomic representation is displayed (Fig. 9).
4. To modify the curve while keeping the atomic representation: set "opacity" to 0.2 in the "light tab" so that the DNA path, the points and the tangents become visible (Fig. 10). In the "tool" tab (Fig. 11), click on "Edit cubic Bézier curves" and move the control point, reorientate the tangents, increase their length to see how the curve behaves. In particular, play with the middle tangent to make the bending less severe if necessary and remove the clashes between atoms that occur at the apex. In the same way, adjust the tangents of the first and last points to remove the structural irregularities occurring at the ends. Repeat **steps 1–4** to create a DNA model wrapping around the gyrase (Figs. 4 and 12).

Making DNA models with various lengths in GraphiteLifeExplorer can help for the interpretation of cryo-EM maps and provide clues for the design of future experiments. Several questions were raised

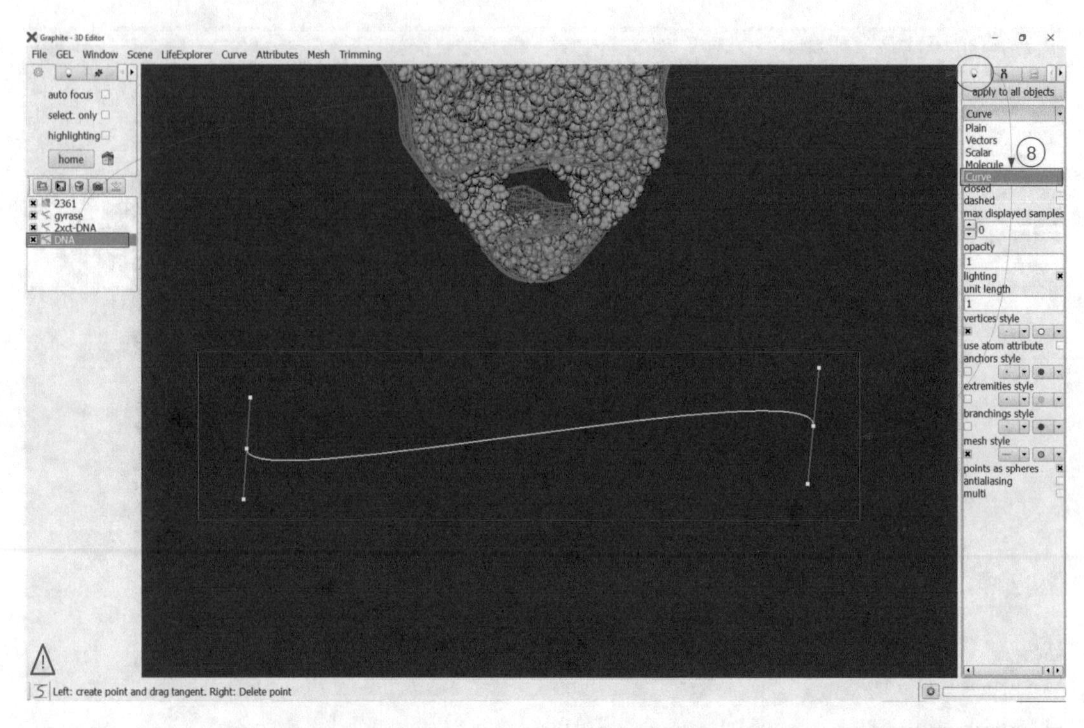

Fig. 7 From the light tab, by clicking on the "curve" representation, the DNA path is displayed (step 8): one clearly notices that the path is not a straight line connecting the two control point but it is tangent to the vertical segments. This property allows to draw complicated shapes with a minimal number of control points

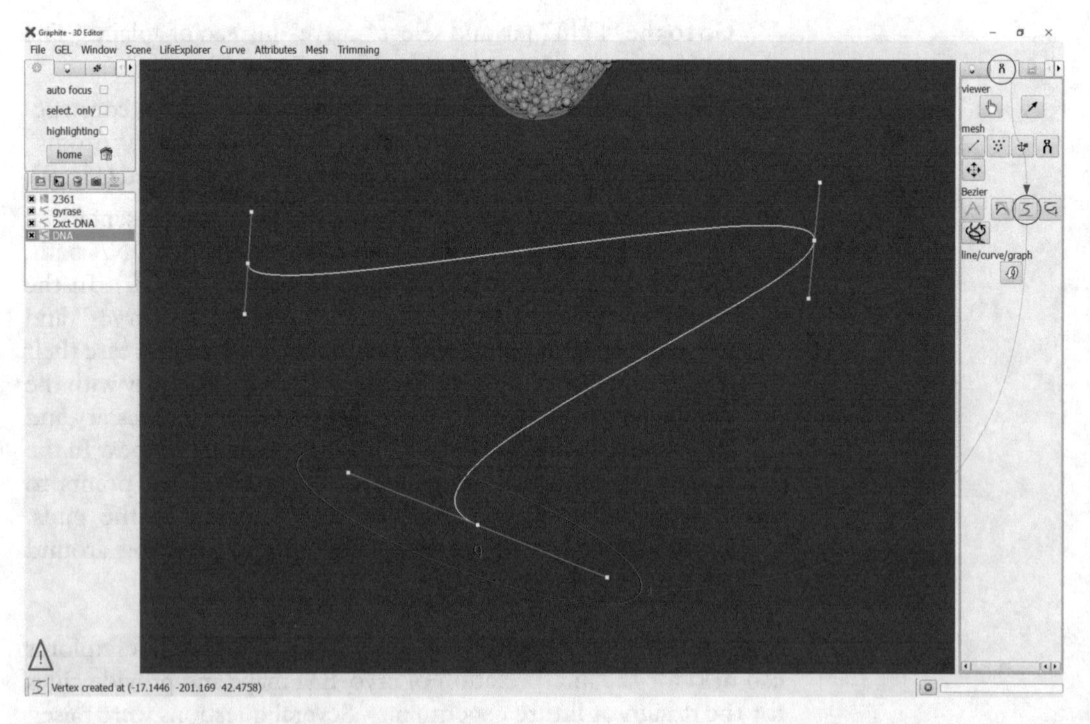

Fig. 8 Once the DNA path is displayed (step 8, Fig. 7) the user goes back to the tool tab, clicks on the first icon of the cubic Bezier modeling tool and draws a third line (step 9)

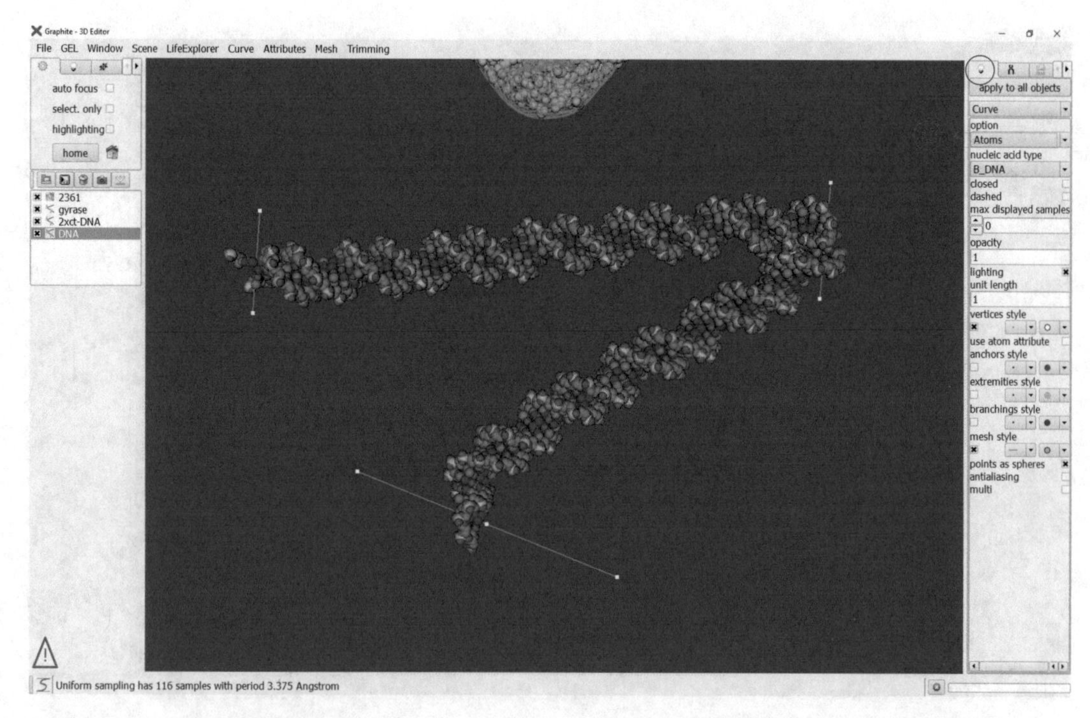

Fig. 9 Once the third line has been added (step 9, Fig. 8), the user goes back to the light tab and selects "Atoms" to obtain an all-atom representation (step 10). DNA must be selected in the outliner to apply any effect on it

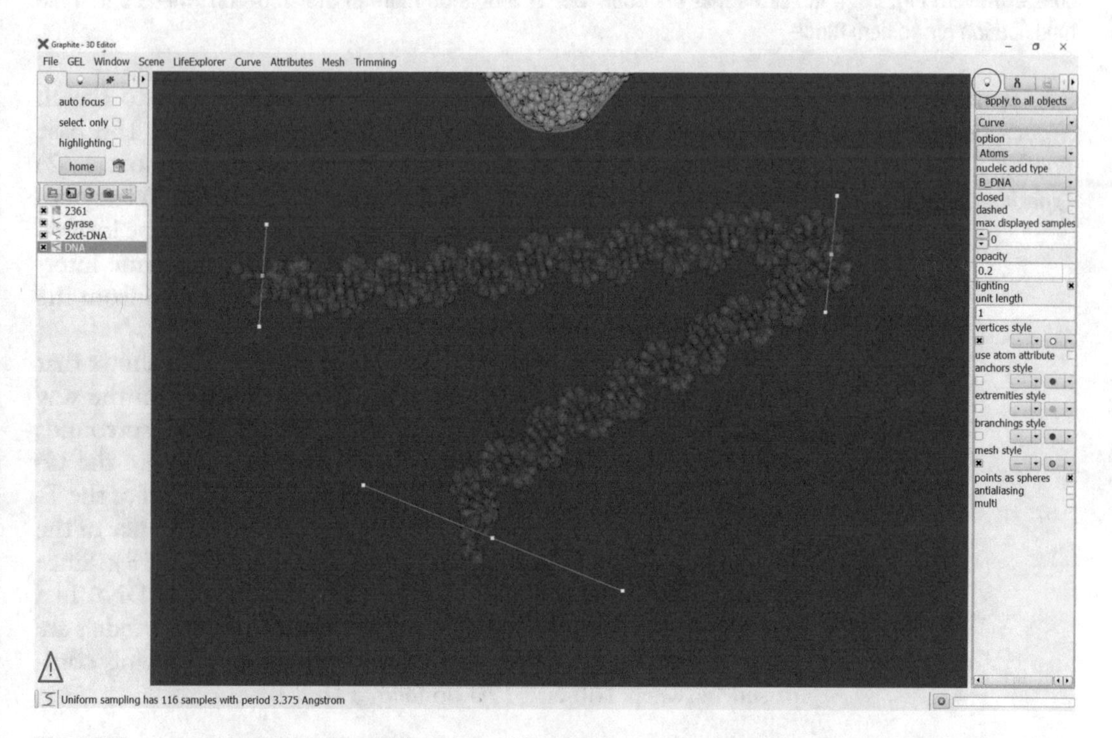

Fig. 10 The user sets opacity of the DNA representation to see control points through the model (step 11). This way the user can modify the model while keeping the atomic representation

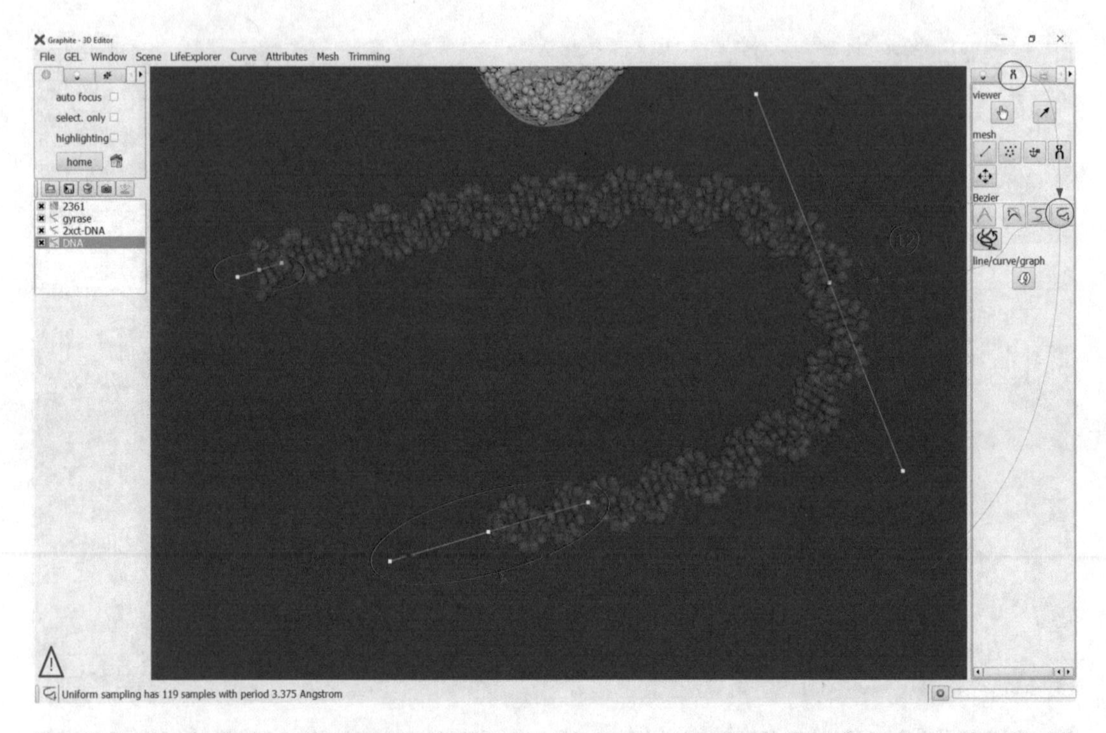

Fig. 11 Back to the tool tab, the user selects the second icon corresponding to the cubic editor and, in the 3D scene, reorientates the tangents, modifies their length to soften the curve and get locally a structurally correct arrangement of atoms (step 12). Notice that the icon in the left lower corner of the screen has changed compared with Fig. 10. It indicates that the cubic Bézier tool is no more in the "creation" mode but in the modification (or edition) mode

3.3 Drawing Hypothesis and Guiding Experimentations

by Papillon and coworkers since they were only able to sketch 130 bp out of 155 bp used in the experiments: (1) Are 155 base pairs really enough to generate experimentally a crossover? (2) What is the path taken by DNA to form a crossover? What does the resulting crossover look like (knowing that it should be located right at the center of the cryo-EM envelope at the enzyme interface, in an area where its density cannot be distinguished from the protein density at the resolution of the current map)?

The visual exploration of the whole gyrase complex shows that the optimal size for a DNA construct clearly depends on the way DNA binds the first disk-shaped pinwheel. If DNA surrounds completely the "disk structure," then a crossing between the G- and the T-segments is not possible using 155 bp. The end of the T-segment from a 155 bp duplex would just reach the entrance of the DNA gate without forming the DNA crossover (Fig. 13). Since GraphiteLifeExplorer allows to draw tortuous shapes of DNA in a few minutes and adjust their length (*see* **Note 8**), several models are easily sketched that rather suggest an optimal length being comprised between 160 and 170 bp (Fig. 14b, c).

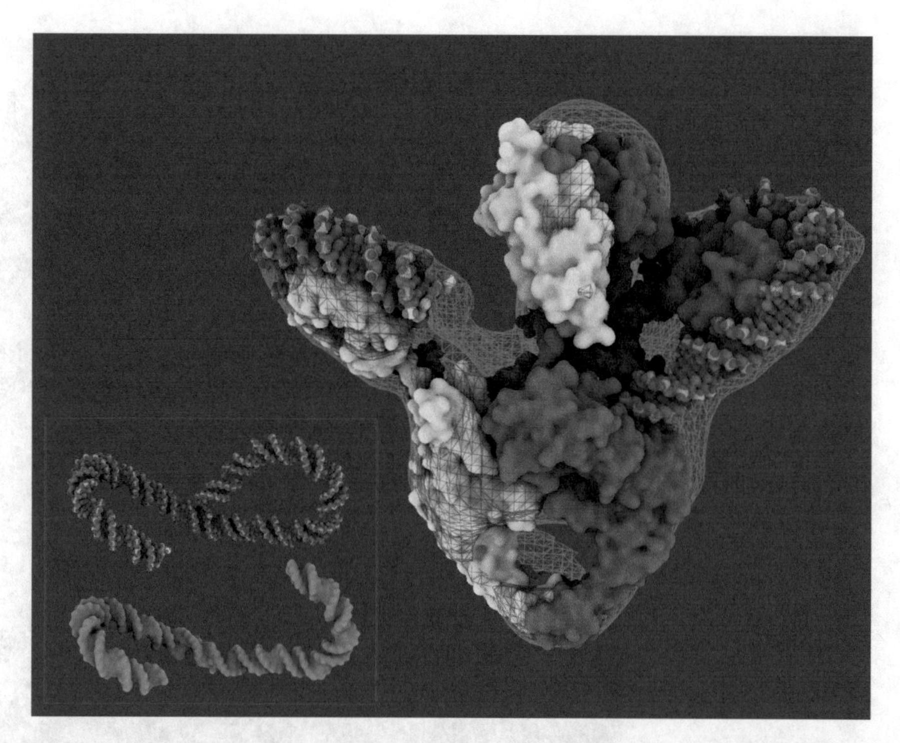

Fig. 12 155 bp DNA model. The protein surface has been colored and various light effects (including ambient occlusion) have been applied to the 3D scene to improve the 3D perception (*see* Subheading 2 for relevant tutorials and *see* **Note 8**). In Medaillon, for comparison, the 155 bp DNA model created with GraphiteLifeExplorer and the 130 bp DNA model obtained by the authors of the cryo-EM map [9] are shown together. The building of the 130 bp model took days to be completed as it is made of short DNA portions docked to the protein and sealed together. The 155 bp DNA duplex has been fulfilled in a few minutes with GraphiteLifeExplorer

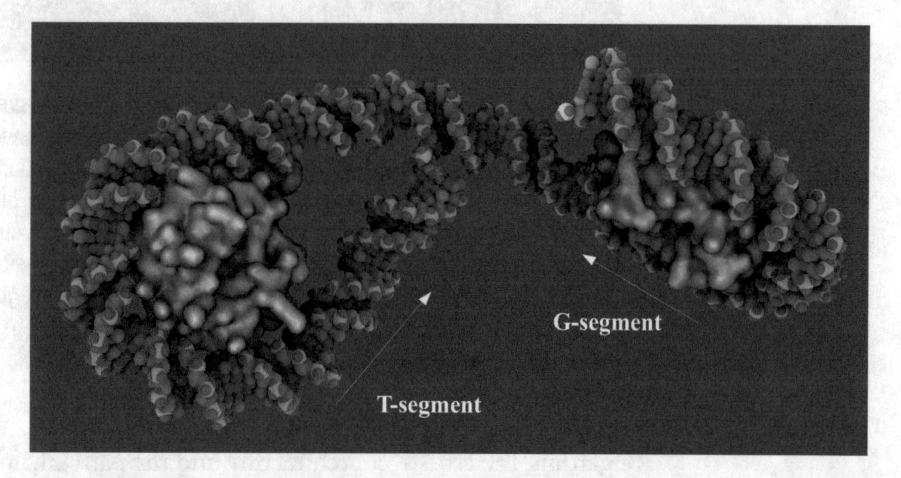

Fig. 13 The G- and T-segments of the contiguous 155 bp DNA are not crossing. We conclude that the initial experimental setup including a 155 bp DNA might not generate a DNA crossover as anticipated while designing the DNA sequence to form a complex with DNA gyrase

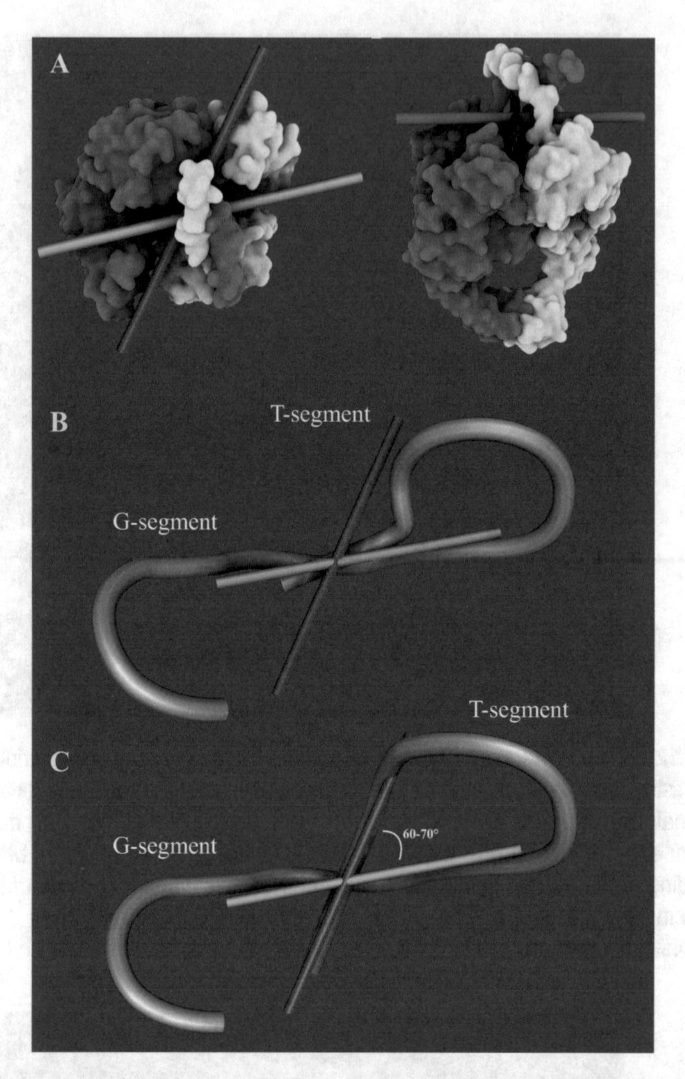

Fig. 14 DNA channels trajectories inside the DNA-binding cleavage domain of DNA gyrase. (**a**) *Top* and *side* views of the domain surface have been represented in the *upper image* (pinwheels and the ATPase domains have been omitted). The *red* and *blue* tubes materialize two grooves in the DNA-binding cleavage domain of DNA gyrase that can accommodate a DNA duplex. (**b** and **c**) Two possible models for the DNA crossover. The G-segment sits in the middle of the protein, always parallel to the *blue* tube. The T-segment, if extrapolated, could potentially follow two trajectories: one is parallel to the *red tube* and the other reenters by the "blue" channel. Both DNA models contain between 160 and 170 bp

Regarding the crossover architecture and the path taken by the DNA, the visual exploration of the gyrase also leads to identify two channels as a possible reentry for the T-segment (Fig. 14a). One channel is larger but forces the DNA to wrap a longer portion of the second pinwheel and to be kinked to reenter through the "blue" channel (Fig. 14b). Also, in this configuration, there is enough room to accommodate the two strands with a correct curvature. However, the G- and the T-segments are superimposed pointing in the same direction, which is inconsistent with the formation of a DNA crossover. The second "red" channel (Fig. 14c) provides enough room for the G- and the T-segments to be superimposed. However, it is narrower and would probably provoke slight rearrangements of the proteic domains in this region as DNA is pinched by the two CTD-domains forming the DNA gate at closure. However, this configuration is in agreement with the existence of a true crossover characterized by an angle of 60–70° between the G- and the T-segments.

3.4 Export Capability

A series of important scientific questions can be studied using molecular dynamics simulations [13]. As an example, the models of the complete gyrase/DNA complex can serve as a starting point to study in atomistic details the conformational stability of the DNA with the gyrase enzyme, and the dynamics of the complex over time. To make this possible, the DNA modeled in GraphiteLifeExplorer with the whole complex can be saved in the PDB file format so that it can be used in a third-party software (*see* **Note 9**). Note that the gyrase/DNA model used in this example consists of roughly 29,000 atoms before including explicit solvation and counter-ions which could translate in a system of more than half a million atoms. For a system of this size, specialized computing hardware (GPU, high performance cluster computing, etc.) is recommended to achieve enough sampling time during MD simulations.

To save the 3D scene in the GraphiteLifeExplorer GSG format, go to the menu File, Save scene, give a name to the scene, for instance *model.gsg* (the suffix *.gsg is mandatory) and press Save (*see* **Note 3**). Besides reading and exporting in the PDB file format, GraphiteLifeExplorer can also read geometrical formats (like surfaces): Wavefront OBJ, 3DS, PLY, and Lightwave LWO. Surfaces created in GraphiteLifeExplorer (from a PDB file for instance) can be exported in the VRML format.

Figure 15 shows a chart synthesizing all steps from the very beginning of the DNA modeling procedure to the generation of a PDB file. The user can load in GraphiteLifeExplorer an interactive model, *model.gsg*, containing the gyrase, the EM map, and one of the two DNA models described at the Subheading 3.3.

4 Notes

1. 3DNA and NAB are popular and powerful packages that can do more than modeling. Both allow to analyze the mechanical parameters of any DNA molecule, be it single or double-

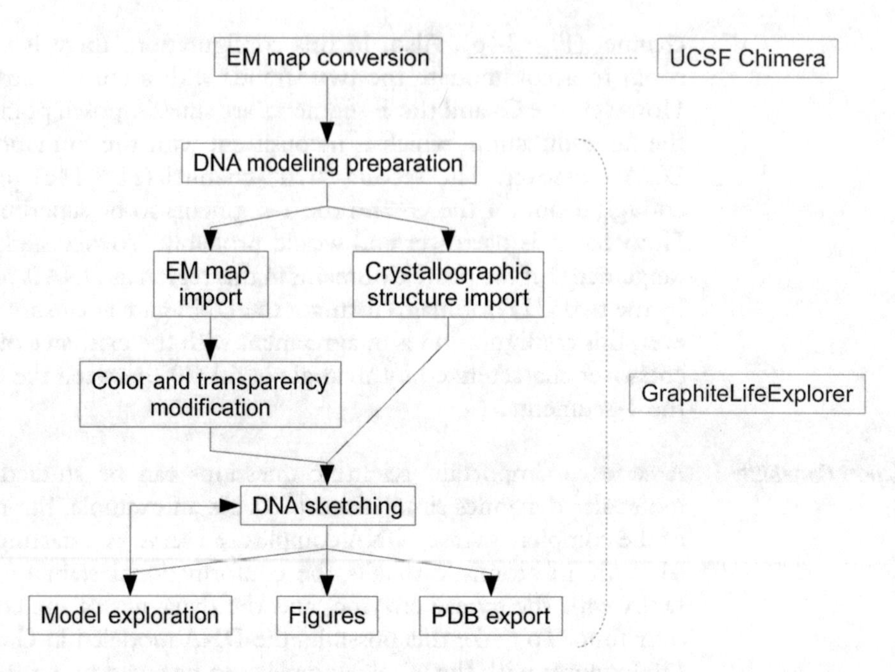

Fig. 15 Summary chart representing the different steps to model a DNA molecule in a protein complex fitted in a cryo-EM map

stranded. NAB, as a part of the Amber suite of programs, also allows to perform molecular dynamics based simulations.

2. mMaya is a free plugin for Autodesk Maya (a commercial product). mMaya is equipped with DNA modeling capabilities that go beyond what GraphiteLifeExplorer offers (custom sequence, animation, etc.). It is accessible at: https://clarafi.com/tools/mmaya/.
3. By default, a scene created and displayed in GraphiteLifeExplorer is saved in a file with the suffix *.gsg. In this format (*G*raphite *S*cene *G*raph), surfaces and DNA models are restored as such. A structure initially opened as a PDB file and part of the scene saved in the gsg file format is restored as a point cloud (all information contained in the PDB file is lost except the coordinates of the atoms). Background colors and light effects are not included in the gsg file format.
4. A Wavefront OBJ file generally contains only polygonal faces describing a surface. It does not contain color definitions that are stored in a separate material library file (recognized by file suffix *.mtl). As a consequence, a second file named 2361.mtl is automatically created by UCSF Chimera. It contains color information associated with the map and specified by the user in UCSF Chimera. This file 2361.mtl is of no importance for GraphiteLifeExplorer which does not read it (in GraphiteLifeExplorer the user can reassign any color to the map which is white by default).

5. This structure (PDB code 2xct) of the DNA-binding cleavage domain in complex with a short DNA duplex had been used by Papillon et al. [9] to generate a complete model of the gyrase fitting in the cryo-EM map. In the text, *2xct-DNA.pdb* refers to the coordinates corresponding to DNA in the crystallographic structure 2xct. We use it to guide the DNA sketching.
6. The DNA editor in GraphiteLifeExplorer proposes two tools: the cubic Bézier tool and the quadratic Bézier tool. The quadratic Bézier tool is intuitive: the user simply draws a line made of control points that form the DNA path (or helical axis). However, the smoothness of the curve cannot be perfectly achieved (contrarily to the cubic mode) and kinks can be created locally. The quadratic mode is ideal to model a straight fiber or a sharply bent DNA. The cubic mode is less intuitive (the user does not create a succession of points but tangents) but allows to sketch perfectly smooth curves with a minimum of control points. Note that "Control" in "control point" means that a point can still be moved or suppressed once created.
7. The icon in the left lower corner shows the current mode: at Fig. 1 for instance, the "hand" indicates the navigation mode. At Fig. 6, it indicates the cubic Bezier "Creation" mode. A frequent mistake consists of willing to fly around the complex while such a modeling mode is turned on. To avoid adding control points we suggest a trick consisting of pressing the Ctrl key for exploring the scene. This way, whatever the modeling mode, the user automatically switches to the navigation mode. The modeling mode is turned on again when we stop pressing the Ctrl key.
8. A tutorial explaining how to fix the length of a DNA model is accessible at http://www.lifeexplorer.info/tutorials/basics/#fixlength. Also, a tutorial showing how to obtain a surface representation from a protein atomic cloud can be found here: http://www.lifeexplorer.info/tutorials/rendering/.
9. A tutorial showing how to export a DNA model in the PDB format is accessible at: http://www.lifeexplorer.info/tutorials/input-output/#pdb. The user can also save in PDB a component previously opened as a PDB file thanks to a function named writePDB. This is useful when the component has been moved with respect to another component. writePDB allows to save the coordinates of all components in the same frame of reference and to have, in a third-party software, the components positioned relatively to each other exactly like in GraphiteLifeExplorer. Note that writePDB does not save all components in a single combined PDB file as Chimera does.

References

1. Glaeser RM (2016) How good can cryo-EM become. Nat Methods 13:28–32
2. Kühlbrandt W (2014) Microscopy: cryo-EM enters a new era. elife 3:e03678
3. Merk A, Bartesaghi A, Banerjee S, Falconieri S, Rao P, Davis MI, Pragani R, Boxer MB, Earl LA, Milne JLS, Subramaniam S (2016) Breaking Cryo-EM resolution barriers to facilitate drug discovery. Cell 165:1698–1707
4. Jolley CC, Wells SA, Fromme P, Thorpe MF (2008) Fitting low-resolution Cryo-EM maps of proteins using constrained geometric simulations. Biophys J 94:1613–1621
5. Raposo AN, Gomes AJP (2016) Computational 3D assembling methods for DNA: a survey. IEEE/ACM Trans Comput Biol Bioinform. 13:1068–1085. doi:10.1109/TCBB.2015.2510008
6. Colasanti A, Lu XJ, Olson WK (2013) Analyzing and building nucleic acid structures with 3DNA. J Vis Exp 74:e4401
7. Macke TA, Case DA (1998) Modeling unusual nucleic acid structures. In: Leontes NB, Jr SLJ (eds) Molecular modeling of nucleic acids. American Chemical Society, Washington, DC
8. Hornus S, Lévy B, Larivière D, Fourmentin E (2013) Easy DNA modeling and more with GraphiteLifeExplorer. PLoS One 8(1):e53609
9. Papillon J, Ménétret J-F, Batisse C, Hélye R, Schultz P, Potier N, Lamour V (2013) Structural insight into negative DNA supercoiling by DNA gyrase, a bacterial type 2A DNA topoisomerase. Nucleic Acids Res 41:7815–7827
10. Fisher LM, Mizuuchi K, O'Dea MH, Ohmori H, Gellert M (1981) Site-specific interaction of DNA gyrase with DNA. Proc Natl Acad Sci U S A 78:4165–4169
11. Pettersen EF, Goddard TD, Huang CC, Couch GS, Greenblatt DM, Meng EC, Ferrin TE (2004) UCSF Chimera—a visualization system for exploratory research and analysis. J Comput Chem 25:1605–1612
12. Goddard TD, Huang CC, Ferrin TE (2007) Visualizing density maps with UCSF Chimera. J Struct Biol 157:281–287
13. Cheatham TE III, Case DA (2013) Twenty-five years of nucleic acid simulations. Biopolymers 12:969–977

Part IV

Imaging the Bacterial Nucleoid

Chapter 16

Multilocus Imaging of the *E. coli* Chromosome by Fluorescent In Situ Hybridization

Bryan J. Visser, Mohan C. Joshi, and David Bates

Abstract

Fluorescence in situ hybridization (FISH) is a widely used technique to detect and localize specific DNA or RNA sequences in cells. Although supplanted in many ways by fluorescently labeled DNA binding proteins, FISH remains the only cytological method to examine many genetic loci at once (up to six), and can be performed in any cell type and genotype. These advantages have proved invaluable in studying the spatial relationships between chromosome regions and the dynamics of chromosome segregation in bacteria. A detailed protocol for DNA FISH in *E. coli* is described.

Key words Fluorescence in situ hybridization, FISH, Bacterial chromosomes, *E. coli* chromosome

1 Introduction

Technological advances using fluorescently labeled DNA binding proteins, such as the fluorescent repressor-operator system (FROS), have revolutionized chromosome dynamics research (reviewed in [1]). While these systems have the clear advantage of allowing the tracking of genetic loci in living cells, they have significant drawbacks. First, for every chromosomal site to be examined (at best two in the same cell) a large array of protein binding sequences must be cloned into the chromosome by homologous recombination—often a difficult procedure because of their repetitive nature (cannot PCR). Second, the resulting large protein complexes often interfere with DNA replication [2] and may even change the cellular localization of the genetic locus being analyzed. Fluorescent in situ hybridization (FISH) requires no cloning, does not involve an in vivo tag of any kind, and can be performed in any strain. Since several genetic loci (up to six) can be analyzed simultaneously, FISH is particularly suited for analyzing the spatial relationship between chromosome regions and was the primary tool used to document chromosome cohesion in *E. coli* [3–5]. Undoubtedly, bacterial FISH is underutilized because it is

Olivier Espéli (ed.), *The Bacterial Nucleoid: Methods and Protocols*, Methods in Molecular Biology, vol. 1624,
DOI 10.1007/978-1-4939-7098-8_16,

technically challenging and most protocols have been optimized for eukaryotic cells. We describe here a highly optimized method that is based on pioneering work by Sota Hiraga [6], and later modified in our *E. coli* chromosome dynamics studies [3, 5, 7]. The methods are applicable to a variety of gram-negative and gram-positive bacteria with minor adjustments to the permeabilization procedure due to differences in cell wall structure.

2 Materials

2.1 Fluorescence Imaging System

As FISH markers contain only a few fluorescent molecules, the primary consideration when choosing an imaging system for FISH is light sensitivity. Sensitivity is determined largely by the imaging sensor or charge-coupled device (CCD), with the most sensitive cameras having very large pixels with a high dynamic range. These cameras generally have electron multiplying sensors (EMCCD) with quantum efficiency (the percentage of photons detected) approaching 100%, thus capable of detecting a single photon. However, with increased pixel size there is an equal reduction in spatial resolution for the object being imaged. While this is usually not a concern with interpretation of FISH signals on eukaryotic chromosome spreads, bacterial applications often require resolution to the light limited value of ~0.2 μm. By sampling theory, the image pixel size must be less than half the desired resolution (0.1 μm). Under 100× magnification, this corresponds to an actual CCD sensor pixel size of less than 10 μm. Note that further reduction in CCD pixel size does not confer any additional gain in resolution because of the light diffraction limit. Cameras, especially EMCCD types, also require an active cooling system (e.g., peltier block or liquid nitrogen) to reduce thermal energy (dark current) that can activate the CCD sensors. With FISH's typical long exposure times, dark current can significantly reduce the ability to detect weaker signals. Lastly, fluorescence excitation and emission filters most useful for multicolor FISH are narrow spectrum bandpass filters designed to reduce bleed-through of fluorescent signals into adjacent filter channels (cross talk; Fig. 1). Because background fluorescence is generally present in all wavelengths, long exposures with a narrow spectrum filter produce a higher signal-to-noise ratio than short exposures with a wide spectrum filter.

1. Fluorescence microscope: the most important features of a fluorescence microscope are high quality (and clean) optics, a high numerical aperture objective, and a multichannel filter wheel or cube turret for illuminating and observing specific fluorescent dyes. Motorized stages and computer-controlled filter selection and image acquisition save time and allow precise three-dimensional imaging.

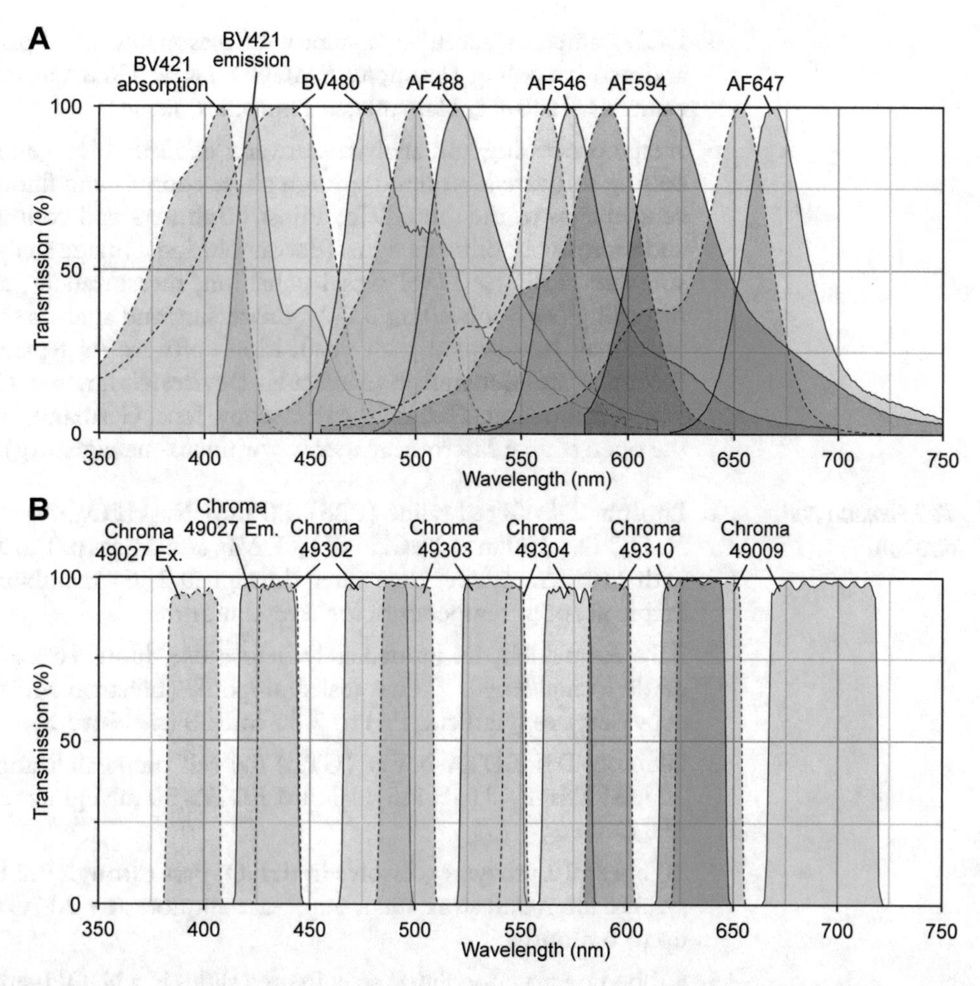

Fig. 1 Fluorescence spectra for six-color FISH. (**a**) Absorption and emission spectra for six fluorescent dyes as indicated; (BV) Brilliant Violet (BD Biosciences, San Jose, CA, USA), (AF) Alexa Fluor (ThermoFisher Scientific). (**b**) Excitation and emission filter sets (Chroma Technology Corp, Bellows Falls, VT, USA) to image six fluorescent dyes with minimal cross talk (dichroic mirror spectra not shown)

2. Illumination source: the most common light source is a high-pressure mercury arc lamp, which has several intensity peaks that correspond to excitation spectra of several fluorescent dyes. An alternative light source is a xenon lamp, which has a more homogeneous spectral coverage and is able to excite far-red dyes better than mercury lamps. Light-emitting diode (LED) illumination technology is rapidly emerging, but may, as of yet, not provide sufficient light output for FISH.
3. Fluorescence bandpass filters: multicolor FISH demands narrow wavelength excitation and emission filters to avoid cross talk between channels. Six-color FISH with minimal cross talk can be performed using the dyes and filter sets shown in Fig. 1.

4. CCD camera: a sensitive camera with reasonably small pixels and active cooling (example, EMCCD 1K × 1K 8 μm pixel camera C9100-02, Hamamatsu Photonics, Japan).
5. Image processing and analysis software: dedicated image processing software is required to align phase contrast and fluorescent images of the same field, adjust brightness and contrast, and remove out of focus signal (deconvolution). Image analysis software facilitates FISH signal detection, measurement, and recording (*see* Subheading 3.3.2). Processing and analysis software may be coupled with acquisition software or separate. Examples are Metamorph (Molecular Devices, Sunnyvale, CA, USA), Axiovision (Carl Zeiss Microscopy, Jena, Germany), and the open source Micro-Manager (www.micro-manager.org).

2.2 Cell Fixation and Hybridization

1. Phosphate-buffered saline (PBS): 80 mM Na_2HPO_4, 20 mM NaH_2PO_4, 100 mM NaCl in dH_2O. Mix and adjust pH to 7.5 with HCl. Autoclave, then filter through 0.2 μm membrane. Stable at room temperature for several months.
2. 2.5% formaldehyde: immediately before use dilute 16% EM-grade formaldehyde from a sealed ampoule (Electron Microscopy Services, Hatfield, PA) to 2.5% in PBS (*see* **Note 1**).
3. Glucose–Tris–EDTA buffer (GTE) for cell permeabilization: 20 mM Tris–HCl (pH 7.5), 10 mM EDTA, 50 mM glucose in dH_2O.
4. 10 mg/ml Lysozyme: dissolve in dH_2O, pass through 0.2 μm syringe filter, and store small single-use aliquots at −20 °C for up to 6 months.
5. Subbed microscope slides: soak frosted slides in a NaOH–ethanol solution (2 M NaOH, 50% ethanol in dH_2O) in a glass staining dish for 1 h, then rinse continually with dH_2O for 10 min (place jar under running water). Transfer the slides (without drying) to a plastic staining dish filled with poly-L-lysine solution (0.01% poly-L-lysine in 0.1× PBS) and soak with gentle shaking for 1 h. Rinse the slides three times in dH_2O and dry with compressed air to avoid streaks. Prepared slides may be stored in a desiccator for several weeks.
6. 20× SSC: 4 M NaCl, 0.4 M sodium citrate in dH_2O, pH to 7.5 with HCl, and autoclave.
7. 50% dextran sulfate: dissolve 2.5 g dextran sulfate in 2.5 ml dH_2O at 60 °C with stirring. Filter through 0.2 μm membrane and store at −20 °C.
8. Denaturing solution: 50 ml 2× SSC, 70% formamide (recently deionized), 0.1 mM EDTA in dH_2O. Verify that pH is 7.0–7.5 with a test strip. Solution may be reused up to 1 month if stored at 4 °C.

9. 1.2× hyb buffer: 60% formamide, 2× SSC, 12% dextran sulfate, 30 mM Na_2HPO_4, 30 mM NaH_2PO_4 in dH_2O. Store small single-use aliquots at −20 °C for up to 1 year.
10. Several glass staining dishes, both upright (~50 ml volume) and horizontal (~200 ml volume) styles. Horizontal staining dishes have removable racks and are convenient for transferring the slides rapidly between successive treatments.
11. 85 °C heating source for denaturing genomic DNA in situ: heated circulating water bath, thermal cycler with in situ PCR attachment, or solid aluminum dry block incubator (*see* Subheadings 3.2.3, 3.2.4)
12. Hybridization chamber: fill an empty tip box (holds four slides) or large Tupperware with inner rack with ~1 cm water. Preheat container to 37 °C before hybridization and cover with lightproof material.
13. Blocking buffer: 1 ml 0.1% Tween 20, 2% BSA in 1× PBS.
14. 10× 4′,6-diamidine-2′-phenylindole dihydrochloride (DAPI): 3 μg/ml in dH_2O.

2.3 Probe Labeling (Nick Translation)

1. Nucleotide stocks: 5 mM each of dATP, dGTP, dCTP, dTTP, and fluorescent or hapten-labeled dUTP.
2. 10× nick translation buffer: 0.5 M Tris–HCl (pH 7.5), 0.1 M $MgSO_4$, 1 mM dithiothreitol (DTT). Store single-use aliquots at −20 °C.
3. 25× enzyme mix: 5 U/μl DNA Pol I and 0.05 U/μl DNase I in storage buffer (25 mM Tris–HCl pH 7.5, 1 mM DTT, 0.1 mM EDTA, 50% glycerol). Store small single-use aliquots at −20 °C.
4. Stop buffer: 0.5 M EDTA (pH 8.0).
5. TE buffer: 10 mM Tris–HCl (pH 7.5), 1 mM EDTA.
6. Sephadex spin columns.
7. Small volume spectrophotometer.

3 Methods

3.1 Probe Preparation

DNA probes can be labeled enzymatically by nick translation, PCR, or random-primed synthesis using a fluorescent nucleotide. Alternatively, a biotin or hapten-labeled nucleotide may be used that is subsequently detected by a fluorescently conjugated streptavidin or antibody. Although indirect labeling methods utilizing secondary fluorescent antibodies produce an amplified signal, they generally have higher background fluorescence and fewer labels can be used in one experiment due to the limited number of available hapten–antibody pairs. Alternatively, fluorophores can be chemically conjugated to dsDNA, with the universal linking system (ULS) being the most common commercially available method (Ulysis Nucleic Acid Labeling Kit, ThermoFisher Scientific, Waltham, MA, USA). Nick translation, our preferred labeling method (described below), is relatively inexpensive, works well with a wide range of fluorescent nucleotides, and has the advantage of producing short DNA fragments without secondary sonication steps. All steps using fluorescent dyes should be performed under low light conditions.

1. Prepare several micrograms of a 2–6-kb DNA fragment at the site of interest by PCR amplification or plasmid purification. Although probes as small as 100 bp can produce detectable signals, longer probes are brighter and less prone to sequence-specific differences in labeling and detection efficiency. It has also been reported that probes prepared from whole plasmids increase FISH signals due to chain hybridization of vector sequences at the site of genomic target [8].
2. Prepare a 50 μl nick translation reaction in a 0.5 ml PCR tube on ice: 5 μl 10× nick translation buffer, 1 μg template DNA (in 1–23 μl H_2O), 5 μl 5 mM dATP, 5 μl 5 mM dCTP, 5 μl 5 mM dGTP, 2.5 μl 5 mM dTTP, 2.5 μl 5 mM fluorescently labeled dUTP, 2 μl 25× enzyme mix, and bring volume to 50 μl with dH_2O.
3. Mix well, incubate at 15 °C for 3 h, and place reaction on ice.
4. Run 5 μl of the reaction on an agarose gel to check for proper fragment length (200–500 bp). If fragment length is >500 bp, add 1 μl of enzyme mix and incubate for an additional 3 h at 15 °C (*see* **Note 2**).
5. Stop reaction by adding 2 μl Stop Buffer.
6. Remove unincorporated fluorescent nucleotides by passing through a Sephadex spin column using the manufacturer's protocol, then precipitate the DNA with ethanol and resuspend in 15 μl TE.
7. (Optional) To calculate labeling efficiency, measure the optical density of the labeled probe using a small volume spectrophotometer at 260 nm and at the excitation maximum of the fluorescent dye and apply the following equation:

$$\text{dye/base ratio} = (\text{Abs}_{\text{dye}} \times \text{EC}_{\text{dsDNA}})/(\text{Abs}_{260\text{corr}} \times \text{EC}_{\text{dye}}),$$

where the extinction coefficient for dsDNA (EC_{dsDNA}) is 6600, and the corrected 260 nm absorbance ($\text{Abs}_{260\text{corr}}$) is calculated from:

$$\text{Abs}_{260\text{corr}} = \text{Abs}_{260} - (\text{Abs}_{\text{dye}} \times \text{CF}_{260}).$$

Extinction coefficient (EC_{dye}) and absorbance contribution at 260 nm (CF_{260}) of the dye are provided by the supplier. Incorporation rates >0.03 (3 dye molecules per 100 bases) are acceptable and are typically 0.05–0.08 in nick translation reactions.

3.2 In Situ Hybridization

3.2.1 Cell Fixation

To preserve chromosome morphology during chemical and temperature treatments of FISH, DNA and protein must be fixed prior to FISH treatment. Aldehyde fixatives, either formaldehyde or glutaraldehyde, physically cross-link proteins and nucleic acid and offer excellent morphological preservation. Glutaraldehyde forms cross-links between more distantly separated molecules and is often used in combination with formaldehyde. Commercial 37% formaldehyde solutions (formalin) often contain up to 10% methanol, which can cause protein clumping, so we recommend using sealed ampoules of formaldehyde that are made by dissolving polymerized formaldehyde (paraformaldehyde) in dH_2O under nitrogen. Alcohol fixatives preserve morphology by precipitating proteins and nucleic acids, and can result in greater accessibility for FISH probes over cross-linking fixatives. However, precipitation causes significant contraction of neighboring molecules and can visibly alter cellular morphology. The addition of acetic acid, which causes cellular swelling, can counteract alcohol induced contraction.

1. Pellet ~2×10^8 cells from an exponentially growing culture (approx. 2 ml at OD = 0.2) by centrifugation at $5000 \times g$ for 3 min at room temperature. This quantity of cells is sufficient for a single FISH experiment including lysozyme optimization.
2. Resuspend cell pellet thoroughly in 1 ml 2.5% formaldehyde in PBS by gently pipetting up and down.
3. Incubate tube at room temperature for 15 min, then on ice for an additional 30 min.
4. Wash cells three times in 250 μl ice-cold PBS by repeated centrifugation ($5000 \times g$ for 2 min at 4 °C). Resuspend the cell pellet in 200 μl ice-cold PBS, and store at 4 °C up to 1 month (cross-links reverse slowly in aqueous conditions).

3.2.2 Lysozyme Treatment

Prior to hybridization, cells must be permeabilized with lysozyme to allow probe entry. This step varies significantly between bacterial species, with gram positives requiring more vigorous treatment. However, even within a species cell wall composition is affected by genotype and growth conditions, and the activity of lysozyme stocks can vary. Overdigestion can undermine the membrane integrity to the point that cells burst, and underdigestion prevents adequate probe penetration. In our experience, improper lysozyme digestion is the most common cause of poor FISH signals, and it is well worth the effort to optimize lysozyme treatment for every experiment.

1. Pellet 200 μl fixed cells (5000 × *g* for 2 min at room temperature) and resuspend in 45 μl GTE.
2. Add 5 μl diluted lysozyme (freshly diluted to 25 μg/ml in GTE from frozen 10 mg/ml stock), place into a 37 °C heated block, and start a timer.
3. At 1-min intervals, transfer 5 μl cells to 45 μl ice-cold GTE. Take a total of six samples. Place all tubes in ice.
4. With the flat side of a pipette tip, spread 5 μl of each sample (keeping the rest of the sample on ice) onto a ~20 × 20 mm area on a poly-L-lysine coated slide and allow to dry 10 min on the bench.
5. Assemble four staining dishes containing room temperature PBS, 70% ethanol (in dH_2O), 85% ethanol, and 100% ethanol. Alcohol baths can be reused in later steps.
6. Place the five slides into a staining rack and immerse into the four dishes in order of increasing ethanol for 1 min each then dry the slides with compressed air.
7. Apply 3 μl PBS with 0.3 ng/μl DAPI, add coverslip, and image cells.
8. A range of cell morphologies should be observed with shortest lysozyme treatments appearing completely intact and the longest treatments appearing burst with DNA outside the cell (Fig. 2). The optimal lysozyme treatment for FISH will have mostly intact cells but a few (1–5%) burst cells. If no burst cells are observed in the longest incubation sample, place the undiluted reaction at 37 °C for additional time.
9. Spread 10 μl diluted cells from the best time point over the entire area (~20 × 40 mm) of a poly-L-lysine-coated slide and allow to dry 20 min on the bench (use a desiccator in humid conditions).

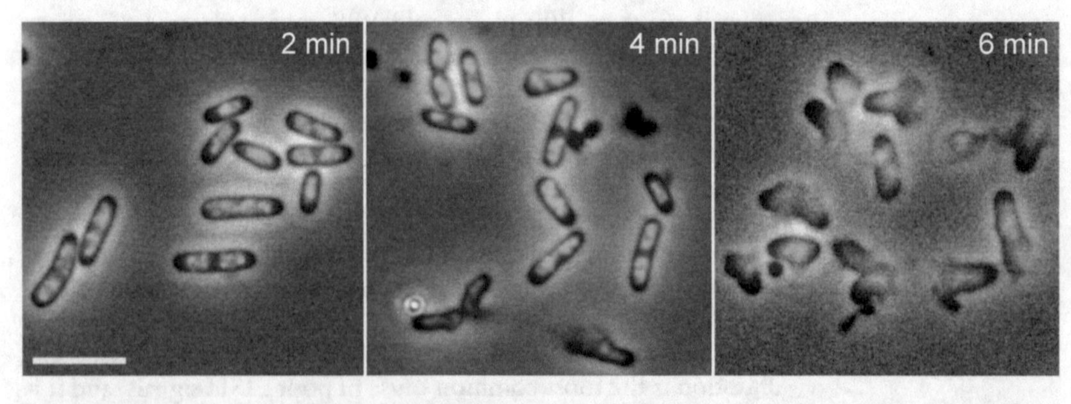

Fig. 2 Lysozyme test slides show underdigested, correctly digested, and overdigested cells. Formaldehyde fixed *E. coli* were permeabilized with lysozyme and incubated for the indicated times and stained with DAPI before imaging with a 100× objective. By 4 min in this example some cells have lost their integrity, but many high-quality cells remain for analysis. Bar is 5 μm

3.2.3 Denaturation and Hybridization, Immersion Method

Traditionally, genomic DNA is denatured in a large volume of hot denaturing solution containing 50–70% formamide, followed by ethanol dehydration and addition of denatured probe DNA in hybridization buffer. An alternative method is provided (Subheading 3.2.4), in which genomic DNA and probe DNA are denatured simultaneously directly in hybridization buffer using a thermal cycler or heated block.

1. Mix 1–2 μl labeled probe DNA (~60 ng/μl in TE) with 10 μl 1.2× hyb buffer. Add dH_2O to 12 μl if necessary.
2. Denature probe DNA for 5 min at 80 °C then place on ice.
3. Wash the slides in fresh PBS for 5 min at room temperature, then in 70%, 85% and 100% ethanol for 1 min each. Dry the slides gently ~2–5 s with compressed air. Chill the 70% ethanol on ice for a later step.
4. Place the slides into 50 ml preheated denaturing solution for 5 min at exactly 75 °C in a heated water bath. Temperature is critical; place a clean thermometer directly in denaturing solution.
5. Dehydrate again by transferring the hot slides into ice-cold 70% ethanol, then room temperature 85% and 100% ethanol for 1 min each. Dry the slides gently with compressed air and proceed immediately to the next step or store the slides in a desiccator.
6. Apply the denatured probe (12 μl) to the center of a 40 mm coverslip and place inverted slide onto the coverslip to avoid air bubbles. Hydrophobic plastic hybridization slips (ThermoFisher Scientific) result in a more even distribution of hybridization buffer over the slide surface.
7. Seal edges of coverslip with rubber cement and place the slides into a prewarmed hybridization chamber at 37 °C. Hybridize in the dark for 12–24 h.

3.2.4 Denaturation and Hybridization, Dry Block Method

1. Immerse the slides in 50 ml fresh PBS for 5 min at room temperature, then in 70%, 85% and then 100% ethanol for 1 min each. Dry the slides gently ~2–5 s with compressed air.
2. Mix 1–2 μl labeled probe DNA (~60 ng/μl in TE) with 10 μl 1.2× hyb buffer. Add dH_2O to 12 μl if necessary.
3. Apply probe mixture to the center of a 40 mm coverslip (plastic or glass) and lower an inverted slide onto the coverslip, and seal the edges with rubber cement.
4. Place the slides into an in situ PCR block, and program the cycler for 80 °C for 5 min then 4 °C for 3 min (temperature ramp times account for slide thickness). Alternatively, two solid

aluminum blocks (dry incubation blocks turned upside down) at 80 °C and 4 °C may be used.

5. Transfer the slides into a prewarmed hybridization chamber and hybridize in the dark at 37 °C for 12–24 h.

3.2.5 Washing and Mounting

To remove unhybridized probe and nonspecific hybrids, a series of washes is performed with increasing stringency. Usually only weakly stringent washes are sufficient to remove nonspecific hybridized probe, but the slides can be rewashed under more stringent conditions if excessive background is present (*see* **Note 3**).

1. Carefully remove rubber cement from the coverslips, but do not remove coverslips from the slides.
2. Immerse the slides in 50 ml room temperature 2× SSC with gentle agitation. Coverslips should come off easily in the buffer. Do not pull, pry, or slide coverslips.
3. Wash the slides in 50 ml 37 °C 2× SSC 50% formamide for 10 min with gentle agitation (rotary platform).
4. Wash the slides again in 37 °C 2× SSC then once in room temperature 2× SSC (10 min each). Drain the slides.
5. (Optional) If using an antibody to detect probe, add 0.1% Tween 20 to above washes. Drain the slides, apply 50 μl blocking buffer, incubate for 15 min at room temperature. Drain the slides, apply 50 μl diluted antibody in blocking buffer, and incubate for 1–6 h in hybridization chamber at room temperature. Wash the slides three times with 500 μl PBS containing 0.1% Tween 20 for 5 min each at room temperature. Repeat blocking, annealing, and washing steps if using a secondary antibody. Drain the slides.
6. Allow the slides to dry 30 min on the bench then add 10 μl mounting medium with 0.3 μg/ml DAPI (optional). To aid spatial alignment of multiple channels (colors) of the same field, add diluted multifluorescent beads (e.g., TetraSpeck Microspheres, ThermoFisher Scientific) in the mounting medium.
7. Seal edges of coverslips with nail polish. The slides may be stored temporarily at 4 °C or for several months at −20 °C.

3.3 Imaging and Analysis

3.3.1 Image Acquisition and Processing

To maximize signal detection, it is important to utilize as much of the dynamic range of your CCD sensor as possible. With typically dim FISH signals, this means taking very long exposures (up to several seconds) with an antivibration table to reduce blurring. It is also important to equalize signals among all colors (channels) by adjusting exposure times and output of the excitation light source. For most applications, increasing sensitivity by pixel binning is not an option with bacterial FISH due to loss of spatial resolution.

Acquisition of multiple images through the *z*-axis allows the capture of signals on the nucleoid periphery that might be outside the focal plane. Subsequent deconvolution to reduce out of focus light can result in a higher signal-to-noise ratio. Pixel shift, or misalignment of phase contrast and fluorescent images taken of a single field, is caused primarily by slight changes in the angles of dichroic mirrors or the filter cubes themselves. Post-acquisition image alignment is facilitated by the addition of multifluorescent beads on the slide.

1. Take several test exposures with each filter set and adjust exposure times such that average signal brightness (pixel grey value) is ~10–50% of the capacity of the CCD sensor. This will allow subsequent 2× bright signals to fall within the range of the camera. Use a nonhybridized slide as a negative control.
2. Take final images using the same optimized acquisition settings for each channel. Scan over the entire slide surface, as some regions will have better signals than others. Be sure that each image contains at least one fluorescent bead.
3. Adjust brightness and contrast settings for each channel as needed to visualize fluorescent signals (e.g., Fig. 3). Importantly, these steps only affect how the 16-bit CCD image (65,536 grey values) is scaled for viewing on an 8-bit monitor (256 grey values), they do not change the raw image data or affect downstream image analysis as long as images are not saved as 8-bit conversions.
4. Align individual channel images using the multifluorescent beads as a reference.

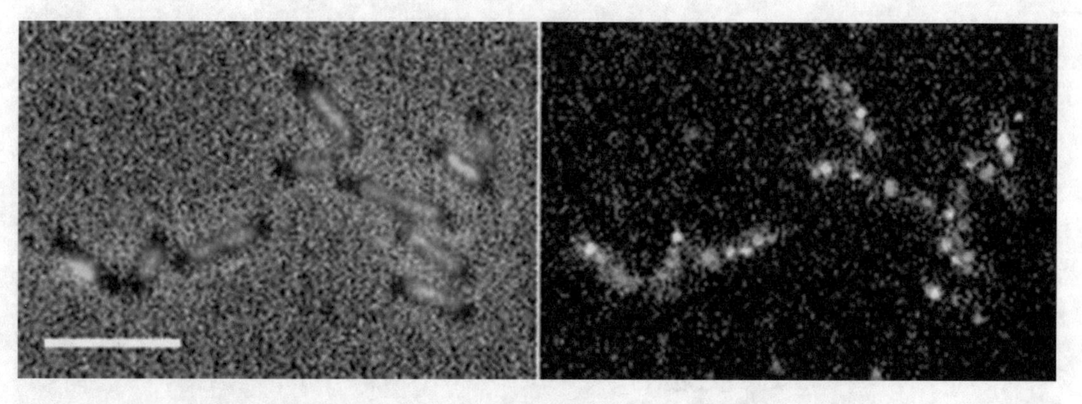

Fig. 3 Five-color FISH with DAPI in *E. coli.* Phase contrast with DAPI fluorescence image (*left panel*). Combined 5-color FISH image (*right panel*). *E. coli* cells exponentially growing in minimal media (containing ≤2 chromosomes) were hybridized with 3 kb probes targeted to five equally spaced (~500 kb) sites on the *E. coli* chromosome. Probes were labeled by nick translation with the following nucleotide base conjugates: digoxigenin-dUTP secondarily detected with anti-digoxigenin BV480 (*blue*), AF488-dUTP (*green*), AF546-dUTP (*yellow*), AF594-dUTP (*red*), and AF647-dUTP (*magenta*). Brightness and contrast of color channels were independently adjusted and merged into a composite image in Axiovision. Bar is 5 μm

5. For 3D applications, obtain three to five z-axis images for each channel. Perform deconvolution (these steps alter the raw image data). Some deconvolution algorithms are additive (increasing overall image brightness), and thus raw images must be well below the bit depth dynamic range to avoid pixel saturation. Proprietary deconvolution algorithms differ widely among and within different software packages, and must be chosen empirically with your particular images.

3.3.2 Signal Calling and Quantification

FISH signals at well-separated loci appear as round, diffraction-limited spots (foci) with a radius (width of the point spread function) determined by the microscope optics and wavelength of light imaged (infra-red foci are approx. twice the size of blue foci). Spot detection programs, such as FocusCounter [9], Spatzcells [10], FindFoci [11], MicrobeTracker [12], and Oufti [13], find and characterize foci by fitting signal intensity profiles to a Gaussian function. Signals resulting from hybridization of probe to target DNA can be discriminated from background signals (free dye molecules, cross-fluorescence, or autofluorescence) by selecting foci with the highest intensity. These programs also automatically detect cell outlines (segmentation) for numeration of foci per cell and localization of foci within cells. As the efficiency of detecting any given locus is always less than 100%, focus counts will under-represent the actual number of spatially separate loci in the cell. When accurate focus counting is important (e.g., using FISH to determine the frequency of segregated loci), measured counts can be adjusted upward by the factor of detection inefficiency, which is calculated from target copy number (e.g., by qPCR or flow cytometry) and the fraction of cells exhibiting zero foci [5].

4 Notes

1. Although paraformaldehyde depolymerizes to formaldehyde in water, some manufacturers may still label solutions as paraformaldehyde. Aqueous formaldehyde decomposes rapidly when exposed to air and heat, precipitating and oxidizing to formic acid, and should be made fresh from a sealed ampoule for every experiment. Formaldehyde is highly toxic and gloves should be worn when handling concentrated solutions under a fume hood.
2. Resulting fragment sizes in nick translation reactions are also dependent on the ratio of DNA Pol I and DNase I, and we have found that an activity ratio of 100:1 (0.6 U/μl/h DNA Pol I and 0.006 U/μl/h DNase I) produces ideal fragment sizes in 3 h. However, variations in enzyme activity of commercially supplied stocks may result in fragment sizes that are over or

under the optimal 200–500 bp range. In these cases, the concentration of DNase I should be increased or decreased, respectively, to produce the desired fragment size. For this reason, new enzyme stocks should be tested prior to use.

3. High background as indicated by abundant signals (foci) in and around cells can be caused by a number of factors including improperly cleaned slides, excessive debris in cell samples, or inadequate washing after hybridization. Signal over the entire nucleoid indicates additional stringent washes are required.

Acknowledgments

We thank Beth Weiner, Nancy Kleckner, Sota Hiraga, and Lucy Shapiro for sharing their detailed FISH protocols, which formed the basis of our methods. We thank Jeff Carmichael (Chroma Technology Corp) for technical assistance with fluorescent filters, and Po J. Chen and Anna K. Barker for comments on the manuscript. All work was supported by NIH Grant GM102679 to D.B.

References

1. Wang X, Montero Llopis P, Rudner DZ (2013) Organization and segregation of bacterial chromosomes. Nat Rev Genet 14 (3):191–203. doi:10.1038/nrg3375
2. Possoz C, Filipe SR, Grainge I, Sherratt DJ (2006) Tracking of controlled *Escherichia coli* replication fork stalling and restart at repressor-bound DNA in vivo. EMBO J 25 (11):2596–2604
3. Bates D, Kleckner N (2005) Chromosome and replisome dynamics in *E. coli*: loss of sister cohesion triggers global chromosome movement and mediates chromosome segregation. Cell 121(6):899–911. doi:10.1016/j.cell.2005.04.013
4. Sunako Y, Onogi T, Hiraga S (2001) Sister chromosome cohesion of *Escherichia coli*. Mol Microbiol 42(5):1233–1241
5. Joshi MC, Bourniquel A, Fisher J, Ho BT, Magnan D, Kleckner N, Bates D (2011) *Escherichia coli* sister chromosome separation includes an abrupt global transition with concomitant release of late-splitting intersister snaps. Proc Natl Acad Sci U S A 108 (7):2765–2770. doi:10.1073/pnas.1019593108
6. Niki H, Hiraga S (1997) Subcellular distribution of actively partitioning F plasmid during the cell division cycle in *E. coli*. Cell 90 (5):951–957
7. Magnan D, Joshi MC, Barker AK, Visser BJ, Bates D (2015) DNA replication initiation is blocked by a distant chromosome-membrane attachment. Curr Biol 25(16):2143–2149. doi:10.1016/j.cub.2015.06.058
8. Weiner BM, Kleckner N (2009) Assaying chromosome pairing by FISH analysis of spread *Saccharomyces cerevisiae* nuclei. In: Keeney S (ed) Meiosis: volume 2, cytological methods. Humana Press, Totowa, NJ, pp 37–51. doi:10.1007/978-1-60761-103-5_3
9. Joshi MC, Magnan D, Montminy TP, Lies M, Stepankiw N, Bates D (2013) Regulation of sister chromosome cohesion by the replication fork tracking protein SeqA. PLoS Genet 9(8): e1003673. doi:10.1371/journal.pgen.1003673
10. Skinner SO, Sepulveda LA, Xu H, Golding I (2013) Measuring mRNA copy number in individual *Escherichia coli* cells using single-molecule fluorescent in situ hybridization. Nat Protoc 8(6):1100–1113. doi:10.1038/nprot.2013.066
11. Herbert AD, Carr AM, Hoffmann E (2014) FindFoci: a focus detection algorithm with automated parameter training that closely matches human assignments, reduces human inconsistencies and increases speed of analysis. PLoS One 9(12):e114749. doi:10.1371/journal.pone.0114749

12. Sliusarenko O, Heinritz J, Emonet T, Jacobs-Wagner C (2011) High-throughput, subpixel precision analysis of bacterial morphogenesis and intracellular spatio-temporal dynamics. Mol Microbiol 80:612–627. doi:10.1111/j.1365-2958.2011.07579.x

13. Paintdakhi A, Parry B, Campos M, Irnov I, Elf J, Surovtsev I, Jacobs-Wagner C (2016) Oufti: an integrated software package for high-accuracy, high-throughput quantitative microscopy analysis. Mol Microbiol 99 (4):767–777. doi:10.1111/mmi.13264

Chapter 17

Imaging the Cell Cycle of Pathogen *E. coli* During Growth in Macrophage

Gaëlle Demarre, Victoria Prudent, and Olivier Espéli

Abstract

The study of the bacterial cell cycle at the single cell level can not only give insights on the fitness of the bacterial population but also reveal heterogeneous behavior. Typically, the DNA replication, the cell division, and the nucleoid conformation are appropriate representatives of the bacterial cell cycle. Because bacteria rapidly adapt their growth rate to environmental changes, the measure of cell cycle parameters gives valuable insights for the study of bacterial stress response or host–pathogen interactions. Here we describe methods to first introduce fluorescent fusion proteins and fluorescent tag within the chromosome of pathogenic bacteria to study these cell cycle steps; then to follow them within macrophages using a confocal spinning disk microscope.

Key words *parS* sites, Microscopy, Fluorescence

1 Introduction

This method is dedicated to the investigation of the bacterial cell cycle during host infection, via the observation by fluorescence microscopy of chromosome tags, nucleoid-associated protein, and FtsZ ring formation. The bacterial cell cycle has been studied for decades in controlled conditions of laboratory growth. Most of the work has focused on growth in the absence of environmental stress. However, in their natural environment bacteria frequently encounter lack of nutrients and oxidative, genotoxic, or acidic stress. The progress of imaging methods allows for the evaluation of the bacterial cell cycle during host cell infection. The present protocol describes classical methods of cell cycle analysis by fluorescence microscopy adapted to the imaging of AIEC (Adherent Invasive *E. coli*) LF82 bacteria [1] during growth inside acidic phagolysosome of macrophages.

FROS (Fluorescent Represssor Operator Sequence) [2, 3] and *parS*/ParB-GFP [4] tags have been used to label specific loci and follow chromosome choreography [5, 6]; the number of foci

Olivier Espéli (ed.), *The Bacterial Nucleoid: Methods and Protocols*, Methods in Molecular Biology, vol. 1624,
DOI 10.1007/978-1-4939-7098-8_17, © Springer Science+Business Media LLC 2017

within a given cell is also a good indication of the availability of nutrients in the environments [7]. The labeling of the chromosome with the nucleoid associated protein HU fused with mCherry [8] is also an indicator of the environment encountered by the bacteria. The shape of bacterial nucleoid changes according to the nutrient used by the bacteria [9] and when genotoxic [10] or oxidative stress [11] is present. Finally cell division can be monitored by the observation of the main septal ring protein FtsZ [12, 13]. FtsZ polymerization is targeted by two cell cycle checkpoints, the SOS regulon and the Nucleoid Occlusion system that would modify the timing of division and provoke cell elongation [14, 15]. Therefore, imaging of the origin of replication, the nucleoid, and the septal ring are good indicators of the bacterial adaptation to an environmental challenge. We selected HU-mCherry, *oriC*::*parS*/ParB-GFP, and FtsZ-GFP fluorescent reporters because they produce robust signals that tolerate fixation and successive illumination during Z stack acquisition and can be accurately scored by image analysis tools such as Oufti [16], Fiji [17], or ObjectJ (https://sils.fnwi.uva.nl/bcb/objectj/index.html).

2 Material

2.1 Plasmids, Strains, and Primers

Human cell line:

THP1 (ATCC® TIB202™).

Bacterial Strains:

LF82 [18] Δ*bla* (a generous gift from Nicolas Barnish).
MG1655 Hu-mcherry [19].

Plasmids:

pKOBEGA (pSC101, amp, TS) [20].
pCP20 (pSC101, amp, TS) [21].
pAD37 (R6k, cm) [22].
pFH2973 [23].
pFWZ5 [24].

Primers:

Construction of the HU-mCherry strain

hupA-up	AAGGATAACTTATGAACAAGACTC
hupA-down	AACAGTAATTGCGAACCTTCGG
hupA verif up	CGTCGCACTCGATGCTTAGC
hupA verif down	CCAGTGGATTTGCTGAAGACC

Insertion *parS*$_{pMT1}$ (pAD37) (*FRT-aph-FRT-parS*$_{pMT1}$).

UpT1	5′<50nt homology to *oriC* region (position-position) >CGG CTG ACA TGG GAA TTA GCC
DownT1	5′<50nt homology to *oriC* region (position-position) >GGT CTG CTA TGT GGT GCT ATC T

2.2 Reagents: Stock Solutions

All solutions must be prepared using ultrapure water (by purifying deionized water, to attain a sensitivity of 18 MΩ-cm at 25 °C).

Prepare the following buffers and stock solutions. Unless otherwise specified, filter solutions using a 0.2 μm low protein binding nonpyrogenic membrane.

10% glycerol in H_2O cold.

Lennox broth (LB).

100 mg/ml ampicillin.

50 mg/ml kanamycin.

30 mg/ml chloramphenicol (pAD37).

20% arabinose.

1M IPTG.

2.3 Cell Culture

800 μg/ml PMA (phorbol 12-myristate 13-acetate).

RPMI 1640 without L-glutamine supplemented with 10% decomplemented fetal bovin serum and 2 mM glutamine (Life Technologies).

Keep at 4 °C no more than 1 month.

1 × phosphate buffer saline (PBS).

24-well plate (1.9 cm^2, no coating).

2.4 Cell Fixation

Cold methanol kept at −20°C.

Fluorescence mounting medium: Dako S3023.

Microscope slides: 76 × 26 mm.

Microscope glass coverslips, 12 mm diameter, no coating.

2.5 Molecular Biology and Microbiology Material

Shaker: Infors Minitron.

Water bath.

Vacuum concentrator: miVac, Genevac.

PCR: C1000 Thermal cycler, Bio-Rad.

Electroporator: Micropulser, Bio-Rad.

2.6 Microscope

Inverted Zeiss Axio Imager with spinning disk CSU W1 (Yokogawa):

100× objective, NA = 1.4, oil, WD = 0.13 (WD: working distance in mm).

Camera ORCA-Flash 4.0 digital CMOS.

Lasers:

490 nm at 150 mW.

561 nm at 100 mW.

642 nm at 110 mW.

Metamorph Premier 7.6 software.

3 Methods

3.1 Strain Construction

AIEC LF82 will be genetically engineered to observe the nucleoid with HU-mCherry, the origin of replication with a $parS_{pMT1}$ tag and the septal ring with a plasmid expressing FtsZ-GFP.

a. Transform AIEC LF82 with the plasmid pKOBEGA

Prepare competent bacteria using a protocol adapted for the strain of interest. For AIEC LF82, a culture of OD = 0.6 is washed 3 times in ice-cold 10% glycerol and the final resuspension volume is 1/250 of the initial volume.

Transform 50 μl of competent bacteria with 30 ng of pKOBEGA.

Plate at 30 °C on ampicillin (100 μg/ml).

b. Construction of AIEC LF82 HU-mCherry strain with λ Red (γ, β, *exo*) recombination [21]

b1. Prepare competent bacteria expressing λ Red (γ, β, *exo*) proteins as follows:

Dilute 500-fold an overnight culture of AIEC LF82 containing pKOBEGA in 50 ml LB (100 μg/ml amp, 0.15% arabinose).

Incubate with shaking at 30 °C up to OD_{600nm} = 0.6.

Incubate at 42 °C for 15 min.

Chill the culture in an ice–water bath for 15 min.

Centrifuge for 10 min at 5000× *g* at 4 °C.

Discard the supernatant and wash cell pellet with 25 ml of 10% cold glycerol.

Centrifuge for 10 min at 5000× *g* at 4 °C.

Wash cell pellet with 12 ml of 10% cold glycerol.

Centrifuge for 10 min at 5000× *g* at 4 °C.

Wash cell pellet with 6 ml of 10% cold glycerol.

Centrifuge for 5 min at 5000× *g* at 4 °C.

Resuspend the cell pellet in 150 μl 10% cold glycerol.

b2. Transform with *hupA-mcherry* PCR product prepared as follows

Perform 10 PCR of 50 μl:

PCR Mix for 50 μl.

5 μl 10× Extaq buffer.

5 μl dNTP mix (2 mM each).

50 ng of genomic DNA (strain MG1655HU-mcherry).

2.5 μl of 10 μM primer hupA-up.

2.5 μl of 10 μM primer hupA-down.

1 unit Extaq Polymerase.

33.7 μl H_2O.

PCR conditions:

Initial denaturation:

98 °C—3 min.

30 cycles:

98 °C—10 s.

57 °C—30 s.

72 °C—1 min/kb.

Final extension:

72 °C—10 min.

Hold:

12 °C.

The expected fragment size is about 3 kb.

Concentration of PCR product:

Use PCR purification kit to clean the DNA (three columns for 10 PCRs, elution with 100 μl elution buffer per column); pool the three elutions in one tube.

Concentrate the DNA with a vacuum concentrator at 45 °C up to a final volume of 30 μl.

Dialyze 30 min on filter membrane (Millipore 0.02 μm) floating on a petri dish filled with ultrapure water.

Electrotransformation of the PCR product:

Preincubate the electroporation cuvette (2 mm) in ice.

In a microcentrifuge tube mix ~5 μg of PCR product with 90 μl of the electrocompetent bacteria.

Transfer the mix to the precooled cuvette. Keep on ice.

Dry the cuvette. Insert it in the electroporator and immediately pulse at 2500 V.

Immediately add 1 ml of LB. Mix thoroughly with the bacteria and transfer to a 2 ml tube.

Incubate for 1 h at 37 °C with shaking.

Plate 100 μl of the transformation on 75 μg/ml kanamycin plate (*see* **Note 1**).

Centrifuge for 2 min the rest of the transformation tube.

Remove the supernatant.

Resuspend the cell pellet with 100 μl of LB, plate it on 75 μg/ml kanamycin plate.

Incubate overnight at 37 °C.

c. Delete the kanamycin gene through the flipase/FRT recombination

When it is required, the kanamycin resistance gene can be removed with Flp recombination using the protocol described in [21]. Choose kn^S and amp^S colonies.

d. Construction of AIEC LF82 HU-mCherry strain with a *parS$_{pMT1}$* tag near *oriC*

Prepare *parS$_{pMT1}$* product by PCR as described in paragraph b2 with 1ng of pAD37 as template and the primers UpT1 and downT1. The size of the expected fragment is about 1.7 kb.

Transform the AIEC LF82 HU-mcherry strain with the pKO-BEGA plasmid.

Prepare AIEC LF82 HU-mcherry pKOBEGA competent bacteria expressing lambda red (beta exo and gam) proteins as described in the paragraph b2.

Transform the strain HU-mcherry *oriC::parS$_{pMT1}$*-FRT-aph-FRT with the pFH2973. Select transformants on 100 μg/ml ampicillin plates.

e. Construction of AIEC LF82 HU-mCherry strain with a plasmid expressing FtsZ-GFP

Transform the AIEC LF82 HU-mcherry strain with the pFWZ5 plasmid.

3.2 Infection

a. Differentiation of THP1 monocytes in macrophages (*see* **Note 2**)

Prepare a 5 × 10^5 cell/ml cell suspension in the cell growth medium.

Just before use, dilute 5 μl of the 800 μM stock of PMA in 195 μl of growth medium (*see* **Note 3**).

Add diluted PMA at a 1000× dilution into the cell suspension.

Introduce a circular coverslip in each well.

Load 1 ml of cell suspension in each well. Remove the bubbles from under the coverslip by softly pressing on the coverslip with a pipette

Incubate for 18 h at 37 °C with 5% CO_2.

b. AIEC LF82 culture

Growth the AIEC LF82 strain of interest overnight in LB supplemented with 100 μg/ml ampicillin and 2 μM IPTG (to induce ParB-GFP from pFH2973). Measure OD_{600nm}, it should be close to 4.

Dilute them in 1× PBS up to an OD_{600nm} of 0.4.

c. Infection

Check the THP1 cells; they should form heaps (*see* **Note 4**).

Wash the differentiated macrophages with 1× PBS. Eliminate the medium by inverting the plate over a trash bin and then above an absorbing paper.

Add 1× PBS drop by drop on the wall of the well.

Mix gently by rotation.

Add 1 ml of growth medium.

Add 78 μl of the bacterial suspension.

Mix gently by rotation.

Centrifuge for 10 min at 900 rpm at room temperature.

Incubate for 10 min at 37 °C with 5% CO_2.

Wash twice with 1× PBS.

After the second wash, remove the last drop of PBS with a micropipette.

Add 1 ml of cell growth medium supplemented with 50 μg/ml gentamycin in each well.

Incubate for 40 min or 24 h at 37 °C with 5% CO_2.

3.3 Cell Fixation and Slide Preparation

a. Fixation of infected cells

At the selected time point the cell should be fixed before imaging.

Wash twice with 1 ml of 1× PBS.

Add 1 ml of 1× PBS and slowly add 1 ml of methanol (kept at −20 °C) (1 Vol methanol/1 Vol PBS) (*see* **Note 5**).

Immediately remove the PBS + methanol mix.

Add very slowly 1 ml of methanol.

Incubate for 5 min on ice.

Keeping the methanol add very slowly 1 ml of 1× PBS.

Remove the PBS + methanol mix.

Add very slowly 1 ml of 1× PBS.

Keep in PBS at 4 °C till mounting.

b. Slide mounting

Load one drop (10 μl) of Dako on the microscope slide.

Remove the coverslip from the plate with a pair of pliers.

Damp the coverslip in fresh 1× PBS.

Drain the PBS for the coverslip holding the coverslip with the pliers.

Dry the coverslip by touching an absorbing paper with the edge of the coverslip.

Transfer the coverslip onto the Dako drop, the side of the coverslip with the cell/bacteria should face the slide.

Incubate overnight at room temperature in the dark.

3.4 Imaging and Image Processing

Imaging is performed on a Zeiss Axio imager microscope equiped with Yokogawa W1 spinning disk confocal head. Routinely, Z stacks of 36 images with a step of 0.4 are recorded to observe the complete macrophage volume. Laser excitations are performed at 488 nm for 600 ms at 80%, 561 nm for 200 ms at 20%, phase contrast for 50 ms. Image acquisition is controlled by Metamorph software.

Images are processed with Fiji for background subtraction, Z projection, Z plane color coding, and 3D reconstructions (Fig. 1). Object counting can be performed in 2D before or after Z projection with Fiji, Oufti, or ObjectJ. Three-dimensional object counting and measurements can be performed after deconvolution (Huygens, Scientific Volume Imaging) with a 3D reconstruction and analysis tool (Imaris, Bitplane).

4 Notes

1. LF82 transformant selection on kanamycin plates required a kanamycin concentration of 75 μg/ml. This concentration may need to be adjusted in order to avoid false positive and to allow the kanamycin resistance from a single chromosomic copy of the kanamycin resistance gene.

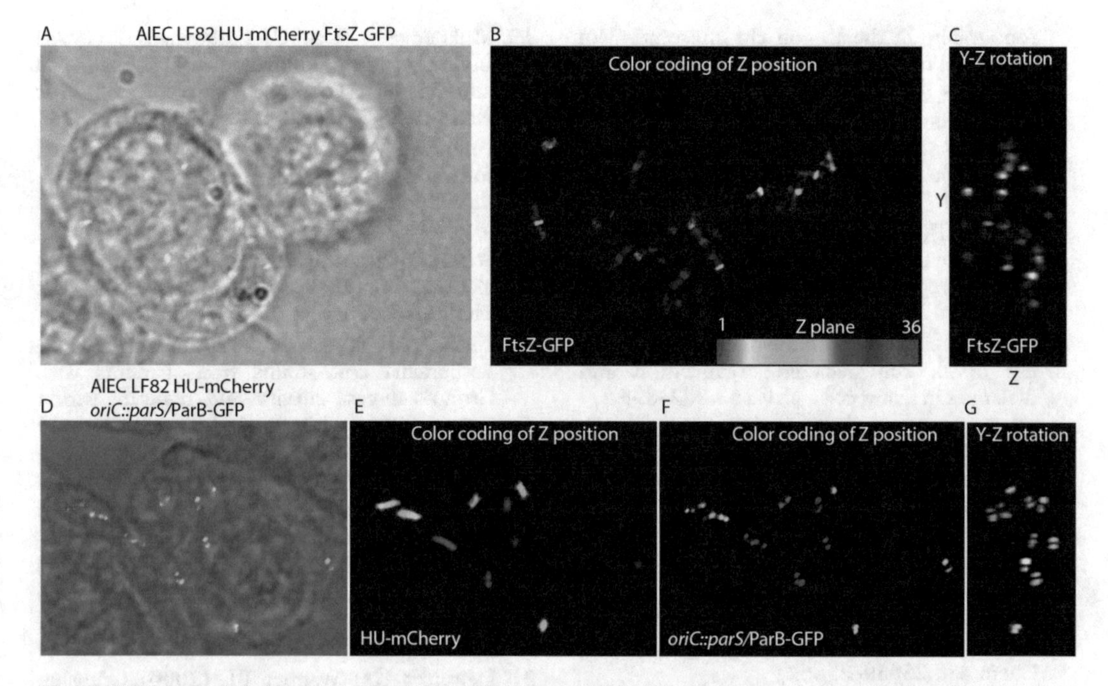

Fig. 1 (**a**) Representative image of AIEC LF82 HU-mCherry FtsZ-GFP, HU-mCherry (*red*), FtsZ-GFP (*green*). The image corresponds to the Z projection (sum of all planes) for the mCherry and GFP signals merged with the best focus plane of macrophage image. (**b**) *Z* plane color coding of the FtsZ-GFP signal. (**c**) 3D reconstruction and rotation along the *YZ* axis of the FtsZ-GFP signal. (**d**) Representative image of AIEC LF82 HU-mCherry *oriC::parS*/ParB-GFP, HU-mCherry (*red*), *oriC::parS*/ParB-GFP (*green*). The image corresponds to the Z projection (sum of all planes) for the mCherry and GFP signals merged with the best focus plane of macrophage image. (**e**) Z plane color coding of HU-mCherry signal. (**f**) Z plane color coding of *oriC::parS*/ParB-GFP signal. (**g**) 3D reconstruction and rotation along the *YZ* axis of the *oriC::parS*/ParB-GFP signal

2. One to three-week-old THP1 monocytes are differentiated within macrophages.
3. Do not introduce a micropipette in the growth medium bottle. Transfer with a pipette a small volume of the medium in an eppendorf tube. Pipette the 195 μl from the eppendorf tube.
4. If the macrophages do not form heaps, they are not adherent and thus not differentiated.
5. Methanol and PBS are added very slowly to avoid detachment of the cells.

References

1. Eaves-Pyles T et al (2008) Escherichia coli isolated from a Crohn's disease patient adheres, invades, and induces inflammatory responses in polarized intestinal epithelial cells. Int J Med Microbiol 298:397–409
2. Gordon GS et al (1997) Chromosome and low copy plasmid segregation in E. coli: visual evidence for distinct mechanisms. Cell 90:1113–1121
3. Lau IF et al (2003) Spatial and temporal organization of replicating Escherichia coli chromosomes. Mol Microbiol 49:731–743
4. Li Y, Youngren B, Sergueev K, Austin S (2003) Segregation of the Escherichia coli chromosome terminus. Mol Microbiol 50:825–834
5. Espeli O, Mercier R, Boccard F (2008) DNA dynamics vary according to macrodomain

topography in the E. coli chromosome. Mol Microbiol 68:1418–1427
6. Youngren B, Nielsen HJ, Jun S, Austin S (2014) The multifork Escherichia coli chromosome is a self-duplicating and self-segregating thermodynamic ring polymer. Genes Dev 28:71–84
7. Nielsen HJ, Youngren B, Hansen FG, Austin S (2007) Dynamics of Escherichia coli chromosome segregation during multifork replication. J Bacteriol 189:8660–8666
8. Fisher JK et al (2013) Four-dimensional imaging of E. coli nucleoid organization and dynamics in living cells. Cell 153:882–895
9. Hadizadeh Yazdi N, Guet CC, Johnson RC, Marko JF (2012) Variation of the folding and dynamics of the Escherichia coli chromosome with growth conditions. Mol Microbiol 86:1318–1333
10. Shechter N et al (2013) Stress-induced condensation of bacterial genomes results in re-pairing of sister chromosomes: implications for double strand DNA break repair. J Biol Chem 288:25659–25667
11. Ushijima Y, Yoshida O, Villanueva MJA, Ohniwa RL, Morikawa K (2016) Nucleoid clumping is dispensable for the Dps-dependent hydrogen peroxide resistance in Staphylococcus aureus. Microbiology 162(10):1822–1828. doi:10.1099/mic.0.000353
12. Ma X, Ehrhardt DW, Margolin W (1996) Colocalization of cell division proteins FtsZ and FtsA to cytoskeletal structures in living Escherichia coli cells by using green fluorescent protein. Proc Natl Acad Sci U S A 93:12998–13003
13. Sun Q, Margolin W (1998) FtsZ dynamics during the division cycle of live Escherichia coli cells. J Bacteriol 180:2050–2056
14. Bernhardt TG, de Boer PAJ (2005) SlmA, a nucleoid-associated, FtsZ binding protein required for blocking septal ring assembly over Chromosomes in E. coli. Mol Cell 18:555–564
15. Mukherjee A, Cao C, Lutkenhaus J (1998) Inhibition of FtsZ polymerization by SulA, an inhibitor of septation in Escherichia coli. Proc Natl Acad Sci U S A 95:2885–2890
16. Paintdakhi A et al (2016) Oufti: an integrated software package for high-accuracy, high-throughput quantitative microscopy analysis. Mol Microbiol 99:767–777
17. Schindelin J et al (2012) Fiji: an open-source platform for biological-image analysis. Nat Methods 9:676–682
18. Glasser AL et al (2001) Adherent invasive Escherichia coli strains from patients with Crohn's disease survive and replicate within macrophages without inducing host cell death. Infect Immun 69:5529–5537
19. Marceau AH et al (2011) Structure of the SSB-DNA polymerase III interface and its role in DNA replication. EMBO J 30:4236–4247
20. Chaveroche MK, Ghigo JM, d'Enfert C (2000) A rapid method for efficient gene replacement in the filamentous fungus Aspergillus nidulans. Nucleic Acids Res 28:E97
21. Datsenko KA, Wanner BL (2000) One-step inactivation of chromosomal genes in Escherichia coli K-12 using PCR products. Proc Natl Acad Sci U S A 97:6640–6645
22. David A et al (2014) The two Cis-acting sites, parS1 and oriC1, contribute to the longitudinal organisation of Vibrio cholerae chromosome I. PLoS Genet 10:e1004448
23. Nielsen HJ, Ottesen JR, Youngren B, Austin SJ, Hansen FG (2006) The Escherichia coli chromosome is organized with the left and right chromosome arms in separate cell halves. Mol Microbiol 62:331–338
24. Wu F, Van Rijn E, Van Schie BGC, Keymer JE, Dekker C (2015) Multi-color imaging of the bacterial nucleoid and division proteins with blue, orange, and near-infrared fluorescent proteins. Front Microbiol 6:607

Chapter 18

Measuring In Vivo Protein Dynamics Throughout the Cell Cycle Using Microfluidics

Roy de Leeuw, Peter Brazda, M. Charl Moolman, J.W.J. Kerssemakers, Belen Solano, and Nynke H. Dekker

Abstract

Studying the dynamics of intracellular processes and investigating the interaction of individual macromolecules in live cells is one of the main objectives of cell biology. These macromolecules move, assemble, disassemble, and reorganize themselves in distinct manners under specific physiological conditions throughout the cell cycle. Therefore, in vivo experimental methods that enable the study of individual molecules inside cells at controlled culturing conditions have proved to be powerful tools to obtain insights into the molecular roles of these macromolecules and how their individual behavior influence cell physiology. The importance of controlled experimental conditions is enhanced when the investigated phenomenon covers long time periods, or perhaps multiple cell cycles. An example is the detection and quantification of proteins during bacterial DNA replication. Wide-field microscopy combined with microfluidics is a suitable technique for this. During fluorescence experiments, microfluidics offer well-defined cellular orientation and immobilization, flow and medium interchangeability, and high-throughput long-term experimentation of cells. Here we present a protocol for the combined use of wide-field microscopy and microfluidics for the study of proteins of the *Escherichia coli* DNA replication process. We discuss the preparation and application of a microfluidic device, data acquisition steps, and image analysis procedures to determine the stoichiometry and dynamics of a replisome component throughout the cell cycle of live bacterial cells.

Key words Single-molecule techniques, Fluorescence imaging, Microfluidics, DNA replication, *Escherichia coli*

1 Introduction

Wide-field fluorescence imaging, based on conventional optical microscopy, can resolve structures in living cells down to approximately 200 nm, due to the diffraction limit of light. Although higher resolution microscopy techniques exist [1, 2], wide-field fluorescence imaging can still provide accurate information on the dynamics of single proteins inside living cells. Moreover, in combination with microfluidics it provides a reliable and statistically sound single-molecule method to observe the motion and

Olivier Espéli (ed.), *The Bacterial Nucleoid: Methods and Protocols*, Methods in Molecular Biology, vol. 1624,
DOI 10.1007/978-1-4939-7098-8_18, © Springer Science+Business Media LLC 2017

stoichiometry of individual proteins inside living cells [3]. To extract this information and subsequently gain insight into the ongoing biomolecular processes, obtained fluorescence images have to be treated with quantitative image analysis procedures.

To investigate dynamic processes during the cell cycle and obtain statistically relevant data it is essential that the environment for each studied cell is equivalent. Microfluidic devices are ideally suited for such studies with time-lapse fluorescence imaging, as they offer well-defined cellular orientation and immobilization, provide a constant flow and interchangeability of nutrients, and allow for high-throughput long-term experimentation of cells. The main advantage of using microfluidics compared to more conventional methods (e.g., the agarose gel [4]) is that, due to the uniform flow of nutrients, the cell's environment can be altered and set equivalent for every measured cell. This environment can also be modified homogeneously by changing the flow through the device (*see* **Note 1**).

This chapter describes the acquisition and post-processing analysis of wide-field fluorescence microscopy images obtained using a microfluidics device (*see* Fig. 1). First, the preparation of *Escherichia coli* (*E. coli*) cells from a genetically engineered strain is described. Second, the fabrication of a microfluidic device loaded with *E. coli* cells that can be readily used for fluorescence imaging is explained.

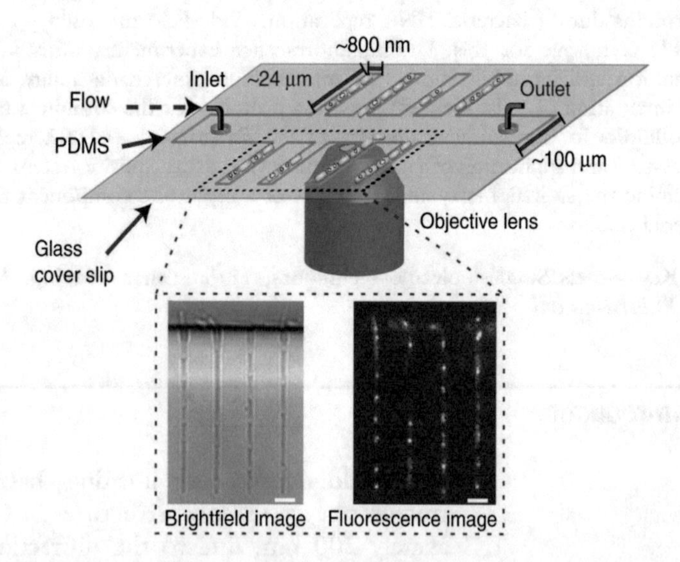

Fig. 1 Wide-field microscopy with a cell-loaded microfluidic device. The microfluidic device used for performing long time-lapse fluorescence microscopy. *E. coli* cells are immobilized in growth channels perpendicular to a main trench through which growth medium is actively pumped. Fifteen channels are monitored in one field of view, with the channel height being 1 µm. The main trench height is ~15 µm. (*inset*) A brightfield image and corresponding YPet-β_2 fluorescence image (80 ms laser light exposure) are acquired every 2.5 min for the duration of the time-lapse experiment

Finally, the operation of the microscope and the custom-built image analysis procedures are discussed in detail.

This protocol is illustrated by using data of fluorescently labeled β_2-sliding clamps (DnaN), a DNA-binding protein with key functions in DNA replication in *E. coli.* They are used as markers to quantify the timing of the DNA replication cycle. A characteristic feature of the β_2-sliding clamp is that it remains DNA-bound during replication (high intensity signal) and diffuses freely when replication ceases (low intensity signal). This allows for the determination of the replication time by observing the evolution of the fluorescence intensity of spots within the cells over time. We recommend the use of the β_2-sliding clamp in addition to other proteins of interest so that processes can be accurately linked to the DNA replication cycle of the cell.

2 Materials

2.1 Microscope

Our optical setup is based on a customized commercial Nikon Ti-Eclipse microscope.

1. Commercial Nikon Ti-Eclipse microscope equipped with a Nikon CFI Apo TIRF 100×, 1.49NA oil immersion objective. The microscope is operated in epifluorescence (EPI) mode.
2. Electron Multiplying CCD (EMCCD) Andor iXon 897 camera operated by a personal computer (PC) running Nikon NIS-elements 4.20.01 software.
3. Cell outlines are imaged using the standard Nikon brightfield halogen lamp and condenser.
4. The fluorescence excitation is performed using an Omicron Laserage LightHUB containing a 514-nm laser.
5. The LightHUB is coupled into a single-mode optical fiber (KineFLEX) connected to the TI-TIRF-E arm input of the microscope.
6. The emission of the fluorescent proteins is projected onto the central part of the EMCCD camera using custom filter sets: Chroma Z514/10×, ET540/30 m, ZT514rdc-tirf (YPet 514 nm).
7. Custom design commercial temperature control housing (Okolabs) enclosing the microscope body.
8. Sample position controlled with a Nikon stage (TI-S-ER Motorized Stage Encoded, MEC56100) with the Nikon Perfect Focus System to eliminate Z-drift during image acquisition.

9. To synchronize camera shutter control with laser exposure, a National Instruments BNC-2115 connector block is used and controlled with NIS-elements.

2.2 PDMS Microfluidic Device Preparation

1. Sylgard 184 Silicone Elastomer Kit to make PDMS in various consistencies.
2. 1H,1H,2H,2H-perfluorodecyltrichlorosilane 97% to make PDMS mold hydrophobic.
3. VWR International 22 × 22 mm coverslips, Thickness No. 1 or 1.5, Cat. No. 631-0124, used for imaging the PDMS device.
4. HelixMark peristaltic pump tubing, Ref 60-825-27 for inlet and outlet.
5. BD 10 mL Syringe Luer-Lok Tip, Ref 300912. + BD Microlance 3 20G 1½″—Nr. 1, 0.9 × 40 mm, Ref 301300 to contain the medium.
6. F560088-90 dispensing needle, standard with thread. Rosa, ø 0.58 mm, 90°, ½″, to connect the tubes with the PDMS device.
7. Harris Uni-Core Multipurpose sampling tool, ø 0.75 mm to puncture holes in PDMS.
8. Harvard apparatus 11 plus syringe pump to control flow.
9. BSA Molecular Biology Grade. B9000S 20 mg/mL Lot: 0051502. Qty: 0.6 mL. Supplied in: 20 mM Tris–HCl, 100 mM KCl, 0.1 mM EDTA and 50% glycerol (pH 8.0 at 25 °C).
10. Customized Okolab Nikon Eclipse-Ti-E Cage Incubator.
11. Fisherbrand Falcon Tubes (50 mL), Cat. No. 06–443-18.
12. Eppendorf Centrifuge 5810R and an Eppendorf IL109 Carrier for the PDMS device.
13. Binder Classic Series B Incubator for PDMS heating.
14. Parafilm (Bemis Company, Inc.).
15. PDMS Mold obtained from Electron Beam Lithography (EBL) imprinted silicon wafer (*see* **Note 2**).
16. Oxygen Plasma-Preen I Barrel Reactor, Plasmatic Systems, Inc.

2.3 Strain and Culture Media

1. *E. coli* strain (YPet-β_2): a derivative of the *E. coli* K12 AB1157 strain.
2. Kanamycin dissolved in dH2O (50 mg/mL as stock solution, filter sterilized 0.22 μm).
3. M9-glycerol media: 1 L of M9 medium contains 10.5 g/L of autoclaved M9 broth (Sigma-Aldrich); 0.1 mM of autoclaved $CaCl_2$ (Sigma-Aldrich); 0.1 mM of autoclaved $MgSO_4$ (J.T. Baker); 0.3% of filter-sterilized glycerol (Sigma-Aldrich) as

carbon source; 0.1 g/L of filter-sterilized "5 amino acids" (L-threonine, L-leucine, L-proline, L-histidine, L-arginine (Sigma-Aldrich)), and 10 μL of 0.5% filter-sterilized thiamine (Sigma-Aldrich).

2.4 Data Analysis and Imaging Software

1. MATLAB (Mathworks, USA) is used for image processing procedures. The Image Processing Toolbox contains useful functions for image operations.
2. ImageJ 1.51a is used for image translation and rolling ball background correction.
3. NIS elements 4.20.01 is utilized to operate and configure the microscope.
4. The Omicron Laserage LightHUB Controller is used to control the laser power digitally.

3 Methods

3.1 Cell Culture

This section refers to the culturing of the YPet-β_2 strain *E. coli* strain (on AB1157 background). This strain is constructed by lambda-red recombination and has been thoroughly described (sequencing, cell morphology, doubling time) previously [3].

1. Inoculate a single colony of the YPet-β_2 strain from an LB-agarose plate (supplemented with the selection antibiotics, kanamycin (50 ng/mL)) and grow it in 5 mL M9-glycerol (supplemented with kanamycin (50 ng/mL)) overnight at 37 °C with 250 rpm shaking.
2. Initiate a secondary culture as a 1:500 dilution from the overnight culture in 5 mL M9-glycerol (supplemented with kanamycin) and incubate at 37 °C with 250 rpm shaking until OD600 0.1–0.3 is reached, to gain an exponential growth phase culture.
3. The secondary culture is used to load the microfluidic device in **step 13** of Subheading 3.2.

3.2 Preparation and Loading of the Microfluidic Device

1. Separate a single microfluidic device from a previously prepared PDMS array (*see* **Note 3**).
2. Place the freshly separated PDMS device into a glass petri dish.
3. Puncture holes at both ends of the main channel with a biopsy needle (this creates the complete main channel that is used to load the device and to provide the nutrient flow).
4. Clean a coverslip by exposing it to a flame, which is a fast and effective method for removing dirt particles from the glass (*see* **Note 4**).

5. Place a clean coverslip (side #1) next to the PDMS device in the glass petri dish.
6. Insert the glass petri dish with the microfluidic device and coverslip into the Plasma-Preen barrel reactor with the channels pointing upward, to activate both surfaces. Vacuum pump the barrel and flow oxygen gas into the chamber for ~10 s. Turn on the microwave at 800 W for 15 s to create oxygen plasma.
7. Flip the PDMS device onto the coverslip and press for full contact (*see* **Note 5**).
8. Heat the PDMS–coverslip construct (in the glass petri dish) for 10 min at 70 °C.
9. Cut roughly 30 cm of tubing for the inlet and outlet of the device. Remove the plastic handles of the two dispensing needles (*see* Subheading 2.2, **item 6**) and insert them in the inlet and outlet.
10. Fill a syringe with 12.5 mL mixture of M9-glycerol (with kanamycin (50 ng/mL)) and BSA (200 ng/mL). Remove air bubbles from the syringe. Insert the syringe needle into one tubing and remove the air by flushing some medium through the tube (*see* **Note 6**).
11. To prepare a mixture for BSA passivation of the PDMS device, add 100 μL of M9-glycerol to 100 μL of BSA in an Eppendorf tube. Create a 200 μL drop of the mixture in a petri dish. Soak up the drop slowly with the hook through the syringe, and avoid suction of air bubbles.
12. Insert the tubing at one end of the PDMS device (via one of the holes created at **step 3**) and slowly flush the channel with the 200 μL mixture, until a drop at the outlet appears (*see* **Note** 7). Keep a potential gradient by virtue of height of the syringe and wait for 45 min, to let the medium reach the end of every channel.
13. Take the second tubing and insert it in the outlet (the other hole, created at **step 3**) of the device. Pump medium through the device until medium appears at the outlet tubing. Disconnect the inlet from the device.
14. To prepare the cells for insertion, pellet 1 mL of the secondary culture by centrifugation (16100 × *g* for 1 min), wash it with 1 mL of fresh M9-glycerol and resuspend the culture in 80 μL M9-glycerol without antibiotics or BSA.
15. Make an 80 μL drop of culture on a petri dish and suck up the drop with the inlet hook connected to the syringe containing the medium. Inject 80 μL of this culture into the device by the inlet tubing until the culture suspension is observed at the outlet tubing. Remove the inlet and outlet tubings.

16. The device with the cells are centrifuged at 2500 × *g*, 15 °C for 10 min (ramp up 5, ramp down 5) to position the cells into the channels (*see* **Note 8**). One side of the PDMS device now contains channels with cells in them. The cells are then incubated for 30 min in the device with M9 drops on inlet and outlet, put within a petri dish containing a water drenched tissue.
17. After incubation, reconnect the inlet and outlet tubings to the device.

3.3 Microscopy

1. Set the temperature of the Okolab Cage to 37 °C.
2. Mount the PDMS device on the microscope. Install the syringe on the syringe pump (input of the device) and lead the outlet tubings coming from the PDMS device (output) into the waste container.
3. Bring the channels into focus with transient light illumination by using the eyepiece. Move the stage towards a position where at least three consecutive channels contain two to three cells (*see* **Note 9**, Fig. 2b). In a typical experiment 100–300 cells are recorded.
4. Remove debris and swimming cells from the main channel by turning on the syringe pump to flush the device at 10.5 mL/h. After the debris is removed, set the syringe pump to operative mode at 0.5 mL/h for the remaining experiment.
5. Turn on the EMCCD camera and set the electron multiplier gain (EM gain) to 0 for taking brightfield images and gain to 300 for taking fluorescence images.
6. Using NIS Acquisition in NIS elements, data acquisition is set to take a brightfield image and a fluorescence image with 80 ms exposure time every 150 s.
7. In the Omicron Control Center (OCC) the 514 nm laser is switched and the power is set to the equivalent of 5 W/cm^2 at the sample height (*see* **Note 10**).
8. The Perfect Focus System (PFS) is used during the measurement to ensure stability of focus along the optical axis during the experiment (*see* **Note 11**).

3.4 Analysis Protocols

1. Correct fluorescence images for nonuniformity by dividing the images with the normalized laser beam image, in ImageJ (*see* Fig. 2). Laser illumination of the sample is nonuniform, due to the Gaussian irradiation profile of the laser beam.
2. Perform rolling ball background subtraction to remove background from the image and to isolate bright spots using the rolling ball background subtraction algorithm in ImageJ, with a rolling ball radius of 10 pixels (0.159 μm/px).

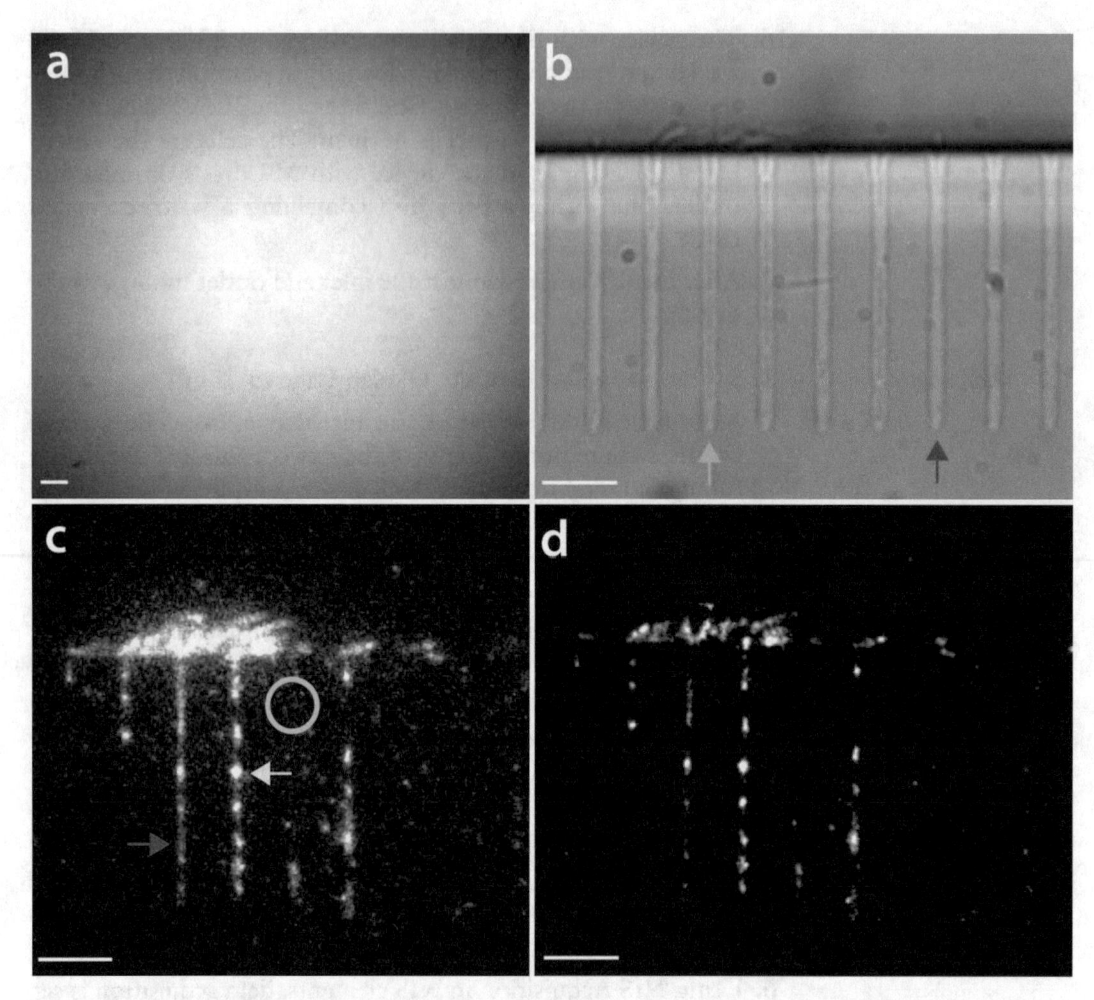

Fig. 2 Examples of images relevant to fluorescence image correction. (**a**) Beam profile of 514 nm laser illumination on an autofluorescent plastic slide (Chroma). *Dark areas* can be observed at the corners of the image, showing its Gaussian illumination profile. (**b**) Brightfield image of the microfluidic device loaded with cells, where the *red arrow* indicates an empty channel, and the *blue arrow* shows a cell-filled channel. (**c**) The raw fluorescence image of the device shown in **b** being illuminated by a 514-nm laser. The *blue area* indicates a region of high background noise, the *yellow arrow* shows a cell with a bright fluorescent spot (actively replicating) and the *red arrow* shows a cell emitting a low uniform intensity (nonreplicating). (**d**) Image **c** with corrected background and laser illumination. The background noise is reduced and the fluorescent spots are clearly visible, including ones on the periphery of the image. *Scale bars* mark 5 μm

3. Align the fluorescence and brightfield images using an X–Y image translation algorithm in ImageJ. The required translation is obtained by performing a calibration measurement with fluorescent beads, 0.5 μm in diameter (*see* **Note 12**).
4. To determine the X–Y drift of the sample during the measurement (*see* Fig. 2d) a high-contrast region is tracked over the time interval of the experiment. Select a high-contrast region of interest (ROI) (e.g., dirt particle or channel edge) within the

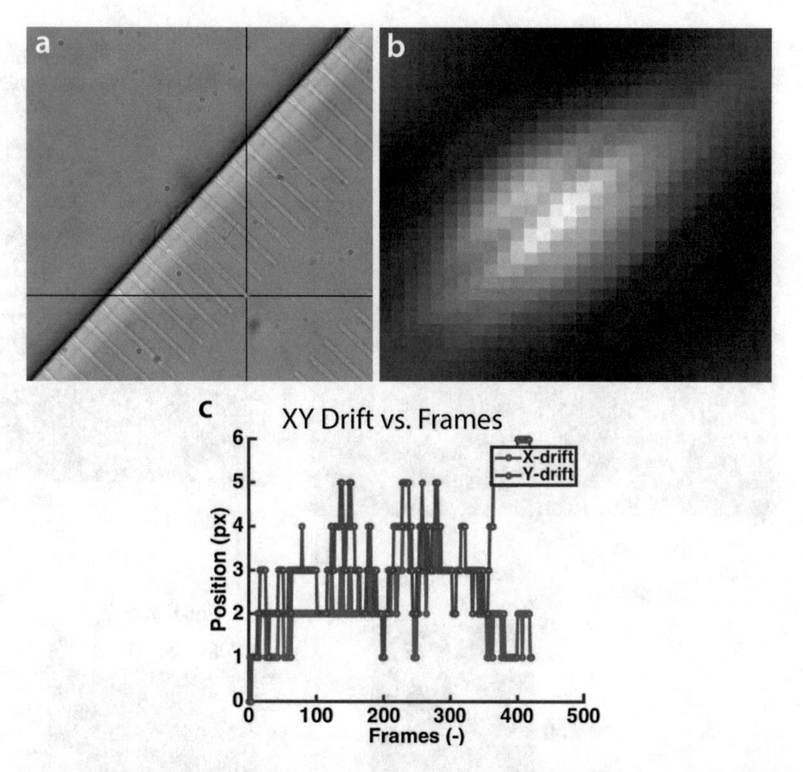

Fig. 3 Determining the drift of the microfluidic device during the time-lapse experiment. (**a**) A high-contrast region of interest (ROI) is selected (*see black crosshair*) in the first brightfield image of the microfluidic device to determine the drift of the sample during the experiment. (**b**) The cross-correlation plot of the ROI compared to the initial ROI image, where the highest intensity value position denotes the shift in position. (**c**) Graph of the resulting translation in *X* and *Y* (in pixels) during the captured frames of the time-lapse measurement. The negatives of the *X* and *Y* in these plots are imposed as translation to the images to counter drift

first image (*see* Fig. 3a). Track the motion of the ROI by using image cross-correlation of the ROI images with the first ROI image (*see* Fig. 3b). Use the resulting X–Y translation vectors (*see* Fig. 3c) to translate the image series correspondingly to stabilize the images.

5. Select the microfluidic channels that are chosen for analysis (*see* **Note 9**) by using a custom made clicking algorithm on a brightfield sample image (Fig. 4a). Clicking parameters (e.g., channel length, channel angle, number of channels, and distance between channels) are specified such that one click is required to select multiple channels. This algorithm makes new time-lapses of individual channels.
6. Generate kymographs for the selected microfluidic channels, for brightfield and fluorescence data (*see* Fig. 4b). The

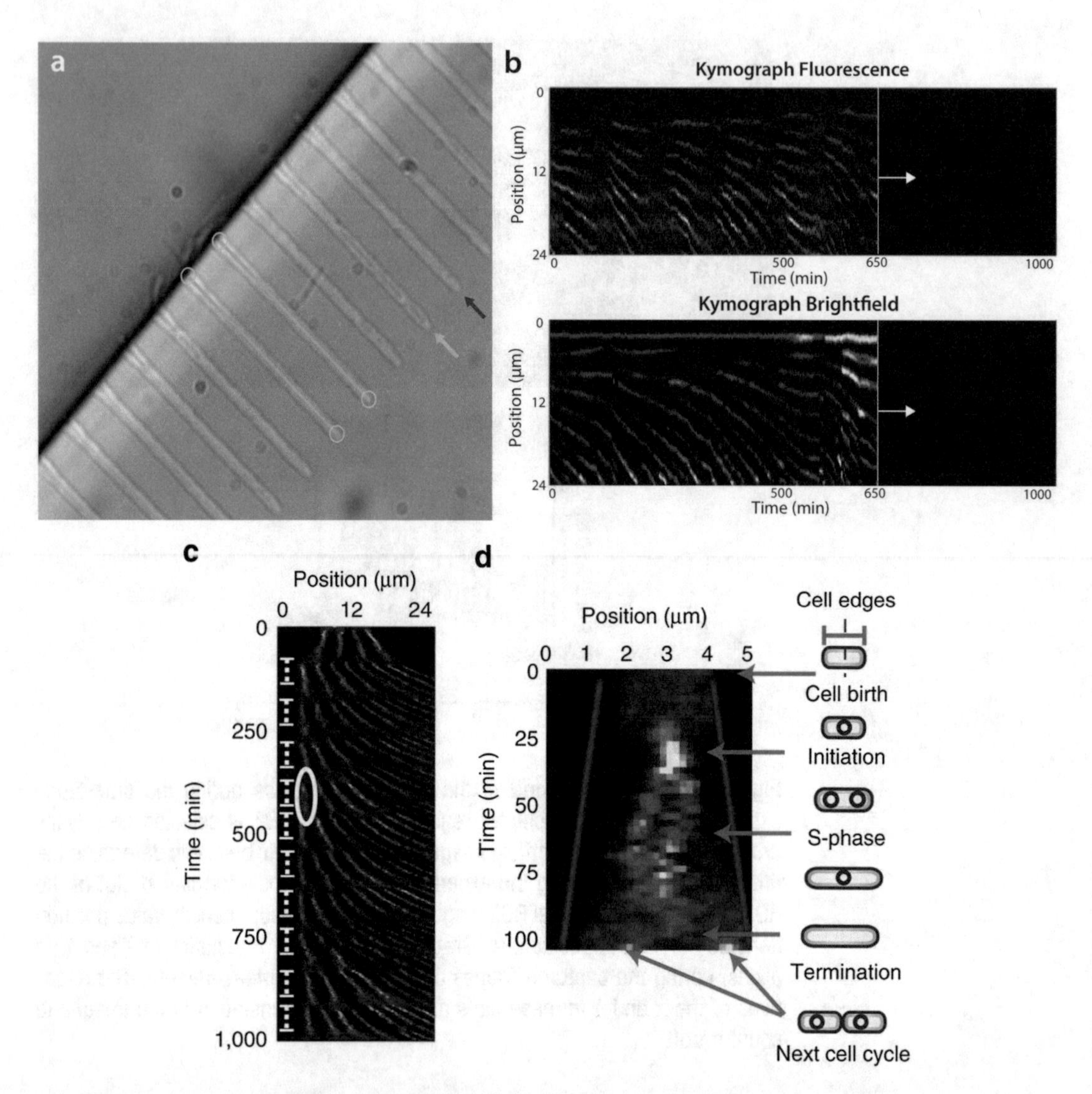

Fig. 4 (**a**) A sample brightfield image of the microfluidic device is used to manually select the channels to make kymographs from. After the user has clicked on the microfluidic channel, *blue lines* show the selection, such that the user can evaluate whether selection is accurately along the channel. A channel containing cells (*green*) and an empty channel (*red*) are indicated by *arrows*. (**b**) Progression of the generation of a fluorescence and brightfield kymograph for one microfluidic channel. The kymograph image is constructed by combining small image strips of the selected microfluidic channel over all time points. In this example the graph is truncated at the image strip corresponding to the 650th minute to show the progressive nature in constructing the kymograph. The curvature of the fluorescence signal is due to the individual cells growing and pushing each other in the direction of the main trench. (**c**) Kymograph of a single growth channel during a time-lapse experiment. Clearly observable diffuse patterns occur at regular intervals, indicating the lack of DNA-bound β_2-clamps before initiation and after termination. The repeating pattern is due to the multiple cycles of replication. Each replication cycle is indicated with a *vertical dashed line* on the *left-hand side* of the image. The *white ellipsoid* indicates an individual replication cycle. (**d**) The detailed kymograph of an individual replication cycle indicated by the *ellipsoid* in **c**. The *blue lines* are the cell boundaries detected from the brightfield field images. The illustrations on the *right-hand side* indicate the different stages of replication that take place during the cell cycle, with *grey arrows* pointing toward the position in the fluorescent data

kymograph image is constructed by combining small image strips of the selected microfluidic channel over all time points. These graphs provide spatial and temporal information of the cell sizes (cell poles) and proteins (fluorescent clusters). The fluorescence kymograph is created by using the corrected fluorescent images, while the brightfield kymograph is constructed by differentiated brightfield images with respect to the direction of the channel. The differentiation is done such that high contrast gradients (space between neighboring cell poles) will show spot-like intensities. Use the differentiated images containing these spots to create the brightfield kymograph.

7. The kymographs allow a high-throughput manual clicking procedure to follow the replication cycle of the cells. Click on the kymograph to select the initiation of replication for the first cells, the first top image strip which has a spot (*see* Fig. 4c, d, replication initiation). Do this step first on a zoomed-in part of the starting section of the kymograph (*see* white circles in Fig. 5a). A larger fraction of the kymograph is then displayed. Click the end of the observed fluorescent spots. Connect these dots to two reappearing fluorescent spots below the termination point (*see* lower white dots in Fig. 5b). This denotes the initiation of the replication cycle of the two daughter cells (*see* Fig. 5b). Click in this manner through every generation.

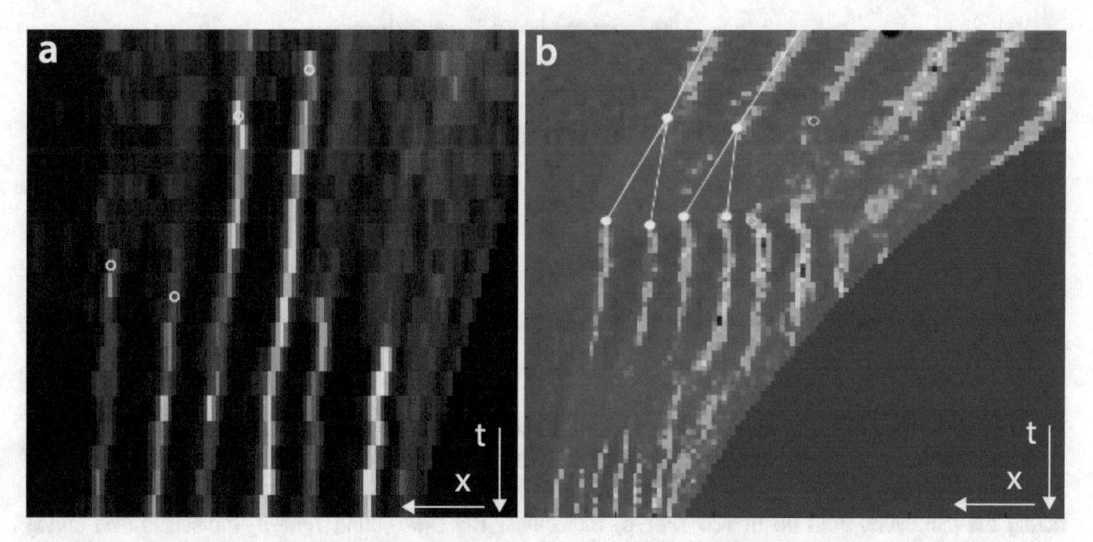

Fig. 5 The kymograph replication cluster clicking procedure. (**a**) A zoomed image of the start of the kymograph is used to click (*white circles*) the beginning of measured fluorescent spots (*bright lines*), indicating the beginning of replication for each cell. (**b**) A larger fraction of the kymograph displayed to click the end of the observed fluorescent spots (replication ceases here), and to connect it to the two reappearing fluorescent spots as the daughter cells start their own replication cycle. Replication profiles are curved in the kymograph as a result of cell growth. This is corrected for to facilitate the user clicking process. The color maps are chosen in such a way that the user can better recognize the cycle events

Different zooms on the kymograph help to complete this procedure for the full time-lapse experiment.

8. Use the clicks created in **step 6** and the brightfield kymograph to determine the replication and division time (*see* Fig. 6a). The division is determined by the appearance of new cell poles in the brightfield kymograph (*see* **Note 13**). The replication time is determined by the distance between the clicks on the fluorescence kymograph. Set intensity thresholds according to the individual intensity value of the protein of interest.

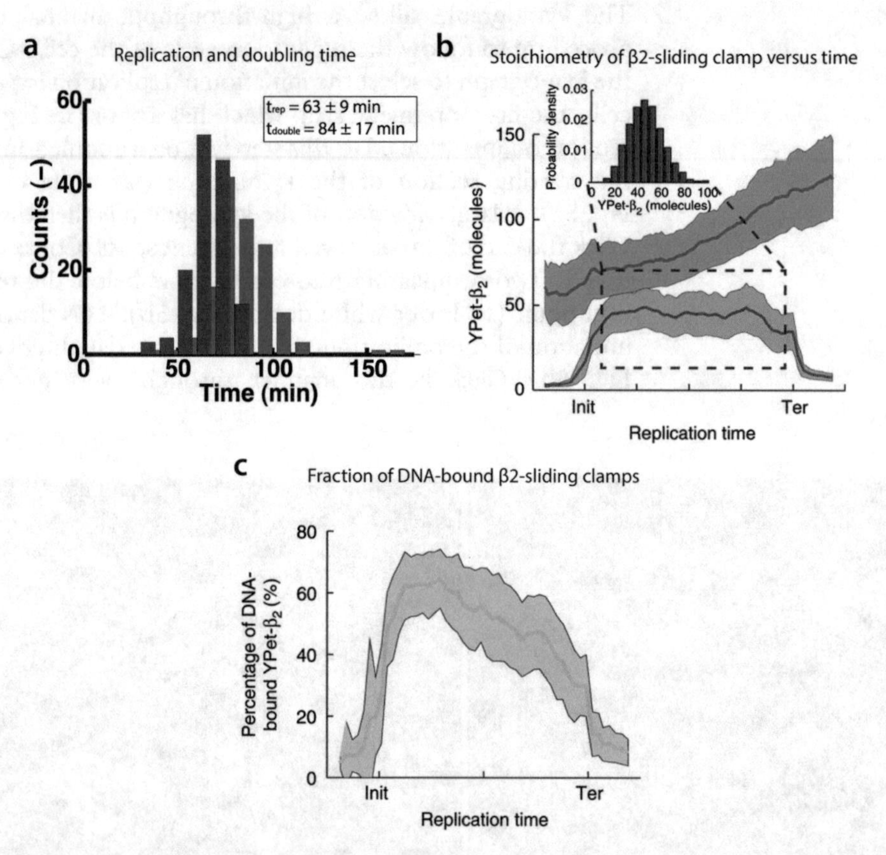

Fig. 6 Quantification of the in vivo β_2-sliding clamp stoichiometry during replication. (**a**) The division (t_{double}, *blue*) and replication (t_{rep}, *red*) time distributions of the YPet-β_2 *E. coli* strain in the microfluidic device during a long time-lapse experiment. (**b**) The YPet–β_2 molecules in the whole cell (*blue curve*) approximately double during the cell cycle, from 60 to 120 YPet–β_2 molecules. The DNA-bound YPet–β_2 molecules (*red curve*) remarkably increase to a mean steady state value of 46 YPet–β_2 molecules following initiation. (*inset*) A histogram of the distribution of number of DNA-bound YPet–β_2 molecules during steady state. (**c**) The fraction of DNA-bound YPet–β_2 molecules varies over the course of the replication cycle. At the midway point of the replication cycle (indicated by a *tick mark*), this fraction equals 45%

9. The output of **step 7** also includes data of nonreplicating or nongrowing cells, so a set of criteria (fluorescence intensity, cell size, and cell cycle time thresholds) is implemented to ensure that the analysis is done only on actively replicating cells.
10. Generate individual cell time-lapse datasets, containing fluorescence image data of single cells, by using the cell pole data from the brightfield kymograph to crop fluorescent images from pole to pole.
11. Apply bleaching correction by measuring the fluorescence intensity per cell length for the individual cells. Correct the fluorescence images for the factor by which the intensity drops during the cell cycle of the cell, to properly account for the bleached molecules (*see* **Note 14**).
12. Analyze the cell time-lapse datasets in terms of spot intensity (stoichiometry, *see* **Note 15**) and the total fluorescence intensity coming from the cell. The spot intensity is properly accounted for by fitting a 2D Gaussian to the spot data, and a weighted integration is done over the area to get the integrated intensity. By knowing the integrated intensity coming from one of the utilized fluorescent proteins, the molecules can then be counted on the observed spot (*see* Fig. 6b). Determine the total cell intensity by summing the intensity of the pixels that make up the cell. After this step it is possible to plot the fraction of molecules in the spot (DNA-bound) versus all the molecules in the cell (DNA-bound and freely diffusing in the cytoplasm, *see* Fig. 6c).

4 Notes

1. There are several alternative and combined applications of the methods for microfluidics [5].
2. If you do not have access to clean room facilities able to do EBL, PDMS molds can be obtained from external sources [6]. Make sure to store the microfluidic channels facing upward (not to damage the protruding details) in a closed container (to protect it from dust contamination).
3. The PDMS array is obtained from a PDMS mold as described previously in detail [7]. PDMS is mixed in a 1:10 ratio, degassed, and poured onto the previously cured PDMS mold. After another degassing step, the device is allowed to cure for 2 h at 85 °C. Once the curing is complete, the PDMS is left to cool down for at least 30 min. Subsequently, the two PDMS layers (the mold and the actual device) are separated from each other. At this point the PDMS mold can be stored for later use.

4. Optionally, sonication can be applied for cleaning the cover glasses; Put the cover glasses in a holder that is submerges into a beaker. Place the beaker into a sonicator (Branson Ultrasonics B1510, USA). Apply 15 min of sonication in acetone (Sigma-Aldrich), then 15 min in isopropanol (Sigma-Aldrich), followed by a 15-min sonication in dH_2O. Blow-dry the cover glasses then with N_2 gas.
5. When attaching the PDMS device to the coverslip, make sure to use the plasma-activated surface of the glass that is facing upward during the oxygen plasma step. Apply some pressure on the PDMS device on the glass to obtain full contact of the surfaces.
6. Before inserting the syringe needle into the tubing, make sure that air bubbles are removed. This is best done by holding the tip upward and tapping it until the bubbles travel close to the tip. Press the end of the syringe until the air is removed.
7. Make sure that you do not introduce air into the device when injecting the M9–BSA mixture.
8. Before centrifuging the cells into the channels, put M9 drops on inlet and outlet to prevent the device to dry. On the coverslip, indicate the direction of centrifugation so that you know what side of the device contains channels with cells.
9. The channels chosen for analysis are the ones that contain more than three cells initially. Cells might stay close to each other even though they are physically separated. This prevents us from detecting the brightfield "spot" in the differentiated brightfield image. However, one could still use the fluorescence images and the clicks generated from the kymographs for determining the division time. As we use minimal medium, initiation of replication in a daughter cell occurs after the cell has divided. Between replication profiles cell division must have taken place. If we are unable to track a new pole between replication cycles, we estimate it to be in the middle of that time interval.
10. The required laser power to reach a specific intensity at the sample height varies from setup to setup [8].
11. Make sure that the Perfect Focus System (PFS) offset is set into the NIS acquisition configuration with command: PFS_define_offset(*offset_value*) (NIS elements). Determine a good offset by looking live at the sample while being laser illuminated, and change the offset such that the fluorescent spots are in focus and the image is sharp.
12. Determine the center position of fluorescent beads (TetraSpeck Fluorescent Microspheres Size Kit, Molecular Probes) for fluorescence images, by fitting a 2D Gaussian function over

the spot profile. The center position in the brightfield image is determined by inverting the image (such that it appears as a light spot) and fitting a 2D Gaussian function over the spot profile. The deviation between the center positions determines the required translation between the fluorescence and brightfield images.

13. It is recommended to judge by eye which channels are best fit for analysis, prior to the experiment. Criteria can be the number of cells in the channel (ideally it should be filled), the fluorescence activity of the cells (a filled channel can contain nonliving cells), the physiological state of the cells (deformed shape and size), or the quality of the channel shape itself (channel can be too wide, judged by mobility of cells within).
14. To correct for photobleaching, we use the following approach. We assume that cell growth and protein copy number in the cell increases linearly from cell birth until cell division. The ratio of these two numbers, as function of time, should remain constant throughout the cell cycle of a cell if there is no photobleaching. However, due to photobleaching, this ratio will decrease as function of time. We fit this curve with a single exponential and multiply the detected fluorescent signal with the appropriate factor as function of time in order to correct for this decline in fluorescence due to photobleaching.
15. To determine the stoichiometry of the spots, one has to know how many photons are excited from the single fluorophore that is used to label the protein. An additional calibration measurement is therefore required to determine the integrated intensity of the spot resulting from a single fluorescent protein. This can be an in vivo photobleaching experiment [9] or an in vitro single-molecule photobleaching assay [10].

Acknowledgments

The authors would like to thank the members of the Nynke H. Dekker laboratory for discussions on the chapter. We thank Sriram T. Krishnan for his active involvement in the experiments discussed here, Jacob W.J. Kerssemakers for contributing to the development of the data analysis procedures, and Margreet Doctor for building and maintaining the experimental setup. The experiments described here were carried out in the Department of Bionanoscience, Delft University of Technology.

References

1. Huang B, Bates M, Zhuang X (2009) Super-resolution fluorescence microscopy. Annu Rev Biochem 78(1):993–1016. doi:10.1146/annurev.biochem.77.061906.092014
2. Schermelleh L, Heintzmann R, Leonhardt H (2010) A guide to super-resolution fluorescence microscopy. J Cell Biol 190 (2):165–175. doi:10.1083/jcb.201002018
3. Moolman MC, Krishnan ST, Kerssemakers JWJ, van den Berg A, Tulinski P, Depken M et al (2014) Slow unloading leads to DNA-bound ß2-sliding clamp accumulation in live Escherichia coli cells. Nat Commun 5:5820. doi:10.1038/ncomms6820
4. Joyce G, Robertson BD, Williams KJ (2011) A modified agar pad method for mycobacterial live-cell imaging. BMC Res Notes 4:73. doi:10.1186/1756-0500-4-73
5. Sia SK, Whitesides GM (2003) Microfluidic devices fabricated in poly(dimethylsiloxane) for biological studies. Electrophoresis 24(21): 3563–3576. doi:10.1002/elps.200305584
6. Microfluidic foundry company website: http://www.microfluidicfoundry.com/moldpdms.html
7. Moolman MC, Dekker NH, Kerssemakers JWJ, Krishnan ST, Huang Z (2013) Electron beam fabrication of a microfluidic device for studying submicron-scale bacteria. J Nanobiotechnol 11:1–10
8. Grünwald D, Shenoy SM, Burke S, Singer RH (2008) Calibrating excitation light fluxes for quantitative light microscopy in cell biology. Nat Protoc 3(11):1809–1814. doi:10.1038/nprot.2008.180
9. Hines KE (2013) Inferring subunit stoichiometry from single-molecule photobleaching. J Gen Physiol 141(6):737–746. doi:10.1085/jgp.201310988
10. Reyes-Lamothe R, Sherratt DJ, Leake MC (2010) Stoichiometry and architecture of active DNA replication machinery in *Escherichia coli*. Science 328:498–501

Chapter 19

Imaging of Bacterial Chromosome Organization by 3D Super-Resolution Microscopy

Antoine Le Gall, Diego I. Cattoni, and Marcelo Nollmann

Abstract

The bacterial nucleoid is highly organized, yet it is dynamically remodeled by cellular processes such as transcription, replication, or segregation. Many principles of nucleoid organization have remained obscure due to the inability of conventional microscopy methods to retrieve structural information beyond the diffraction limit of light. Structured illumination microscopy has recently been shown to provide new levels of spatial details on bacterial chromosome organization by surpassing the diffraction limit. Its ease of use and fast 3D multicolor capabilities make it a method of choice for imaging fluorescently labeled specimens at the nanoscale. We describe a simple high-throughput method for imaging bacterial chromosomes using this technique.

Key words Bacterial chromosome, Nucleoid, Structured illumination microscopy, Super-resolution microscopy, High-throughput

1 Introduction

How chromosomes are organized within the cell is a fundamental problem in biology. Most bacteria possess a single circular chromosome, which must be tightly compacted to fit within a micrometer sized volume, called the nucleoid. Sophisticated regulatory mechanisms ensure the packaging and dynamic organization of the DNA content within the nucleoid at different hierarchical levels, from the nanometer to the micron scale. Traditional ensemble approaches in biochemistry, genetics or molecular biology have provided invaluable insights into this organization. Of particular interest, bacterial chromosomes have been shown to be structured by several hierarchical levels of organization, including topological domains (10–20 kb) [1, 2], chromosome interaction domains (50–300 kb) [3–5] and macrodomains [6]. In particular, it is not yet understood how these different levels of structural organization are orchestrated at the single cell level, coordinated with the cell

Olivier Espéli (ed.), *The Bacterial Nucleoid: Methods and Protocols*, Methods in Molecular Biology, vol. 1624,
DOI 10.1007/978-1-4939-7098-8_19,

cycle, or coupled to DNA management processes such as replication.

Fluorescence microscopy enables the in vivo observation of DNA while permitting the localization of proteins at the single cell level with high specificity. However, the maximum resolution attainable by conventional microscopy approaches is intrinsically restricted to the detection of spatial scales above the limit imposed by the diffraction of light (typically 200–300 nm and 700–900 nm in the lateral and axial directions, respectively). As most hierarchical organization levels of the bacterial nucleoid lie in the 10–500 nm scale, they cannot be accessed by conventional microscopies. In recent years, a plethora of techniques using various strategies to overcome the intrinsic microscope optical resolution have been developed [7–10]. Among these techniques, three-dimensional structured illumination microscopy (3D-SIM) enables a twofold increase in both lateral and axial resolution as compared to diffraction-limited microscopies [11]. 3D-SIM relies on the modulation of the frequency content of the specimen using an excitation grid pattern that is shifted or rotated during image acquisition. Reconstruction procedures permit high-frequency information to be extracted to obtain a 3D super-resolved image. Using this technique on the *Bacillus subtilis* and *Escherichia coli* chromosomes, we revealed the existence of a new higher-order level of organization of the nucleoid that was not visible by diffraction-limited fluorescence microscopy methods [4, 12]. Nucleoids in those species appear structured by High-density DNA Regions (HDRs). Besides circumventing the diffraction limit of light, 3D-SIM offers an ease of use and fast 3D multicolor capabilities that will likely ensure it to become a method of choice for imaging fluorescently labeled specimens at the nanoscale. However, like any super-resolution technique, its use, especially on live biological specimens, requires careful data acquisition and processing procedures to avoid reconstruction artifacts.

Here, we describe a method to image the bacterial nucleoid and DNA-binding proteins simultaneously at super resolution and at high throughput. We provide a step-by-step protocol describing the cell culture, sample preparation, and acquisition and reconstruction of multicolor super-resolution images on a commercially available 3D-SIM platform (GE Healthcare OMX). Specifically, we demonstrate our method by imaging the nucleoids of *Bacillus subtilis* (*B. subtilis*) or *Escherichia coli* (*E. coli*) using protein fusions or chemical labeling, and simultaneously detect the nuclear distribution of ParB, a protein that assembles at centromeric sequences to form a mitotic-like apparatus assuring the inheritance of duplicated genetic material before cell division.

2 Materials

All solutions are prepared using ultrapure water and analytical grade reagents when possible. Prepare and store all stock reagents at 4 °C. Solutions percentage are weight/volume (w/v %) unless otherwise indicated.

2.1 Bacterial Strains

1. *Bacillus subtilis* strains: wild-type *B. subtilis* (PY79 background), chromosomal ParB-GFP strain (HM671, PY79 background), with resistance to kanamycin.
2. *Escherichia coli* strain: HU-mCherry labeled nucleoids and plasmidic ParB-mVenus (DLT3053/pJYB234), with resistance to chloramphenicol.

2.2 Growing Media

1. Luria–Bertani (LB) medium: 4 g NaCl, 4 g tryptone, 2 g yeast extract, bring volume to 400 mL with water. For *E. coli* culture add 4 mg of thymine.
2. Minimal media (M9, 10×): 60 g of Na_2HPO_4, 30 g of KH_2PO_4, 5 g of NaCl, 10 g of NH_4Cl. Bring to 1 L with H_2O.
3. *B. subtilis* minimal media (M9bs): 50 μL of 2 mg/mL Fe-NH_4-citrate, 0.6 mL of 1 M $MgSO_4$, 0.1 mL of 100 mM $CaCl_2$, 0.2 mL of 65 mM $MnSO_4$, 0.1 mL of 1 mM $ZnCl_2$, 0.1 mL of thiamine 2 mM, 2 mL of 5% glutamate, 7.5 mL of 20% succinate. Bring volume to 100 mL with M9 (1×) solution. Store at 4 °C.
4. *E. coli* minimal media (M9ec): 2 mL of 20% glucose, 100 μL of 1 mM $MgSO_4$, 100 μL of 0.1 M $CaCl_2$, 100 μL of 1 mg/mL vitamin B1, 2 mL of 2 mg/mL thymine, 200 μL of 10 mg/mL leucine. Bring volume to 100 mL with M9 (1×) solution. Store at 4 °C.
5. Casamino acids (CAA) supplement: 20%. Store at 4 °C.
6. Antibiotics, final concentrations: chloramphenicol, 10 μg.mL^{-1}, kanamycin, 2 μg.mL^{-1}.

2.3 Sample Preparation

1. Rectangular coverslip (#1.5H, Deckgläser, 24 × 60 mm).
2. 1 mm thick glass slide (SuperFrost, Ultra Plus, 25 × 75 mm).
3. Acetone.
4. Bunsen burner.
5. Homemade Teflon rack.
6. Double-side adhesive tape.
7. Razor blade.
8. Fluorescent beads (40 nm, Trans FluoSpheres and 100 nm Tetraspeck, Invitrogen).

9. Agarose (A4804, Sigma-Aldrich).
10. Minimal medium (M9bs/M9ec).
11. Dry bath heater.
12. Optional: plasma cleaner Femto (Diener Electronic).

2.4 Structured Illumination Microscope and Software

Instrumental setup (GE Healthcare OMX version 3).

- 405/488/561 nm excitation lasers.
- DAPI filter: 419–465 nm.
- A488 filter: 500–550 nm.
- A561 filter: 581–618 nm.
- Objective 100×, NA 1.4.
- Nanometric piezo-stage.
- Four electron-multiplying charge-coupled device (EM-CCD) cameras (Photometrics Evolve).

Software

- OMX acquisition software.
- SoftWoRx version 5.
- ImageJ (Bioformat, SIMcheck plugins).
- Icy/Imaris/Matlab.

3 Methods

3.1 Growing Cells

Bacterial strains are stored in a mix of LB and glycerol (50%) at −80 °C. All cell culture steps should be carried out under flame or in a hood.

1. Streak bacteria from −80 °C glycerol stock onto solid (plate) Luria–Bertani (LB) medium (i.e., agar plate) complemented with appropriate antibiotics.
2. Culture agar plates 24 h at 37 °C.
3. Select a single colony from plate and perform at least five serial dilutions with M9bs/M9ec supplemented with CAA 20 μg/mL in sterile 50 mL falcon tubes (1/10 each in 5 mL final volume) and grow overnight at 30 °C with agitation at 200 rpm.
4. The next morning measure the optical density (OD) at 600 nm and select dilution of cells exponentially growing (OD ~0.3–0.5). Further dilute to OD ~0.05 in M9 (M9bs or M9ec) supplemented with CAA 20 μg/mL (25 mL final volume in 250 mL flasks for optimal aeration) and incubate at 30 °C with agitation at 200 rpm.

5. Measure OD at 1 h intervals and when reaching OD ~ 0.3–0.4, take 1 mL of bacterial suspension and place it into a 1.5 mL Eppendorf tube.
6. If necessary, add 7.5 μL of a 0.3 μM solution of DNA dye (DAPI, Invitrogen) to 1 mL of bacterial suspension and incubate for 5 min.
7. Spin down cells in a bench centrifuge at 1500 × *g* for 4 min for *B. subtilis* or 7000 × *g* for 1.5 min for *E. coli*.
8. Resuspend in 1 mL of M9bs or M9ec.
9. Repeat **steps** 7 and **8** two more times.
10. Discard supernatant and resuspend pellet in 20 μL M9bs or M9ec.
11. Add 1 μL of 1/10 dilution of 40 nm fluorescent beads to the suspension and gently mix (*see* **Note 1**).

3.2 Mounting Agar Pads

1. Rinse microscope coverslips and glass slides with acetone and dry them over an open flame to eliminate any remaining fluorescent contamination (*see* **Note 2**).
2. Place microscope coverslips and glass slides on a Teflon rack (homemade). Plasma-clean it for 15 min with O_2 gas at 0.4 mbar at 70% power (*see* **Note 3**).
3. Deposit a piece of double-side adhesive tape on a glass slide and extrude a ~ 5 mm V-shaped channel from its center using a razor blade (Fig. 1, *see* **Note 4**).
4. Spread 15 μL of 2% melted agarose (diluted in M9bs or M9ec media, melted at 90 °C) on the center of the glass slide and cover right away with a second glass slide to ensure a flat agarose surface (Fig. 1).
5. Let agarose cool down to room temperature for ~2 min at a horizontal position with external pressure to ensure a flat agarose surface.
6. Carefully remove the top glass slide by sliding it down towards the bottom of the V-shape channel to ensure the agarose pad remains on the bottom glass slide.
7. Deposit ~4 μL of vegetatively growing bacterial resuspension (*see* Subheading 3.1) onto the agarose (*see* **Note 5**) and let settle for 2 min.
8. While carrying out **step** 7, remove the second face of the double side tape then gently seal the pad with a clean coverslip by pressing its top surface (*see* **Note 6**).
9. Store your agar pad in a sealed box with water to prevent sample from drying.

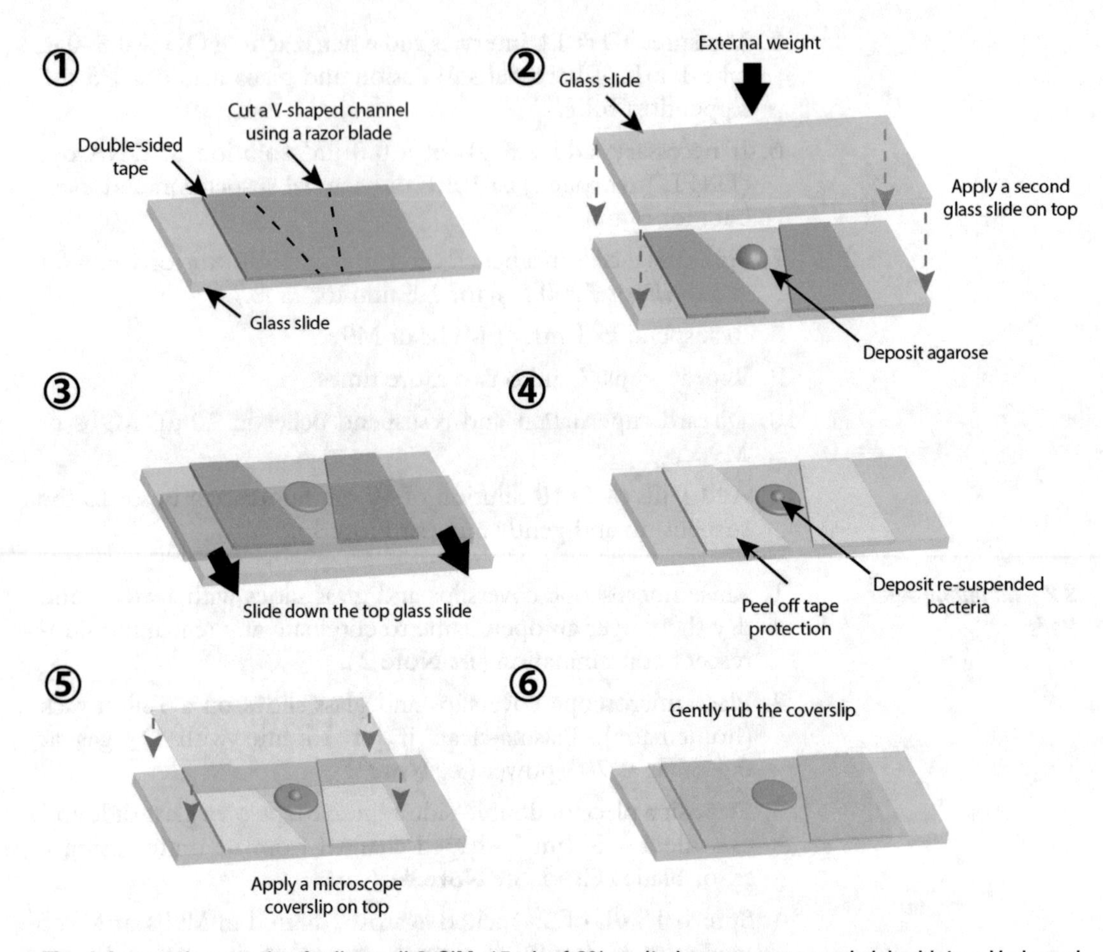

Fig. 1 Agar pads mounting for live-cell 3-SIM. 15 μL of 2% melted agarose are sandwiched into a V-shaped channel between two 1-mm thick glass slides. Once cooled to room temperature under external pressure (*see* main text), 4 μL of resuspended bacteria are deposited onto the agarose before sealing the sample with a microscope coverslip

3.3 Imaging

This section describes initial acquisition conditions and optimizations for imaging bacterial chromosomes in 3D-SIM. Before starting high-throughput acquisitions, additional controls described in Subheading 3.5 should be performed.

1. Set software to the 100× objective settings and Z scan acquisition mode (center of the stack).
2. Mount slide with agarose pad onto the microscope and find the focus on the agar pad–coverslip interface using brightfield illumination.
3. Under brightfield illumination, explore sample and while verifying its integrity find a region of the pad with a dense monolayer of bacteria (~300–500 bacteria per field of view can be found in optimal conditions) (*see* **Note 7**).

4. Assign the appropriate fluorescent filters to the cameras for ParB (GFP or mVenus labeled) and DNA (DAPI-stained or Hbs-mCherry) emission and excitation wavelengths (*see* **Note 8**) and switch to structured illumination mode (*see* **Note 9**).
5. Readjust the focus to find the optimum axial position between the two fluorescence emission channels for ParB and DNA imaging (*see* **Note 10**).
6. Adjust laser excitation intensities and camera integration times (*see* **Note 11**) and gains to exploit as much as possible the dynamic range of the camera (>10^4 levels of grey for a 16-bit camera) while minimizing photobleaching and phototoxicity. Anticipate a few hundreds images using your current imaging settings for volumetric SIM acquisition, *see* **Notes 12** and **13**)
7. Set the Z step between each slice to 125 nm (*see* **Note 14**) and the number of slices to 15–17 to anticipate the chromatic shift between emission channels and any slight tilt of the sample plane by including information above and below signal's plane of focus.
8. Move to a nearby nonilluminated region of the pad (~100 μm), let the sample stabilize for 10–15 s and launch the acquisition of a 3D-SIM stack.
9. Open the acquired 3D-SIM stack in a data inspection software (e.g., ImageJ/Icy/Matlab) for a rapid visual inspection of your data.
10. Inspect your data for minimum fluorescence intensity variations during acquisition (Fig. 2, *see* **Note 15**), good modulation contrast, and absence of artifactual signatures (Fig. 2). Check for presence of first and second modulation peaks and absence of atypical structures in the Fourier transform of the raw data. A more thorough assessment of image acquisition will be performed in Subheading 3.5.
11. Optimize imaging conditions by repeating **steps 5–10** until you get satisfying imaging conditions.

3.4 Data Reconstruction and Alignment

Reconstruction of 3D-SIM raw data is performed using softWoRx v5.0 (Applied Precision, *see* **Note 16**). Channel specific Optical Transfer Functions (OTFs) are used for reconstruction and are computed from an experimental point spread function (PSF). This experimental PSF is obtained from 3D stack of images of 100 nm multicolor labeled FluoSpheres (Life Technologies) provided as input to softWoRx. For all emission channels, set the reconstruction filter (Wiener filter) settings and background subtraction value to optimal values (0.002 and 65, respectively, under the imaging conditions of this report, *see* **Note 17**). These settings will depend on your image background and signal-to-noise ratio and may be adjusted depending on your imaging conditions.

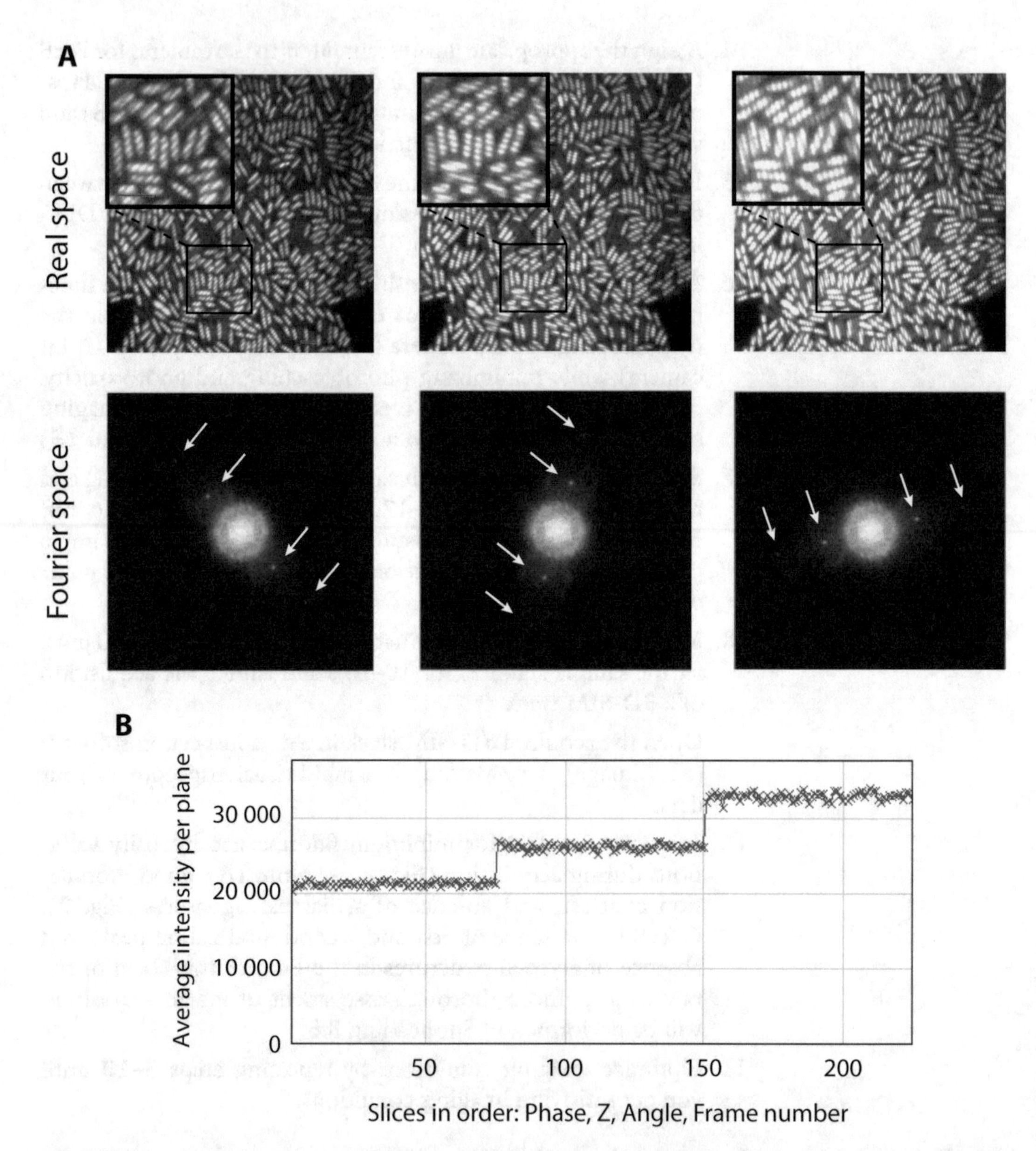

Fig. 2 Visual inspection and optimization of 3D-SIM acquisition. (**a**) Single plane images of a 3D-SIM raw acquisition of *B. subtilis* nucleoids stained with DAPI, showing the illuminated sample with three excitation pattern orientations. *Top panels* show the illuminated sample in the real space with insets showing a higher magnification in which the illumination pattern is clearly visible, while bottom panels display their respective Fourier transform. White arrows in bottom panels point toward the first and second diffraction order of the structured illumination ensuring optimal modulation of the frequency content of the sample with the structured illumination pattern. (**b**) Average intensity for each imaging plane (in the order: Phase, Z, Angle, Time). This allows to ensure minimum fluorescence intensity variations during acquisition due to photobleaching, intensity differences between angles (as illustrated with the increasing fluorescence signal on the above figure), and illumination flickering amongst other measures

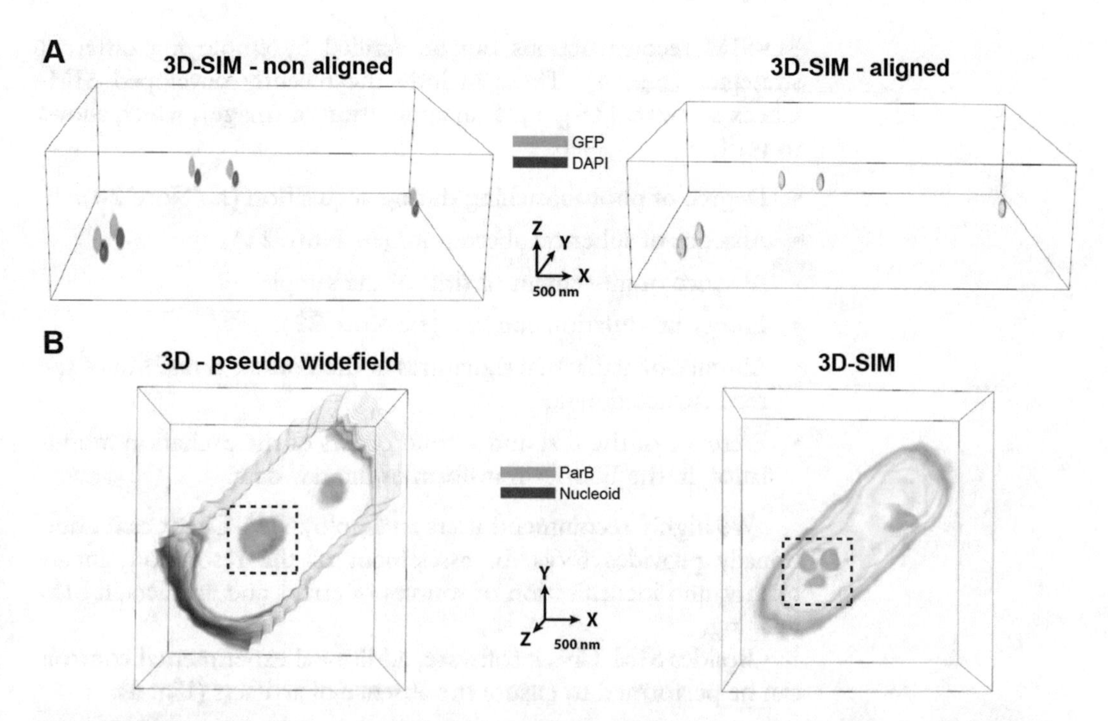

Fig. 3 3D-SIM alignment and reconstruction. (**a**) 3D-SIM images of beads imaged in GFP (*green*) and DAPI (*red*) channels before and after alignment (*left* and *right panels*, respectively). (**b**) Top view of the same nucleoid (*red*) and ParB foci (*green*) from *B. subtilis* in 3D epifluorescence (pseudo-widefield reconstruction from 3D-SIM raw data without channel alignment, *see* **Note 18**) and reconstructed from 3D SIM raw data (with channels alignment). *Dotted square* in *left* and *right* images highlights the same ParB focus composed of three ParB subclusters only resolvable when using 3D-SIM

For multicolor acquisitions, images from different emission channels are aligned to each other to get the most accurate overlay of fluorescent structures (Fig. 3). This step is critical, particularly on the GE Healthcare OMX platform since emission on each channel is recorded on a distinct camera. Multichannel alignment calibration is typically performed using fluorescent beads bound to a coverslip as alignment markers, but other strategies such as nanostructures consisting of arrays of pinholes can be used. The 3D coordinates of each alignment marker then provide references to correct for X–Y–Z translation, distortion, magnification, and rotation of each fluorescent channel. Alignment markers must be immersed in the same media used for experiments, as optical aberrations depend on the refractive index of the sample media. OMX Editor performs automated extraction of the coordinates of alignment references and efficiently applies the obtained correction parameters to align multichannel 3D-SIM acquisitions.

3.5 Data Acquisition and Reconstruction Validation

Inaccurate instrument settings and suboptimal imaging conditions can cause artifacts that are often difficult to distinguish from relevant structural features (*see* **Note 19**). Absence of artifacts in

3D-SIM reconstructions can be verified by employing different strategies (Fig. 4). These include the recently developed SIM-Check software [13], a plugin application of ImageJ, which allows to verify:

- Degree of photobleaching during acquisition (*see* **Note 20**).
- Absence of spherical aberrations (*see* **Note 21**).
- Absence of movement or drift of the sample.
- Proper modulation contrast (*see* **Note 22**).
- Absence of artifactual signatures in the Fourier transform of the reconstructed data.
- Presence of the first and second orders of the excitation modulation in the Fourier transform of the raw data.

We highly recommend users to employ SIM-Check as it additionally provides tools for assessment of the resolution, image quality, and identification of sources of errors and artifacts in SIM imaging.

Besides SIM-Check software, additional experimental controls can be performed to ensure the absence of artifacts (Fig. 4):

- Imaging cytosolic GFP (or other fluorescent markers with similar spectral properties) on anucleated bacterial cells (obtained by treatment with mitomycin at 50 ng/mL during 30 min) using the same imaging conditions as those for whole labeled chromosomes. High-frequency signal should be exclusively distributed at the frontier between the cytosol and membrane where the fluorescence intensity signal drops to zero.
- 3D-SIM time-lapse acquisition of labeled nucleoids. The overall nucleoid morphology should be conserved between the images and only small dynamical changes between acquisitions should be apparent. To minimize dynamical changes during sequential acquisitions as well as photobleaching, the total number of Z slices can be reduced to ~12.
- Two-color 3D-SIM imaging of nucleoids simultaneously labeled with spectrally distinct probes (for example Hbs-GFP and DAPI). Nucleoid structures should appear similar in both colors and only small dynamical changes between acquisitions should be apparent.

3.6 Reconstruction Visualization

Volumetric visualization and inspection of structures of interest can be challenging due to mixture of both low and high spatial frequencies in three dimensions. Exploration of 3D-SIM datasets can be performed using freely available software packages such as Icy, which permits fast and handy representations of your data, or using more sophisticated or automated solutions such as Imaris or Matlab

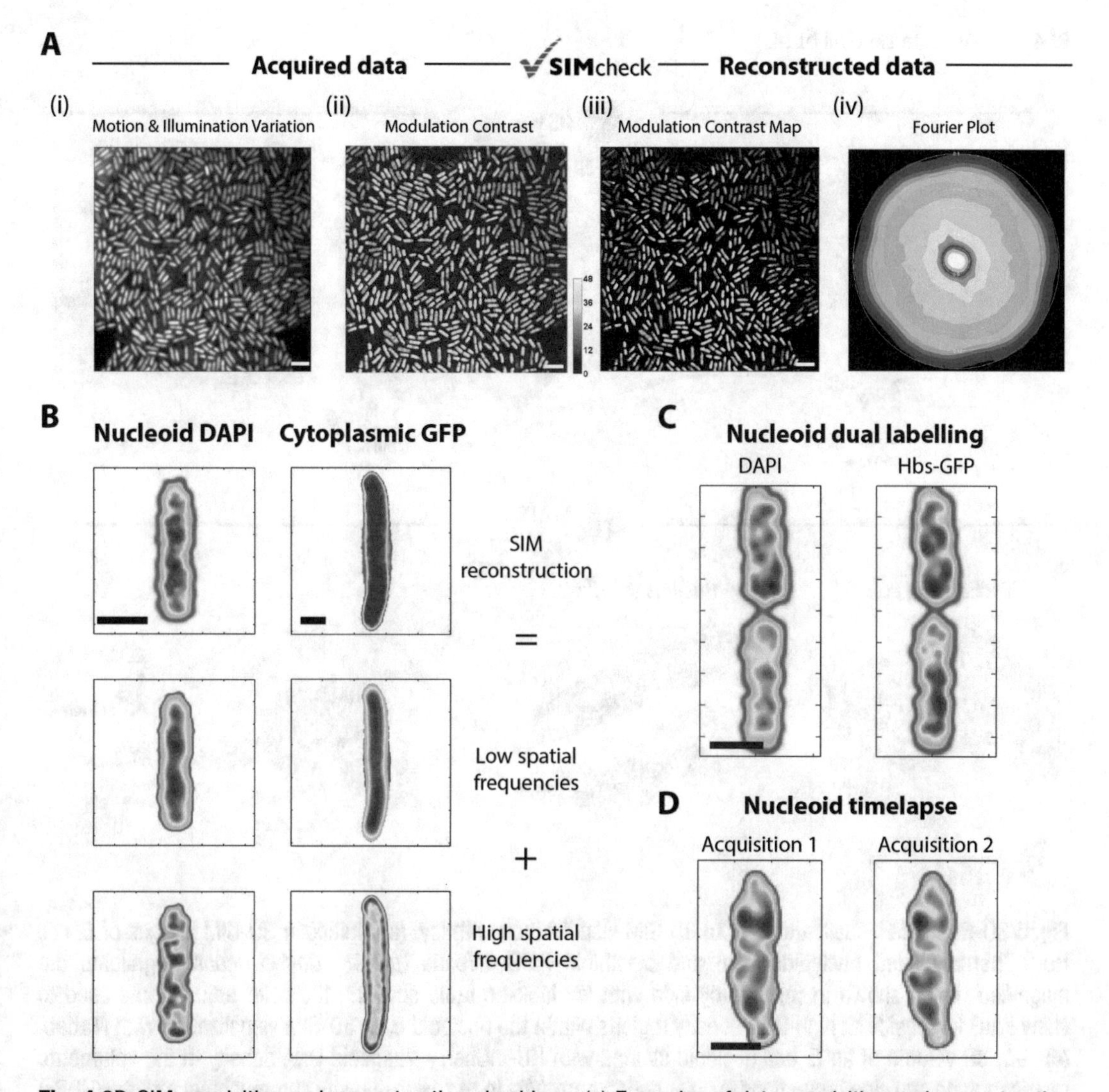

Fig. 4 3D-SIM acquisition and reconstruction checkings**.** (**a**) Examples of data acquisition and reconstruction checkings using SIMCheck on *B. subtilis* chromosomes stained with DAPI. Image (*i*) assesses motion or illumination variations of the sample during the acquisition by averaging and then assigning false colors to images corresponding to the same illumination angle (*see* **Note 23**). Image (*ii*) maps the adequacy of the modulation contrast in the image for reconstruction (*see* **Note 24**). Color code for structures reconstruction should read as inadequate, adequate, good, very good, and excellent for *purple, red, orange, yellow*, and *white*, respectively. Image (*iii*) maps the reliability of the structures in the reconstructed data using the same color code of ii (*see* **Note 25**). Image (*iv*) represents the lateral Fourier transform of the full reconstructed 3D-SIM dataset to assess the resolution of the reconstructed data as well as potential reconstruction artifacts. *Scale bar* is 5 μm. (**b–d)** Experimental controls to assess the absence of artifacts in 3D-SIM nucleoids images of *B. subtilis*. (**a**) Maximum intensity projections of spatial frequencies decomposition of bacterial nucleoids (chromosomes stained with DAPI, *left panels*) and cytosolic GFP (*right panels*, anucleated cells, *see* Subheading 3.5) imaged in 3D-SIM using similar imaging conditions. High spatial frequencies corresponding to high DNA density regions should solely be observed in fluorescently labeled nucleoids images and be absent in cytosolic GFP images. (**c, d**) Comparison of 3D-SIM maximum intensity projections of chromosomes imaged in two colors (panel **c**, DAPI and Hbs-GFP labeling) or in time lapse (panel **d**, DAPI labeling). Overall nucleoid morphology and high DNA density regions should be conserved between images and only small dynamical changes between acquisitions should be apparent. Color codes represent DAPI/GFP fluorescence intensity increasing from *blue* to *red*. *Scale bar* is 1 μm

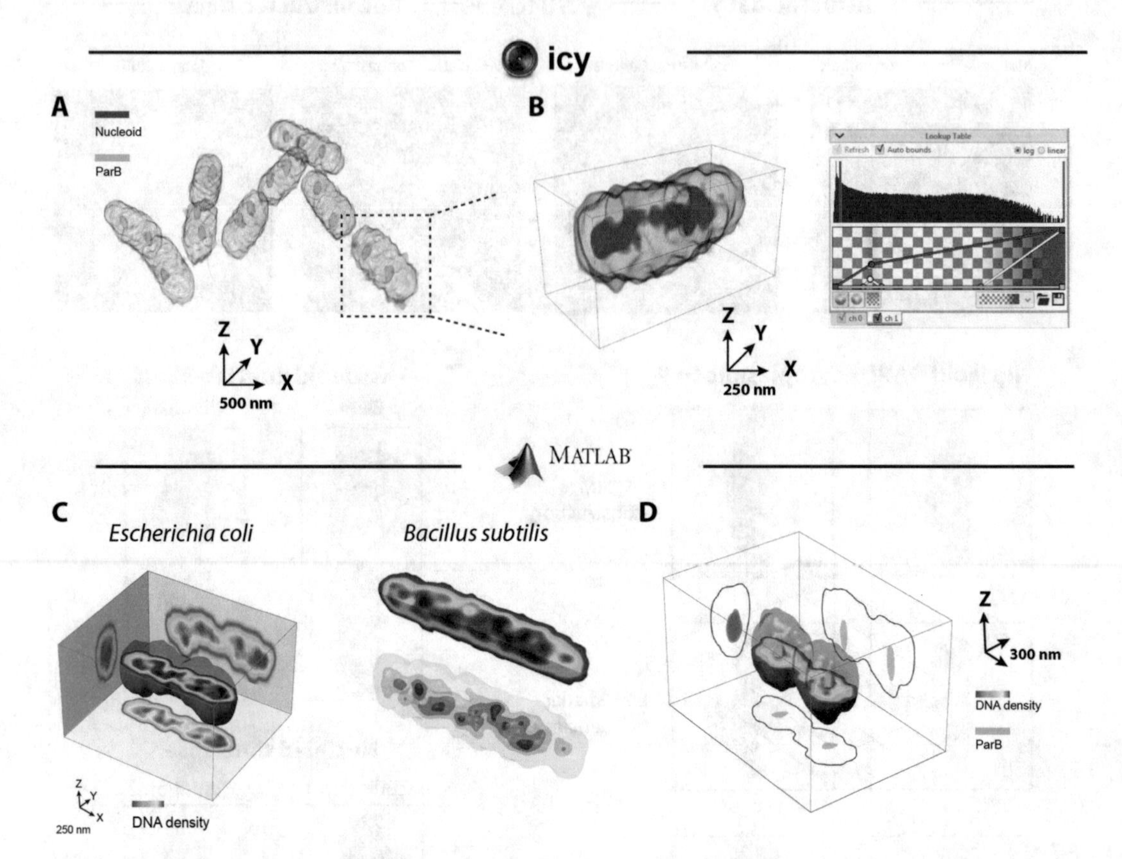

Fig. 5 3D-SIM data visualization. (**a, b**) 3D-SIM visualization with Icy. (**a**) Multicolor 3D-SIM images of *E. coli* HU-mCherry labeled nucleoids (red) and plasmidic ParB-mVenus (*green*). *Dotted square* highlights the magnified region shown in (**b**) side by side with Icy look-up table showing the color adjustments used to show ParB foci reside at high DNA density regions within the nucleoid **c, d**. 3D-SIM visualization with Matlab. (**c**) *Left*: 3D volume of an *E. coli* nucleoid stained with HU-mCherry. Nucleoid DNA density in the volumetric representation and side projections is color coded from *blue* to *red*. High-density chromosomal regions (HDRs) stretch from pole to pole and are joined by a semicontinuous filamentous density. *Right top panel*: 3D-cut of a DAPI-stained *B. subtilis* nucleoid (*top*) allows to see the nucleoid interior. *Right lower panel*: same nucleoid from top represented using Iso-levels (*Light-gray* to *red color-scale* represents low to high DNA densities, respectively). (**d**) 3D volume of an *E.coli* nucleoid stained with HU-mCherry (*solid red*) and ParB-mVenus foci (*green*) with orthogonal 2D projections of nucleoid (*red contour*) and ParB (*green spot*) densities. Figures from panels **a**, **c**, and **d** were reproduced with permission from [4, 12]

(Fig. 5). We here give a step-by-step guide to represent your reconstructed nucleoids in 3D using Icy:

1. Open your reconstructed and aligned 3D-SIM data in Icy.
2. Define a Region of Interest (ROI) containing your selected nucleoids of choice (*see* **Note 26**).
3. Rescale your ROI by doubling its dimensions (number of rows, columns and slices) in order to attenuate aliasing effects of the voxel size.

4. In the drop-down menu of your active sequence, select 3D representation
5. If your computer is equipped with a compatible graphic card, tick the OpenGl option on the right-side panel of the software.
6. Still on the right-side panel of the software, select the fluorescent channel corresponding to your nucleoid data and pick the colormap of your choice.
7. Adjust the minimum and maximum intensity values for each color of the selected colormap to display your features of interest (Fig. 5, panels a, b)

4 Notes

1. Fluorescent beads may have a tendency to adhere to bacteria cells. To circumvent this, beads can be stuck to the microscope coverslip prior to the agar pad assembly (*see* Subheading 3.2).
2. Beware not to deform microscope coverslips as you pass them under the flame. Deformed coverslips introduce non-even flatness of the sample as well as optical aberrations significantly degrading the quality of the data.
3. The plasma cleaning step can be substituted by incubating the microscope glass slides and coverslips in a KOH bath at 1 M, under sonication for 15 min, followed by repeated washes with water.
4. More than one sample can be extruded on the same slide. By adjusting the channels widths and the amount of agarose, up to three samples can be prepared in the same coverslip without cross-contamination.
5. In case the agarose pad is in contact with the adhesive tape, extrude carefully furrows using a clean razor blade prior to depositing bacteria to prevent bacterial suspension to spread on the adhesive tape.
6. Adding beads to the coverslip surface prior to final pad mounting provides a means of ensuring final multichannel 3D-SIM alignment and reconstruction process described in Subheading 3.4. This also provides the means for assessing the final resolution of the reconstructed data.
7. Use the "mosaic" acquisition option of the OMX acquisition software to find a dense region of bacteria faster.
8. After assigning the fluorescent filter set to the camera, the imaging order for each fluorescent channel can be selected when employing sequential acquisition mode. The most

energetic excitation wavelength should then be the last to avoid excitation cross talks and limit phototoxicity of the sample during acquisition.

9. GE Healthcare OMX platform offers a fixed grid spacing. Note however that other commercially available SIM platforms such as Elyra (Zeiss) offer the possibility of adjusting this spacing depending on the fluorescent channel.
10. Due to inherent microscope chromatic aberrations the optimum focus will be a compromise between each fluorescent channel.
11. Typical acquisition parameters used are: DAPI, 5–10 ms exposure with 10% transmission of the 405 nm excitation line; ParB-mVenus, 10 ms exposure with 31.3% transmission of the 488 nm excitation line and HU-mCherry, 10 ms exposure with 31.3% transmission of the 568 nm excitation line.
12. Imaging bacterial chromosomes stained with DAPI requires careful controls of imaging conditions to avoid photodamage. In our hands, extensive illumination with UV light on DAPI-stained *B. subtilis* nucleoids leads to artifactual expansion followed by shrinking of the nucleoid volume.
13. Three different angles (−60°, 0°, and +60°) as well as five phase steps are used to reconstruct each 3D-SIM image plane. Thus, imaging 17 sections of your sample requires a total of 255 images.
14. Other commercially available SIM platforms may offer the possibility to adjust the Z step size depending on the fluorescence wavelength to fulfill the Nyquist sampling criterion.
15. Fluorescence intensity variations between frames may arise from different sources with the most commons being photobleaching (if possible keep photobleaching under <~30%) and intensity differences between illumination angles (*see* Fig. 2, panel b).
16. Open-source SIM packages can be found elsewhere [14–16].
17. Wiener filtering settings can be adjusted to attenuate reconstruction artifacts. It permits a smoother final image reconstruction but at the cost of a lower resolution.
18. SIMcheck utilities permit to reconstruct pseudo widefield volumes from raw 3D-SIM data by averaging images of all phases and angles from the same imaging planes.
19. Artifacts in reconstructed images may arise from different sources that include sample photobleaching, refractive index mismatch between immersion oil, coverslip or immersion media, or insufficient modulation contrast.

20. Photobleaching can be minimized by changing imaging conditions. This includes lowering the number of Z slices or balancing excitation intensities with camera integration times.
21. Spherical aberrations can be caused by refractive index mismatch. They can be attenuated by optimizing the immersion oil refractive index to reduce the haloing (asymmetric axial shape) of the point spread function of fluorescent beads. In our hands, oils with refractive index of ~1.510 give the best results when imaging nucleoids stained with DAPI or HU-GFP; however, bear in mind that this optimization is wavelength dependent.
22. Poor modulation contrast may be caused by low signal-to-noise ratios, diffusing samples or high background masking the excitation modulation.
23. SimCheck Motion and Illumination Variation checking: All images corresponding to the same illumination angle are averaged and intensity-normalized. The three resulting images are then assigned to distinct false color (Cyan, Magenta and Yellow) and merged into a single output CMY-merged image. Any colorization of the output image indicates potential motions of the sample or uneven illumination during the acquisition. Sample motion can be minimized by reducing the number of optical sections of the sample. Also, ensure that the acquisition mode is set to "all Z then channels" and not "all channels then Z."
24. SimCheck Modulation Contrast (MCN) checking depicts the ratio of the modulation amplitude to noise amplitude in the acquired data.
25. SimCheck Modulation Contrast Map displays the ratio of the modulation amplitude to noise amplitude in the acquired data multiplied with the intensity for each pixel in the reconstructed data, normalized to the maximum intensity in the image.
26. The size of the ROI will depend on the configuration of your computer. We recommend 64-bit operating systems equipped with multicore processors and 8 GB of RAM.

Acknowledgments

This research was supported by funding from the European Research Council under the 7th Framework Program (FP7/2010-2015, ERC grant agreement 260787) and from the Agence Nationale pour la Recherche for projects HiResBacs (ANR-15-CE11-0023) and IBM (ANR-14-CE19-0025-02).3D-SIM experiments were performed at Montpellier RIO imaging. We acknowledge support from France-BioImaging (FBI, ANR-10-INSB-04).

References

1. Postow L, Hardy CD, Arsuaga J, Cozzarelli NR (2004) Topological domain structure of the *Escherichia coli* chromosome. Genes Dev 18:1766–1779
2. Deng S, Stein RA, Higgins NP (2005) Organization of supercoil domains and their reorganization by transcription. Mol Microbiol 57:1511–1521
3. Le TBK, Imakaev MV, Mirny LA, Laub MT (2013) High-resolution mapping of the spatial organization of a bacterial chromosome. Science 342:731–734
4. Marbouty M et al (2015) Condensin- and replication-mediated bacterial chromosome folding and origin condensation revealed by hi-C and super-resolution imaging. Mol Cell 59:588–602
5. Wang X et al (2015) Condensin promotes the juxtaposition of DNA flanking its loading site in Bacillus subtilis. Genes Dev 29:1661–1675
6. Espéli O, Boccard F (2006) Organization of the *Escherichia coli* chromosome into macrodomains and its possible functional implications. J Struct Biol 156:304–310
7. Betzig E et al (2006) Imaging intracellular fluorescent proteins at nanometer resolution. Science 313:1642–1645
8. Hess ST, Girirajan TPK, Mason MD (2006) Ultra-high resolution imaging by fluorescence photoactivation localization microscopy. Biophys J 91:4258–4272
9. Rust MJ, Bates M, Zhuang X (2006) Sub-diffraction-limit imaging by stochastic optical reconstruction microscopy (STORM). Nat Methods 3:793–795
10. Heilemann M et al (2008) Subdiffraction-resolution fluorescence imaging with conventional fluorescent probes. Angew Chem Int Ed 47:6172–6176
11. Gustafsson MGL et al (2008) Three-dimensional resolution doubling in wide-field fluorescence microscopy by structured illumination. Biophys J 94:4957–4970
12. Le Gall A et al (2016) Bacterial partition complexes segregate within the volume of the nucleoid. Nat Commun 7:12107. doi:10.1038/ncomms12107
13. Ball G et al (2015) SIMcheck: a toolbox for successful super-resolution structured illumination microscopy. Sci Rep 5:15915
14. Lal A, Shan C, Xi P (2016) Structured illumination microscopy image reconstruction algorithm. IEEE J Sel Top Quantum Electron PP:1–1
15. P. Křížek, T. Lukeš, M. Ovesný, K. Fliegel, G. M. Hagen (2015). SIMToolbox: a MATLAB toolbox for structured illumination fluorescence microscopy. Bioinformatics 32:318–320
16. Müller M, Mönkemöller V, Hennig S, Hübner W, Huser T (2016) Open-source image reconstruction of super-resolution structured illumination microscopy data in ImageJ. Nat Commun 7:10980

Chapter 20

Sequential Super-Resolution Imaging of Bacterial Regulatory Proteins, the Nucleoid and the Cell Membrane in Single, Fixed *E. coli* Cells

Christoph Spahn, Mathilda Glaesmann, Yunfeng Gao, Yong Hwee Foo, Marko Lampe, Linda J. Kenney, and Mike Heilemann

Abstract

Despite their small size and the lack of compartmentalization, bacteria exhibit a striking degree of cellular organization, both in time and space. During the last decade, a group of new microscopy techniques emerged, termed super-resolution microscopy or nanoscopy, which facilitate visualizing the organization of proteins in bacteria at the nanoscale. Single-molecule localization microscopy (SMLM) is especially well suited to reveal a wide range of new information regarding protein organization, interaction, and dynamics in single bacterial cells. Recent developments in click chemistry facilitate the visualization of bacterial chromatin with a resolution of ~20 nm, providing valuable information about the ultrastructure of bacterial nucleoids, especially at short generation times. In this chapter, we describe a simple-to-realize protocol that allows determining precise structural information of bacterial nucleoids in fixed cells, using direct stochastic optical reconstruction microscopy (*d*STORM). In combination with quantitative photoactivated localization microscopy (PALM), the spatial relationship of proteins with the bacterial chromosome can be studied. The position of a protein of interest with respect to the nucleoids and the cell cylinder can be visualized by super-resolving the membrane using point accumulation for imaging in nanoscale topography (PAINT). The combination of the different SMLM techniques in a sequential workflow maximizes the information that can be extracted from single cells, while maintaining optimal imaging conditions for each technique.

Key words Super-resolution microscopy, Single-molecule imaging, Bacterial nucleoid, Protein quantification, Bacterial regulatory proteins

1 Introduction

In contrast to earlier assumptions, bacteria do not represent unorganized entities, whose existence is merely based on entropy and randomly occurring events. In fact, a great deal of organization is required to orchestrate processes such as transcription,

The original version of this chapter was revised. An erratum to this chapter can be found at DOI 10.1007/978-1-4939-7098-8_25

Olivier Espéli (ed.), *The Bacterial Nucleoid: Methods and Protocols*, Methods in Molecular Biology, vol. 1624,
DOI 10.1007/978-1-4939-7098-8_20,

translation, replication, signal transduction, and many more in such a small volume of only a few femtoliters. Adaption to unfavorable conditions is achieved within minutes, facilitating survival even in hostile environments [1]. The investigation of these processes remains challenging, because the temporal resolution of many commonly used techniques is too low to catch the fast dynamics of cellular processes in bacteria, especially at the single cell level [2]. In order to explore bacterial homeostasis, fluorescence microscopy became a widely used tool to visualize various cellular processes. However, conventional light microscopy is limited by the diffraction of light which restricts the achievable lateral resolution to 200–300 nm. Super-resolution microscopy circumvents this limitation either by structural remodeling of the fluorescence signal or its separation in time [2]. The latter is used by single molecule localization microscopy (SMLM) techniques which rely on the detection of separated signals emitted from single molecules and their precise localization with accuracies of typically 10–20 nm (*see* Fig. 1) [3, 4].

Photo-modulatable fluorescent proteins can be genetically fused to the protein of interest, and together with sparse photoactivation allow for quantitative super-resolution imaging using photoactivated localization microscopy (PALM, Fig. 1B) [5]. This procedure is not limited to fixed cells and also allows tracking single proteins over time in living cells (single particle tracking PALM, sptPALM), delivering valuable information both about the spatial distribution and diffusive behavior [6, 7]. A useful extension is to use lipophilic dyes that dynamically bind to cell membranes which allow visualizing the bacterial morphology via point accumulation for imaging in nanoscale topography (PAINT, Fig. 1C) [8, 9]. Since the localization accuracy in SMLM is dependent on the number of photons of single emitter signals (*see* Subheading 3.6.2) [10], bright organic dyes are used in order to further improve the achievable resolution. These dyes can be transferred into a long-living dark state and stochastic reactivation of a subset of molecules is used in stochastic optical reconstruction microscopy (*d*STORM, Fig. 1D) [11]. Recent studies introduced functionalized nucleosides, which can be specifically labeled with organic dyes using click chemistry [12–15]. Their uptake and incorporation into nascent DNA during replication therefore offers the possibility to label the bacterial nucleoid and to generate highly resolved snapshots of the chromosomal ultrastructure in chemically fixed cells (Fig. 1E) [16]. Fixation compensates for the poor temporal resolution provided by SMLM, which is typically in the range of minutes and which would otherwise blur the obtained super-resolution image due to the fast dynamics occurring in living cells.

In this chapter, we describe a sequential imaging workflow which allows (1) extracting spatial and quantitative information of

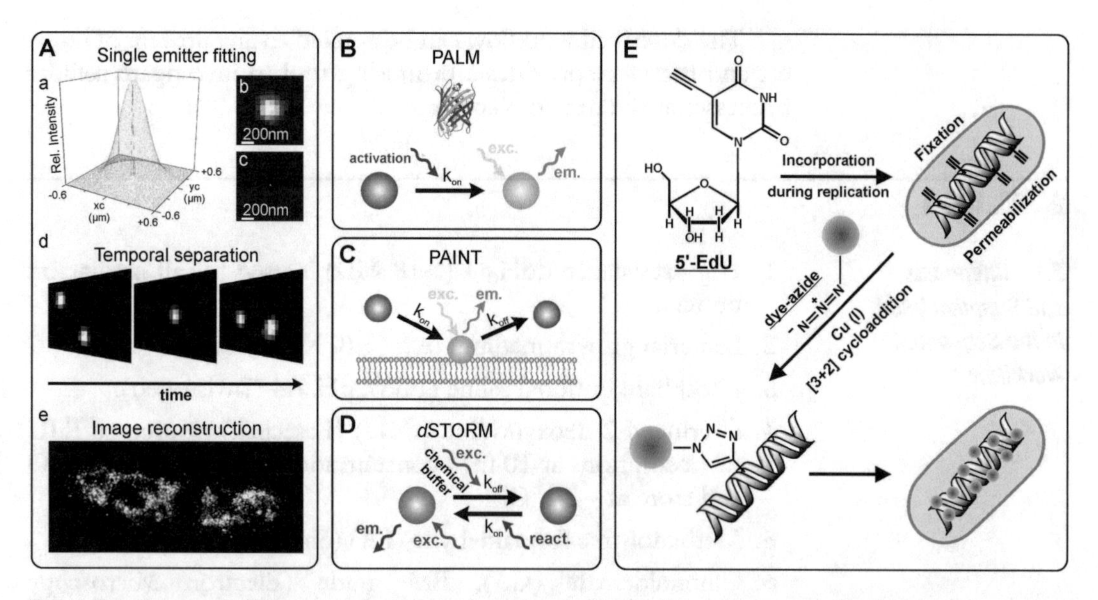

Fig. 1 Principles and different modes of single molecule localization microscopy. (**A**) A single point spread function (*a*, *b*) can be precisely localized by applying a Gaussian fit function to determine the signal centroid with a precision of typically 10–20 nm (*c*). Temporal separation in time (*d*) and subsequent fitting of single emitters facilitates the reconstruction of an artificial, super-resolved image of the labeled structure (*e*, shown: bacterial nucleoid). (**B**) PALM imaging of genetically attached photo-modulatable fluorescent proteins is achieved by iteratively photo-converting or -activating small subsets of fluorescent proteins over time. In the illustrated case, an initially nonfluorescent protein (e.g., PAmCherry1) is converted into its excitable form using UV irradiation. Upon excitation with the readout laser, the fluorophore bleaches irreversibly facilitating quantitative PALM measurements (*see* Subheading 3.6.3). (**C**) Membranes can be highly resolved using PAINT by adding appropriate concentrations of lipophilic dyes to the imaging buffer. Upon binding, the quantum yield of the dye is enhanced and the emitted signal can be analyzed according to (**A**). Optimal concentrations are given by the ratio of k_{on}/k_{off}, which can also be regulated by the intensity of the excitation light source. (**D**) In *d*STORM imaging, the initially fluorescent organic dyes are transferred into a stable, nonfluorescent dark state using reducing buffer conditions and laser excitation. Single fluorophores stochastically return into the excitable state or can be actively reactivated by laser light with shorter wavelengths (*purple arrow*). (**E**) Labeling scheme for specific labeling of bacterial DNA using click-chemistry. The thymidine analog 5′-ethynyl-2-deoxyuridne (EdU) is supplied to the growth medium and incorporates into nascent DNA during replication. After fixation and permeabilization, azide-functionalized organic dyes can be covalently attached to the alkyne group of EdU using a copper-catalyzed cycloaddition

proteins using PALM, and (2) correlating this information to the ultrastructure of the bacterial nucleoid visualized by *d*STORM imaging of click-labeled DNA [17, 18]. The exact positioning of proteins and nucleoids within the same single cell is facilitated by PAINT imaging of the bacterial membrane.

The positions of the imaged bacteria are saved relative to prominent landmarks within the sample (e.g., corners of the chamber). After removal of the sample buffer, these positions can be easily repositioned using a motorized stage. This way, each SMLM technique can be performed at optimal imaging conditions and prevents possible interferences that would occur during parallel imaging.

The described workflow can be applied to any protein of interest and therefore provides a promising tool to investigate cellular processes at the near-molecular level.

2 Material

2.1 Chemicals and Supplies Used in the Sequential Workflow

1. High resistance ddH_2O (>18 MΩ) is used for all media and buffers.
2. Bacterial growth medium (e.g., LB, M9 minimal).
3. Phosphate buffered saline (PBS), pH 7.4 (Invitrogen).
4. 5′-ethynyl-2-deoxyuridine (EdU) (Baseclick) ➔ prepare EdU stock solutions at 10 mM concentration in DMSO or ddH_2O and store at −20 °C.
5. Methanol-free formaldehyde (FA) (Sigma-Aldrich).
6. Glutaraldehyde (GA), EM grade (Electron Microscopy Sciences).
7. 200 mM sodium phosphate buffer, pH 7.5 (NaH_2PO_4 and Na_2HPO_4, both Sigma-Aldrich).
8. NH_4Cl (Sigma-Aldrich).
9. $NaBH_4$ (Sigma-Aldrich).
10. 8-well-chamber slides (Sarstedt).
11. KOH (Sigma-Aldrich).
12. Poly-L-lysine (Sigma-Aldrich).
13. Extracellular matrix (ECM) gel from Engelbreth-Holm-Swarm murine sarcoma (Sigma-Aldrich) (optional).
14. Nile Red or Rhodamine 6G (Sigma-Aldrich).
15. Triton-X100 (Sigma-Aldrich).
16. Tris (Sigma-Aldrich) ➔ prepare 1 M stock solution pH 8.0.
17. $CuSO_4$ (Sigma-Aldrich) ➔ prepare 100 mM stock solution in ddH_2O.
18. Alexa 647azide (Thermo Scientific) ➔ prepare azide stock solution at 5 mM concentration in anhydrous DMSO and store at −20 °C in the dark.
19. Ascorbic acid (Sigma-Aldrich).
20. β-mercaptoethylamine (MEA) (Sigma-Aldrich).
21. Sodium chloride (Sigma-Aldrich).
22. Immersion oil (Merck, catalog number 115577).
23. Mowiol 4-88 (Sigma-Aldrich).
24. TetraSpeck Beads, 0.1 μm (Thermo Fisher Scientific, catalog number T7279).

2.2 Bacterial Strains

For our studies, we used *Escherichia coli* K12 MG1655, which is engineered to express the protein of interest from its native promoter fused to PAmCherry1 on the chromosome [18, 19]. However, our approach is not limited to chromosomal expression and can be performed on plasmid-based expression systems.

2.3 Experimental Setup

This section provides an overview about the essential microscope components and required devices used in our sequential SMLM studies.

1. Our custom-built setup for single-molecule detection is based on an inverted microscope body (Nikon Eclipse Ti), mounting a high NA TIRF objective (100× Apo TIRF oil immersion objective, NA 1.49). The microscope body includes a TI-S-ER motorized stage, which is essential for easy repositioning of the imaged ROIs during the sequential workflow (*see* Subheading 3.5.1). The illumination mode can be changed from widefield to TIRF illumination by manual adjustment of an external TIRF mirror.
2. As a laser source, we use a Coherent Innova 70C Spectrum argon/krypton mixed gas laser, which emits multiple laser lines. Alternatively, single-line light sources such as diode lasers can be used instead of the multiline laser, which typically provide intensities in the range of 100–500 mW for a specific wavelength. The important parameter, however, is the irradiation density at the sample, which depends on the laser beam expansion and optical components. We use irradiation densities of 1–1.5 kW/cm^2 for PALM and PAINT and ≥2 kW/cm^2 for *d*STORM imaging. Inappropriate laser densities might lead to insufficient signal separation and therefore imaging artifacts may occur, especially in densely labeled samples [20, 21]. The desired wavelength is selected by an acousto-optical tunable filter (AOTF; AA Opto Electronic). A Coherent Cube UV diode laser (404 nm/100 mW output power) is additionally coupled into the microscope. UV laser intensity can be manually adjusted using an optical density wheel.
3. Appropriate filter sets are used for each channel. The 568 nm and 647 nm excitation laser lines pass a multiband cleanup filter (AHF, catalog number F69-647). Excitation and fluorescence emission lights are separated using a multiband dichroic mirror (AHF, catalog number F73-888). Fluorescence emissions of PAmCherry1/Nile Red and Alexa Fluor 647 are filtered using a 610/60 (AHF, catalog number F47-616) and 700/75 (AHF, catalog number F47-700) bandpass filter, respectively.
4. The fluorescence signal is detected on an Andor Ixon Ultra EMCCD chip (DU-897U-CS0-#BV; Andor), using a preamplifier gain of three, an EM-gain of 200 and active frame transfer for PALM, PAINT, and *d*STORM measurements.

2.4 Software

1. μManager [22] is used for camera, motorized stage, and UV laser control.
2. Data analysis is performed using rapi*d*STORM v3.31 [23].
3. The super-resolved images are processed and registered using the open-source image processing package Fiji [24].
4. Determination of the experimental localization precision of SMLM techniques and quantitative analysis of PALM data is carried out using LAMA [25].

3 Methods

3.1 Cell Culture

Inoculate bacterial cultures from glycerol stocks onto a plate and then place a single colony into the desired growth medium supplemented with appropriate antibiotics and incubate overnight at 32 °C with 200 rpm shaking speed for optimal aeration. Incubation at 37 °C might introduce artifacts arising from improper folding of the fusion protein. Suitable controls (growth curves, optical inspection of cell morphology) should be performed. Inoculate the working culture to a starting OD_{600} of ~0.02–0.05. Determine the OD_{600} every 30 min to obtain the doubling time and to check for culture viability. EdU is added to an end-concentration of 10 μM at OD_{600} ~0.25 for 40 min when working in LB-medium, which is also the time required for chromosome replication (C-period) [26]. In general, the EdU exposure time depends on the culture mass doubling time in the respective medium (*see* **Note 1**). In minimal medium, the EdU concentration might be changed to prevent cytotoxic effects (*see* **Note 2**).

3.2 Slide Preparation

We typically immobilize the cells on 8-well chamber slides. Initially, clean the glass surface from dust and other particles using 3 M KOH for 30 min. Remove the solution and wash the chambers once with ddH_2O, once with PBS for 5 min, again twice with ddH_2O and let the surface dry. Add 0.01% poly-L-lysine solution for 10 min to coat the glass surface. Remove the poly-L-lysine solution, wash the chambers twice with ddH_2O and let them dry completely for 2 h in a dust-free environment.

3.3 Fixation and Immobilization

After the indicated time of EdU exposure, cells are fixed in solution using crosslinking reagents such as formaldehyde (FA) or glutaraldehyde (GA). We usually use 1–2% methanol-free FA (final concentration) for quantitative PALM measurements. Prepare the fixation mixture containing 300 μl of 200 mM sodium phosphate ($NaPH_4$) buffer and 225 μl 8%/16% FA (for 1% and 2% final concentration, respectively). Add 2 × 638 μl bacterial suspension to the fixation mix and immediately invert the reaction tube several times. Incubate 30 min at room temperature. Afterward, pellet the cells 3 min at 5000–6000 × *g*. Wash the cells once with buffer containing 50 mM NH_4Cl in order to quench excess formaldehyde. Resuspend the pellet in ~200 μl buffer and add the solution onto prepared 8-well chamber slides for 10 min. Remove the cells and wash the chambers two times thoroughly. *Optional*: Add fiducial markers as drift indicators. We usually use a mixture of 60 and 80 nm colloid gold particles (e.g., Sigma-Aldrich). These particles are visible in the PAmCherry1/Nile Red channel and some particles are also visible in the AF647 channel. When FA is used for fixation, all buffers should be adjusted to the osmolality of the growth medium, in order to prevent changes in cell morphology during the sequential workflow (*see* **Note 3**).

3.4 Optional: Embedding in a Hydrophilic Gel Matrix

To prevent cells from moving and detaching from the glass surface, a hydrophilic gel matrix may be applied to the sample [27]. This is not necessary, but is advised if the cells are used over several days. Therefore, thaw ECM gel from Engelbreth-Holm-Swarm murine sarcoma on ice. Dilute the ECM gel with ice-cold ddH2O using prechilled pipet tips to a final protein concentration of 3.5–4.5 mg ml^{-1} (*see* **Note 4**). Add 30 μl of diluted gel to each chamber of the 8-well chamber slide and incubate for 5 min on ice to provide proper sample coverage. Incubate at 37 °C for 30–40 min in a humid environment to mediate gel polymerization. Note: the gel should not dry completely! The gel will liquefy again at low temperatures, but can be post-fixed using 1% FA in sodium phosphate buffer or PBS for 15 min. Quench excess formaldehyde using buffer supplemented with 50 mM NH_4Cl.

3.5 Sequential Imaging

The following section describes the sequential workflow presented in Fig. 2. In order to ensure optimal imaging conditions for each SMLM technique, we recommend following the presented order (*see* **Note 5**).

3.5.1 PALM and PAINT Imaging

1. Start the microscope setup and let it stabilize. Cool the camera to −70 °C. Set the preamplifier gain to three and activate frame transfer. Position the sample on the microscope. The chamber slide should be placed tight against the border of the sample holder. This prevents rotation and is important for reliable repositioning of the cells at later time points. Set the origin to ($X = 0$ and $Y = 0$) at a prominent landmark within the sample. We usually take one corner of the chamber. Search for regions of interest and save the positions relative to the origin. Also save the position and dimensions of the ROIs by adding them to the ROI-manager in μManager.
2. Change the position to cells of no interest and adjust the focus and laser beam angle. We usually measure in HILO (highly inclined and laminated optical sheet) mode (*see* **Note 6**).

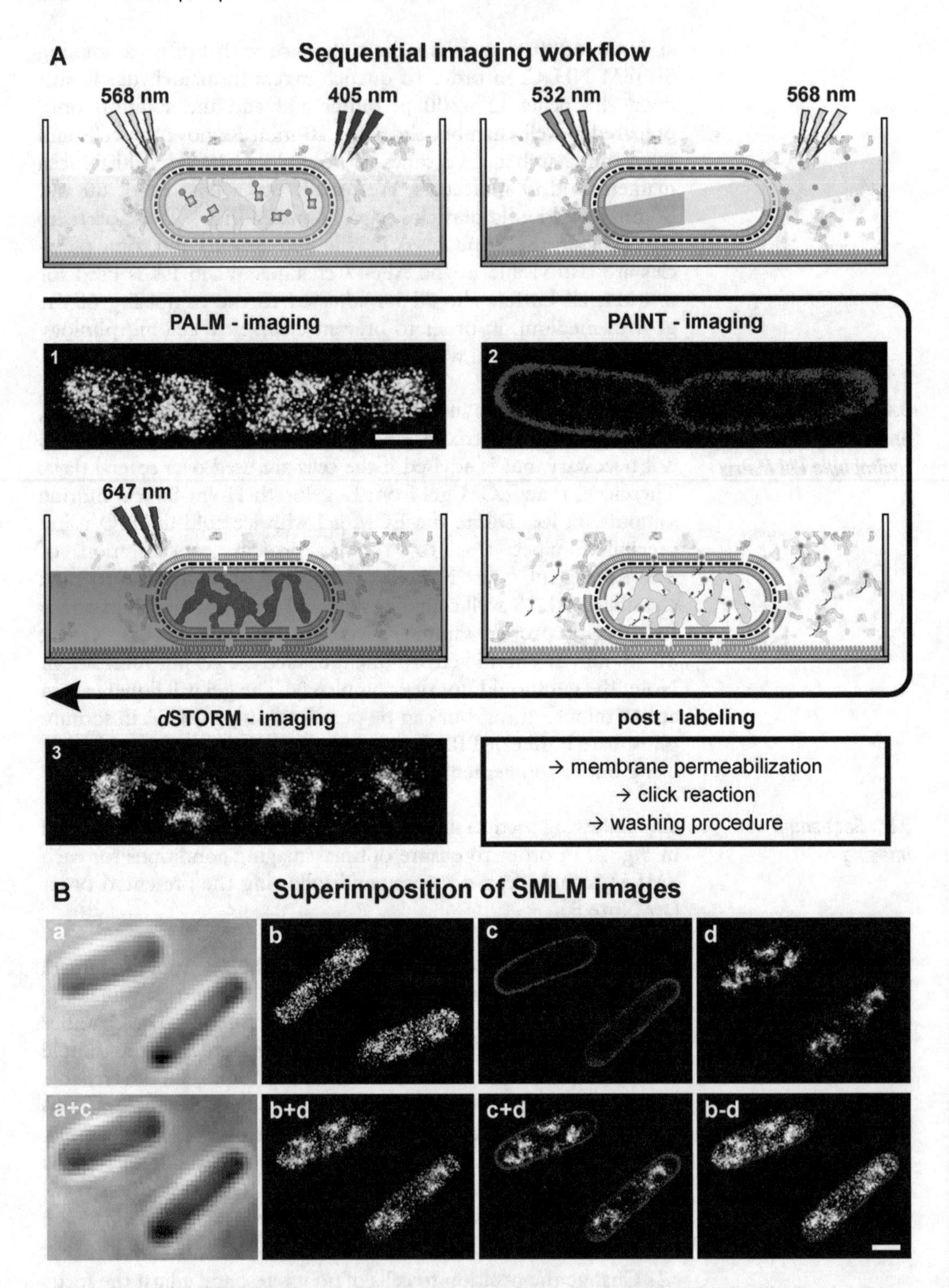

Fig. 2 Sequential SMLM imaging of different molecule classes in single *E. coli* cells. (**A**) Fixed cells are immobilized on poly-L-lysine coated chamber slide systems. The sample can be coated with a hydrophilic gel matrix in order to prevent movement or detaching of bacteria during the imaging workflow (*see* Subheading 3.4).

Set the EM-gain to 200 for higher sensitivity in PALM, PAINT and *d*STORM measurements.

3. Return to the region of interest and run acquisition of the PALM measurements. Start illuminating the sample with 568 nm excitation laser light while the acquisition is already running to also detect preactivated PAmCherry1 molecules. We use laser densities of ~1.3 kW/cm^2 and frame rates of 10–15 Hz for PALM measurements. Single signals should be mostly present in only a single frame to ensure a good signal to noise ratio.
4. Add UV activation light in the regime of a few W/cm^2 and adjust the intensity during acquisition to maintain optimal fluorophore density (typically 1–2 molecules per cell in each camera frame). The optimal number of recorded camera frames depends on the copy number of the protein of interest. For RpoC-PAmCherry1, which is present at ~3600 molecules per cell in LB, up to 10,000 frames are required to read out all the molecules.
5. After imaging all regions of interest, remove the sample from the microscope and apply PBS containing 200–400 pM Nile Red for PAINT imaging of the bacterial membrane. Place the sample back on the microscope and search for the origin which is set in the PALM measurement. Again, set the origin to $X = 0$, $Y = 0$. The relative ROI positions saved during PALM imaging should facilitate easy repositioning. Apply the ROI, which is saved in the ROI-manager and position the cells similarly as in the PALM measurement.
6. Illuminate the sample with low 568 nm laser intensity until the outlines of the membrane-labeled bacteria are visible. Adjust the focal plane and laser beam angle. The correct focal plane is indicated by intense cell envelopes, and a less intense cytosol of the bacteria. Change to a frame rate of 33–50 Hz, depending on the laser densities used for PAINT imaging. For cells grown in LB, well separated signals can be observed using a laser

Fig. 2 (continued) First, the fusion protein RpoC-PAmCherry1 is imaged in PBS buffer using PALM (*1*). The sample buffer is then removed and lipophilic dyes (e.g., R6G, Nile Red) are added to the buffer at suitable concentrations (typically 200–400 pM) for PAINT imaging of the bacterial membrane (*2*). Afterward, the sample is permeabilized and incorporated EdU is labeled with Alexa Fluor 647 via click-chemistry. A thiol-containing reductive buffer system is added to the sample and the ultrastructure of the nucleoids is finally visualized using *d*STORM (*3*). (**B**) The brightfield image (*a*) and the super-resolved SMLM images of the fusion-protein (*b*), cell membrane (*c*) and nucleoids (*d*) can be superimposed in order to calibrate the brightfield-image (*a+c*), correlate the protein distribution to the nucleoid ultrastructure (*b+c*), investigate the positioning of the nucleoids within the cell envelope (*c+d*), or visualize all super-resolved channels at once (*b–d*). Scale bars are 1 μm. Adapted from [17],

density of ~1–1.5 kW/cm^2 at a frame rate of 33 Hz and a Nile Red concentration of 400 pM. Higher laser densities lead to faster photobleaching and allow higher frame rates, but the signal intensity might drop due to the faster photobleaching kinetics. Typically 6000 frames are enough to visualize the membrane as a continuous structure. After PAINT imaging of all the regions of interest, remove the sample from the microscope and proceed with post-labeling of the EdU containing DNA.

3.5.2 Labeling of Chromosomal DNA

1. Wash the chamber twice with PBS to remove Nile Red and permeabilize the cells for 30 min using 0.5% Triton X-100 in PBS. Meanwhile, prepare the click-reaction-buffer according to [28]. The buffer consists of 100 mM Tris pH 8.0, 1 mM $CuSO_4$, 5 μM AF647 azide, and 100 mM ascorbic acid. Prepare fresh ascorbic acid stock solution before usage. Add the components in the described order to ddH_2O. Adjust the osmolality to match the growth medium using NaCl in order to prevent shrinking of FA-fixed cells.
2. Wash the permeabilized cells twice with PBS and add the click-reaction buffer. We typically add 300 μl click reaction buffer to each chamber for 30 (uncoated cells) or 60 min (gel-coated cells). After incubation, rinse the sample three times in PBS to remove the majority of unbound azide molecules. Wash the cells 2–3 times for 5 min (uncoated cells) or up to 1 h (coated cells) and store the sample in PBS with adjusted osmolality.

3.5.3 dSTORM Imaging

1. Prepare the imaging buffer for *d*STORM measurements. Our imaging buffer consists of 100 mM Tris or 1×PBS, supplemented with 50–100 mM MEA. The choice of the imaging buffer depends on the osmolality of the growth medium. For cells grown in minimal medium or LB Lennox (containing 5 g NaCl), we use 100 mM Tris buffer containing 50 mM MEA and 10 mM NaCl. PBS containing 100 mM MEA (adjust pH value to 8.0 using KOH after addition of MEA) is used for cells grown in LB-Miller (containing 10 g NaCl). Due to the high label density, we do not recommend using an oxygen-scavenger system, since signal separation might not be possible due to the altered blinking statistics of AF647. Also, the required glucose also changes the osmolality of the imaging buffer, which might lead to changes in cell morphology.
2. Add the imaging buffer to the sample and place it on the microscope. Relocate the same regions of interest and adjust the focal plane and laser beam angle while illuminating the sample with low 647 nm laser densities. Switch the frame rate to 33–50 Hz and increase the laser density to ~2 kW/cm^2 in order to drive the majority of AF647 molecules into the dark

state. Higher laser densities can be applied to facilitate higher frame rates. Start the acquisition when the optimal fluorophore density is reached (~2 molecules per frame per cell). We typically record 10,000–15,000 camera frames, while maintaining fluorophore density using either 488 nm or 405 nm reactivation. This number of frames is usually sufficient to visualize the bacterial chromosome as a continuous structure (Nyquist–Shannon criterion).

3. Single SMLM channels are spatially shifted due to the imaging procedure and therefore have to be superimposed during the analysis procedure (Fig. 3A) (see Subheading 3.6.3).

3.5.4 STED Imaging of Click-Labeled *E. coli* Nucleoids

1. To demonstrate the versatility of the described click-labeling approach, click-labeled nucleoids of *E. coli* are imaged with STED super-resolution microscopy (Fig. 4E) [30]. Please note that well-performing dyes differ between SMLM and STED and need to be chosen carefully for each method. In this work, Atto565 is used for STED and AF647 for SMLM. STED microscopy has a lower nominal resolution (<40 nm) than SMLM (<20 nm), but we consider it to be a bona fide complementary approach: STED imaging on a well-aligned STED microscope can be considered less prone to imaging artifacts for novice super-resolution microscopists compared to SMLM. Whereas SMLM can miss parts of the structure or even create artifacts if nonoptimal imaging strategies are applied (highlighted in [21]), STED microscopy is more robust. If the STED imaging fails, e.g., due to a misaligned microscope (see below), the STED-image appears less resolved than the confocal image. Although STED can not necessarily reveal all details of a structure, the comparison between SMLM and STED can be employed to confirm the absence of the most common, non-reagent derived artifacts in super-resolution imaging.

2. STED images are acquired with an inverted TCS SP8 3× microscope (Leica Microsystems, Mannheim Germany) equipped with a 100×/1.4 NA oil objective (Leica HC PL APO CS2—STED White). Samples are fixed using 2% formaldehyde and labeled with Atto565 as described in Subheading 3.5.2, embedded in Mowiol and imaged in line sequential mode alternating between standard confocal imaging and STED super-resolution imaging. Stimulated emission depletion is conducted with a doughnut-shaped 2D-depletion beam using a 660 nm CW–laser (power ~400–600 mW). Excitation is performed with 574-nm laser light derived from a white light laser (Leica Microsystems, Mannheim Germany). The power of the 574-nm laser line is increased for the super-resolution image to compensate for the smaller PSF and

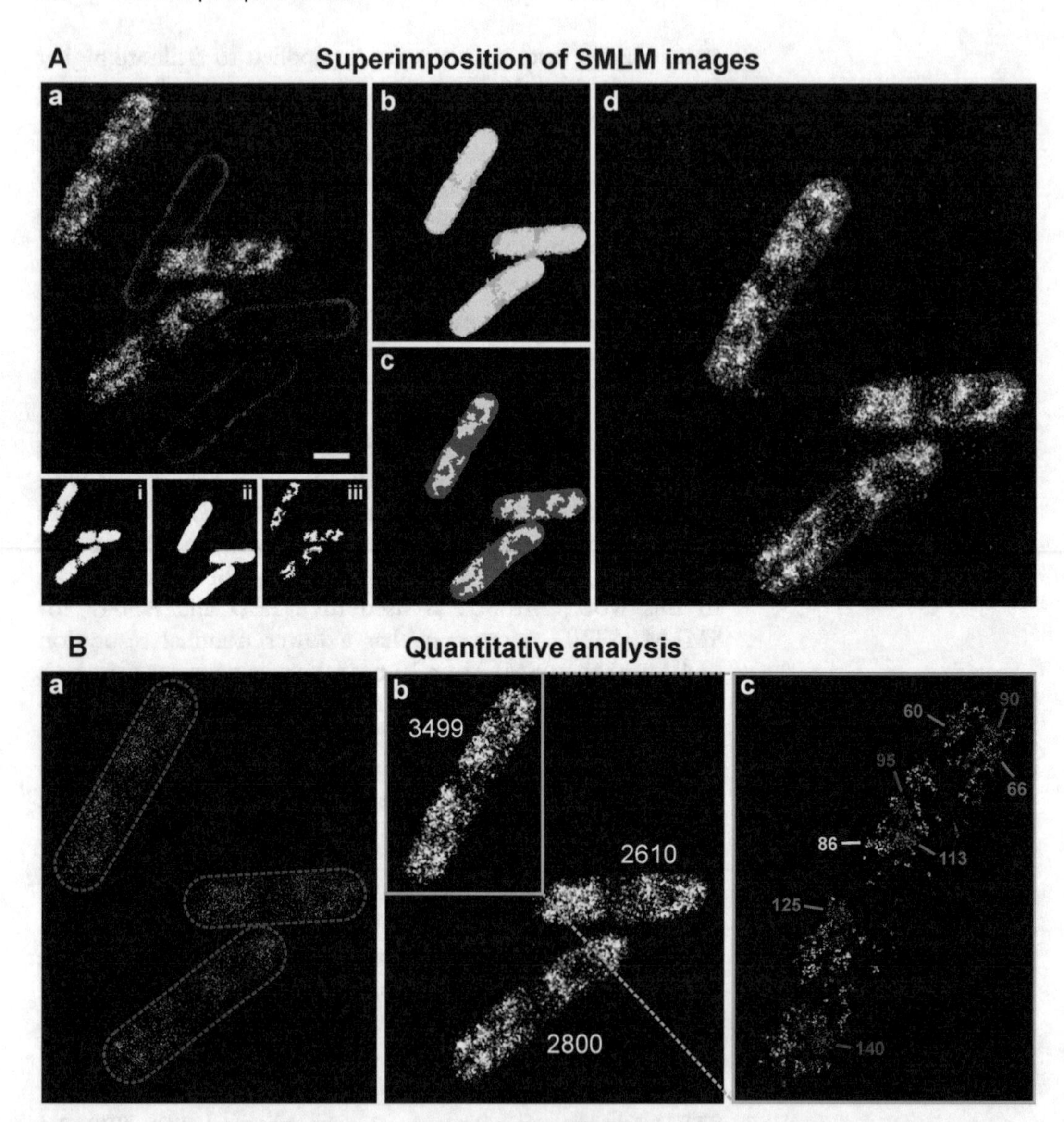

Fig. 3 Superimposition of different SMLM channels obtained using the sequential imaging workflow and quantitative analysis of PALM data. (**A**) Sample removal during the workflow ultimately leads to translationally shifted channels (*a*). The PAINT channel is false-colored in *red*, the PALM image of RNAP in *yellow* and chromosomal DNA in *cyan hot*. Masks of the respective channels are created using Fiji (*i–iii*) and are used for Fourier Transformation based registration (*see* Subheading 3.6.3). The maximum overlap between the PALM and membrane masks (*b*, *yellow and green*), as well as between the DNA and membrane masks (*c, yellow and red*) are calculated. The translation matrix is applied to the SMLM images, resulting in the registered triple color SMLM image (*d*). Adapted from [17], © IOP Publishing. Reproduced with permission. All rights reserved. (**B**) The quantitative nature of PALM imaging is conserved in the sequential workflow and allows the determination of protein copy-numbers per cell. Therefore, the SMLM image is rendered according to the number of localizations using LAMA [25] (*a*). The cell outlines extracted from the PAINT image (*red dashed lines*) can be used to measure the integrated intensity of the rendered image, which correspond to the number of localizations (*b*). PALM data can also be processed using a density-base clustering algorithm (DBSCAN) in order to identify subclusters [34] (*c*). Colored numbers indicate the number of localizations which are present in the respective subclusters. Scale bar in (**A**) is 1 μm

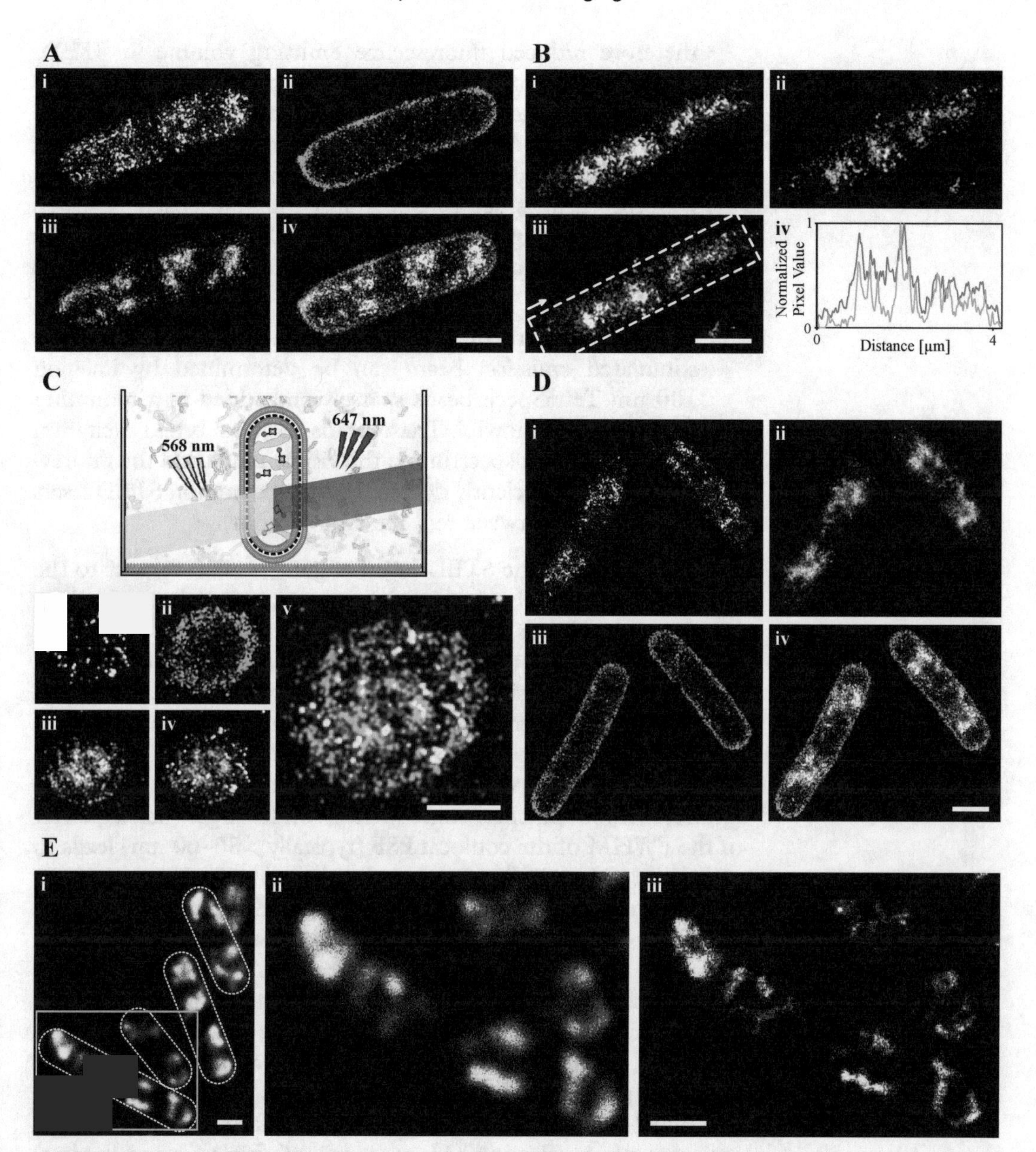

Fig. 4 Sequential SMLM imaging of bacteria grown in different media. (**A**) *E. coli* MG1655 derivative KF26, which chromosomally expresses RpoC-PAmCherry1 at its native locus, was grown in LB Miller at 37 °C. Overlaying the PALM (*i*), PAINT (*ii*) and *d*STORM images (*iii*) indicates that most RNAP molecules are transcribing active genes at the nucleoid surface (*iv*) [18]. (**B**) As expected, the nucleoid-associated protein H-NS (*i*) populates the same area as the nucleoid (*ii*). The intensity plot of the registered channels (*iii*) can be used to analyze the distribution of H-NS and DNA. Cells were grown in M9 minimal medium. (**C**) Distribution of the transcriptional regulator fusion protein OmpR-PAmCherry1 in acidic minimal medium (pH 5.6). Embedding in the hydrophilic gel matrix sometimes results in vertically fixed cells. This allows visualizing the cross-section of the bacterial cell cylinder. OmpR-molecules (*i*) are preferentially located close to the plasma membrane (*ii*). The nucleoid is a condensed ring-like structure in the center of the cross-section (*iii*). DNA filaments seem to link the nucleoid to the cell membrane and OmpR populates these linkages (*iv*). The overlay

therefore reduced fluorescence emitting volume in STED-mode by otherwise constant imaging parameters. The detectors (HyD, Leica Microsystems, Mannheim Germany) are gated with 2.5–6 ns with respect to the 574 nm line of the pulsed white light laser with 80 MHz repetition rate [31]. Detector gain is set to 300% and the scan speed is set to 600 Hz resulting in a dwell time of 600 ns. The pixel size is set to 19 nm, the pinhole to 0.8 AU and a 32× line average is applied.

3. The alignment of the excitation beam and the donut-shape stimulated emission beam can be determined by imaging 100 nm TetraSpeck beads sparsely embedded in a mounting medium, e.g., Mowiol. These beads display a broad excitation and fluorescence spectrum in the visible range and the fluorescence can be efficiently depleted with all common STED lasers (in the range between 592 nm and 775 nm).

The quality of the STED beam alignment with respect to the confocal excitation beam (or vice versa) can be determined by imaging single beads using standard confocal imaging and STED imaging in parallel, using a pixel size of 20 nm and creating an overlay. The smaller signal of the super-resolved beads should lay in the center of the same bead in the confocal image. In practice, a slight misalignment displayed by a decentered STED image (<20 nm) leads to a very mild decrease in fluorescence and can be ignored without practical consequences. Shifts larger than a quarter of the FWHM of the confocal PSF (typically >50–60 nm) leads to an increasing loss of fluorescence signal (and apparent resolution), which can in extreme cases result in a STED image considered worse in terms of resolution and contrast than the confocal image. A realignment of the laser beams is mandatory in these cases. If the beads deviate from a round shape (in x/y) or a z-stack reveals a severely tilted PSF or otherwise distorted PSF deviating from the typically "cigar"-like shape, a major realignment should be performed in order to avoid images with poor resolution, low signal-to-noise ratio or in rare cases, artifacts. Before starting a realignment, confirming (1) an image of single (monodisperse)

Fig. 4 (continued) of all channels is shown in (*v*). Adapted from [18] with permission from The Royal Society of Chemistry. (**D**) PAINT imaging of DNA using a novel fluorogenic dye JF646 (*i*) [29] and *d*STORM imaging of click-labeled nucleoids (*ii*) both facilitate super-resolution imaging of the bacterial nucleoid (*iv*). The membrane can still be imaged using PAINT after permeabilization of the cells (*iii*). (**E**) STED imaging of Atto565 labeled nucleoids. Confocal imaging of a region of interest (*i*) already indicates the highly structured nucleoids of cells grown in LB at 32 °C. (*ii*) Magnified view of the marked area (*green rectangle*) in (*i*). STED imaging greatly enhances the resolution, revealing the highly condensed filaments of the nucleoid (*iii*). Scale bars are 1 μm in (**A**), (**B**), (**D**), (**E**) and 0.5 μm in (**C**)

beads, (2) testing multiple positions on the test samples, avoiding singular effects including air bubbles in the mounting medium, (3) inspecting the sample for scratches in the glass, and (4) avoiding air bubbles in the immersion medium (e.g., immersion oil) can save time to avoid unnecessary alignment efforts.

3.6 Data Analysis

This section describes the analysis of SMLM measurements. The proposed parameters are specific to our microscopic setup and may be altered in order to achieve similar results for measurements recorded on other systems.

3.6.1 2D Reconstruction of SMLM Images

All SMLM measurements have been analyzed using rapi*d*STORM v3.31 [23]. The most relevant parameters are discussed in this section. The pixel size has to be determined for each optical setup (e.g., by measuring a grid with defined distances; 158.3 nm per pixel in our case). The full-width-at-half-maximum (FWHM) of the fluorescent signal can be measured experimentally using the output module "Estimate PSF form" in rapi*d*STORM. Single PSFs can be manually selected from preselected PSFs from the measurement and the mean FWHM is determined as the average value of typically 40 PSFs. We determined a PSF FWHM of 320 nm and 380 nm for PAmCherry1/Nile Red and AF647, respectively. Due to the low background and high signal-to-noise ratio, signals are selected using a fixed global threshold and filtered by the number of emitted photons subsequently (>100 photons for Nile Red, >150 photons for PAmCherry1 and >350 photons for AF647). Signals spreading over multiple frames are tracked using a spatiotemporal Kalman filter with a distance threshold of 90/60 nm (PAmCherry1 and AF647, respectively) and a dark time of two frames.

3.6.2 Determination of the Localization Precision

A good measure for SMLM imaging quality and maximum achievable resolution is the obtained localization precision, which depends on many parameters such as the optical properties of the microscope, imaging conditions or fluorophore brightness. Theoretical estimations [10, 32] define the lower boundary of the localization precision, which is usually not reached under experimental conditions. We therefore decided to apply an experimental approach, which determines the nearest neighbor distances of fluorophores in adjacent frames [33]. The statistical distribution of these nearest neighbor distances can be mathematically fitted, providing the experimental localization precision as the standard deviation of the first Gaussian part in the fit function. Using this approach, we determined our experimental localization precision to be 12.0 ± 0.3 nm, 22.1 ± 1.1 nm, and 12.6 ± 0.5 nm for PALM, PAINT, and *d*STORM measurements, respectively [17]. These values can be used to apply a Gaussian filter to the respective SMLM images, with a σ corresponding to the determined experimental localization precision.

3.6.3 Drift Correction and Channel Alignment

Due to the low temporal and high spatial resolution, SMLM techniques are quite prone to sample drift during acquisition. Various variants of drift correction have been developed, some of them rely on fiducial markers as drift indicators, and others use defined structures for feature-based drift correction. The latter case is predominantly used for correction of linear drift, while the use of fiducial markers also allows for nonlinear drift correction and can be further used to correct for chromatic aberrations between different channels. For our studies, we used a drift correction approach based on fast-Fourier-transformation (FFT), in order to correct for linear drift. Since linear drift introduces a directed pattern to the image, this pattern is visible in the Fourier transformed image, which allows the extraction of the dimensions of drift in X and Y directions. Note that this approach works well for medium/high abundant proteins (1000+ localizations per cell), but fails when facing low copy numbers. In this case, we suggest adding gold beads to the sample (mixture of 60 nm and 80 nm colloidal gold particles) as fiducial markers. Due to the sample removal during the sequential workflow, the different SMLM channels are inevitably shifted and have to be registered accurately. The size, shape, and orientation of the bacteria can be used as an intrinsic feature for the superimposition of single SMLM channels (Fig. 3A). Binary masks of the signal are created using Fiji. The translational shift between the masks can be determined using the Fiji plugin "2D stitching," which calculates the maximal overlay between the respective binary masks. The accuracy of this approach increases with the number of bacteria in the region of interest.

3.6.4 Quantitative Analysis of PALM Data

One great advantage of the presented sequential workflow is the possibility of super-resolving three different classes of molecules, while conserving the quantitative nature of PALM imaging (Fig. 3B) [17]. However, applying optimal imaging conditions, using the appropriate fitting parameters and thresholds are just a few of the crucial parameters for reliable quantitative PALM measurements. Signals that emit over multiple frames have also to be tracked to a single localization in order to prevent overestimation of the copy number. The study of Endesfelder et al. [19] highlights the correct choice of the mentioned parameters. Protein copy-numbers in single cells can easily be determined using LAMA [25]. SMLM localization files can be converted into images, in which every localization of a fluorophore adds one gray value to the respective pixel (Fig. 3Ba). Therefore, the number of localizations per cell is represented by the integrated intensity within the cell cylinder (Fig. 3Bb). This number can be converted to the protein copy number, e.g., by correcting for fluorescent protein blinking and maturation time. The PAINT image provides the sharp outlines of the bacterial cells, which are used to measure the area and number of PALM localizations in each single cell. In addition to copy-numbers and molecular densities, density-based clustering algorithms such as DBSCAN [34] can be applied to PALM measurements to identify protein subclusters, e.g., presumed transcription hot spots of clustered RNA polymerases (Fig. 3Bc) [19]. DBSCAN cluster analysis can also be performed using LAMA [25]. In this example, molecules with at least four neighbors in a radius of 30 nm are considered to be clustered.

4 Conclusions and Perspectives

Due to their small size, bacteria are excellent targets for single molecule localization microscopy studies. The presented sequential workflow combines different SMLM techniques under optimal imaging conditions and thus facilitates the localization and quantification of proteins of interest via PALM and correlation of their position with respect to the nucleoid and cell envelope. This approach is applicable to different *E. coli* strains and growth media (Fig. 4). We confirmed that the majority of RNA polymerase molecules transcribe active genes at the surface of the highly condensed and structured nucleoids during fast growth (Fig. 4A, LB Miller, 37 °C) [16]. Less structuring of the nucleoid is observed in M9 minimal medium (Fig. 4B). The nucleoid associated protein H-NS co-localizes with the chromosome, illustrating its role in chromosome organization. In addition, low-abundant signaling proteins such as OmpR can be visualized and its position with respect to the chromosome and cell membrane determined. We discovered that OmpR was located near the plasma membrane under acidic conditions (Fig. 4C, 100 mM MES buffer, pH 5.6), while the chromosome was condensed in a ring-like structure along the bacterial length axis [18]. These examples illustrate that our approach is applicable over a broad range of targets and conditions. Nucleoids can also be visualized using novel fluorogenic dyes which allows PAINT imaging of unmodified DNA (Fig. 4D) [18, 29] and will facilitate sequential imaging under nonreplicating growth conditions (e.g., in stationary phase). Moreover, super-resolution imaging of click-labeled nucleoids is not restricted to SMLM and can be performed using STED microscopy [30] (Fig. 4E). 2D-STED microscopy delivers a lower nominal resolution than SMLM, but can be exploited as a bona fide control experiment delivering artifact-free super-resolution images and the z-stacking capability of confocal microscopy. 3D-STED microscopy [35], resulting in an approximately isotropic resolution of ~100 nm and comparably shorter imaging times (<1 to a few minutes), can be beneficial, if throughput is more important than maximum resolution and protein counting capabilities.

With the development of further specific labeling strategies, the presented sequential workflow can be extended to multiple proteins of interest, which will be extremely useful to investigate protein interactions within DNA-associated processes such as replication, transcription, nucleoid structuring, or signal transduction.

5 Notes

1. The exposure time of EdU to the bacterial cultures should be varied according to the mass doubling time in the respective medium. For this purpose, we always record a growth curve by measuring the OD_{600} (here referred to as A) of each culture every 30 min. The OD_{600} is plotted against the time (t) and the mass doubling time (t_d) can be extracted by using Eqs. 1 and 2. We typically incubate the cells in EdU-containing medium for 1.5 mass doubling times, which is ~40 min in rich medium and ~66 min in minimal medium (M9 supplemented with 0.1% casamino acids) in order to reliably label the entire chromosome.

$$A(t) = A_0 \times e^{k \times t} \tag{1}$$

$$t_d = \frac{\ln 2}{k} \tag{2}$$

2. The concentration of EdU should be carefully determined according to the growth medium. 10 μM EdU works well for growth in rich medium. However, cytotoxic effects cannot be excluded in minimal media. Here, we recommend growing cells in the respective medium, with and without addition of EdU, fixing the cells and comparing the cell and nucleoid morphology using Nile Red and DAPI, without performing the click reaction. If the cell or nucleoid morphology is impaired due to EdU incorporation, lower EdU concentrations, or the addition of native nucleosides to the growth medium can help to reduce cytotoxic effects.
3. We apply low concentrations of FA (1–2%), since high concentrations might denature fluorescent proteins impairing quantitative imaging [19]. However, FA fixed cells are still (at least to some extent) osmotically active. Therefore, buffers with the osmolality adjusted should be used to prevent changes in cell shape and/or size. The addition of glutaraldehyde (GA) renders the cells osmotically inactive (e.g., 1% FA, 0.1–0.2% GA). We usually incubate for shorter times with an FA/GA mixture (10–15 min). Excess glutaraldehyde has to be quenched by a freshly prepared solution of 0.2% sodium borohydrate in PBS

for 3 min, followed by three washing steps with PBS. Note that you will lose a significant amount of photoactivatable proteins, and we therefore suggest using FA–GA mixtures only on high-abundance proteins, which are not analyzed quantitatively.

4. Coating with extracellular matrix is not essential, but recommended when the samples are stored for longer times. Note that the protein concentration in the ECM may differ between suppliers and lot numbers. High concentrations might affect post-labeling, while low concentrations might lead to insufficient polymerization. The sample should be kept in a humid environment during the polymerization process to prevent evaporation and drying out.
5. The described order in the sequential workflow cannot be changed without introducing unfavorable conditions. We strongly suggest performing PALM imaging first in order to facilitate quantitative imaging. Performing the copper-catalyzed click reaction prior to PALM imaging leads to irreversible destruction of fluorescent proteins [36]. Furthermore, the photoactivation cycle of PAmCherry1 requires molecular oxygen [37], which is mostly depleted in switching buffers for *d*STORM. Nile Red emits fluorescence in the same channel as PAmCherry1, requiring PALM imaging to be performed prior to PAINT imaging, since Nile Red might not be washed out completely. However, *d*STORM imaging could be carried out before PAINT imaging. Although the cells have to be permeabilized prior to click labeling, PAINT imaging is still possible after permeabilization and *d*STORM imaging. However, we prefer the suggested order of PALM, PAINT, and *d*STORM imaging due to the ease of sample handling.
6. HILO illumination applies an inclined laser beam that generates a laminal sheet of illumination. This confinement of the excited volume in the z-direction reduces out-of-focus fluorescence, which would impair image quality in widefield imaging [38]. The advantage over total internal reflection microscopy (TIRF) is the increased penetration depth, which allows imaging of the bacterial "mid plane" with an increased signal-to-noise ratio.

Acknowledgments

C.S., M.G., and M.H. acknowledge funding by the German Science Foundation (DFG, grant CEF 115). The authors are grateful to Luke Lavis for kindly providing the Hoechst-JF646 dye. LJK is supported by VA IBX-000372 and NIH AIR21-123640 grants and an RCE in Mechanobiology from the Ministry of Education, Singapore.

References

1. Rojas E, Theriot JA, Huang KC (2014) Response of *Escherichia coli* growth rate to osmotic shock. Proc Natl Acad Sci U S A 111 (21):7807–7812
2. Turkowyd B, Virant D, Endesfelder U (2016) From single molecules to life: microscopy at the nanoscale. Anal Bioanal Chem 408 (25):6885–6911
3. Furstenberg A, Heilemann M (2013) Single-molecule localization microscopy-near-molecular spatial resolution in light microscopy with photoswitchable fluorophores. Phys Chem Chem Phys 15(36):14919–14930
4. Heilemann M (2010) Fluorescence microscopy beyond the diffraction limit. J Biotechnol 149(4):243–251
5. Betzig E, Patterson GH, Sougrat R et al (2006) Imaging intracellular fluorescent proteins at nanometer resolution. Science 313 (5793):1642–1645
6. Manley S, Gillette JM, Patterson GH et al (2008) High-density mapping of single-molecule trajectories with photoactivated localization microscopy. Nat Methods 5 (2):155–157
7. Stracy M, Lesterlin C, Garza de Leon F et al (2015) Live-cell superresolution microscopy reveals the organization of RNA polymerase in the bacterial nucleoid. Proc Natl Acad Sci U S A 112(32):E4390–E4399
8. Sharonov A, Hochstrasser RM (2006) Wide-field subdiffraction imaging by accumulated binding of diffusing probes. Proc Natl Acad Sci U S A 103(50):18911–18916
9. Lew MD, Lee SF, Ptacin JL et al (2011) Three-dimensional superresolution colocalization of intracellular protein superstructures and the cell surface in live Caulobacter crescentus. Proc Natl Acad Sci U S A 108(46): E1102–E1110
10. Thompson RE, Larson DR, Webb WW (2002) Precise nanometer localization analysis for individual fluorescent probes. Biophys J 82 (5):2775–2783
11. Heilemann M, van de Linde S, Schuttpelz M et al (2008) Subdiffraction-resolution fluorescence imaging with conventional fluorescent probes. Angew Chem Int Ed Engl 47 (33):6172–6176
12. Kolb HC, Finn MG, Sharpless KB (2001) Click chemistry: diverse chemical function from a few good reactions. Angew Chem Int Ed Engl 40(11):2004–2021
13. Salic A, Mitchison TJ (2008) A chemical method for fast and sensitive detection of DNA synthesis in vivo. Proc Natl Acad Sci U S A 105(7):2415–2420
14. Ferullo DJ, Cooper DL, Moore HR et al (2009) Cell cycle synchronization of *Escherichia coli* using the stringent response, with fluorescence labeling assays for DNA content and replication. Methods 48(1):8–13
15. Raulf A, Spahn CK, Zessin PJM et al (2014) Click chemistry facilitates direct labelling and super-resolution imaging of nucleic acids and proteins. RSC Adv 4(57):30462–30466
16. Spahn C, Endesfelder U, Heilemann M (2014) Super-resolution imaging of *Escherichia coli* nucleoids reveals highly structured and asymmetric segregation during fast growth. J Struct Biol 185(3):243–249
17. Spahn C, Cella-Zannacchi F, Endesfelder U et al (2015) Correlative super-resolution imaging of RNA polymerase distribution and dynamics, bacterial membrane and chromosomal structure in *Escherichia coli*. Methods Appl Fluoresc 3(1):14005
18. Foo YH, Spahn C, Zhang H et al (2015) Single cell super-resolution imaging of *E. coli* OmpR during environmental stress. Integr Biol (Camb) 7(10):1297–1308
19. Endesfelder U, Finan K, Holden SJ et al (2013) Multiscale spatial organization of RNA polymerase in *Escherichia coli*. Biophys J 105(1):172–181
20. Burgert A, Letschert S, Doose S et al (2015) Artifacts in single-molecule localization microscopy. Histochem Cell Biol 144 (2):123–131
21. Endesfelder U, Heilemann M (2014) Art and artifacts in single-molecule localization microscopy: beyond attractive images. Nat Methods 11(3):235–238
22. Edelstein A, Amodaj N, Hoover K et al (2010) Computer control of microscopes using micro-Manager. Curr Protoc Mol Biol. Chapter 14: Unit14.20
23. Wolter S, Schuttpelz M, Tscherepanow M et al (2010) Real-time computation of subdiffraction-resolution fluorescence images. J Microsc 237(1):12–22
24. Schindelin J, Arganda-Carreras I, Frise E et al (2012) Fiji: an open-source platform for biological-image analysis. Nat Methods 9 (7):676–682
25. Malkusch S, Heilemann M (2016) Extracting quantitative information from single molecule super-resolution imaging data with LAMA – localization microscopy analyzer. Sci Rep 6:34486

26. Michelsen O, de Mattos T, Joost M, Jensen PR et al (2003) Precise determinations of C and D periods by flow cytometry in *Escherichia coli* K-12 and B/r. Microbiology 149(Pt 4):1001–1010
27. Zessin PJM, Krüger CL, Malkusch S et al (2013) A hydrophilic gel matrix for single-molecule super-resolution microscopy. Opt Nanoscopy 2(1):4
28. Qu D, Wang G, Wang Z et al (2011) 5-Ethynyl-2′-deoxycytidine as a new agent for DNA labeling: detection of proliferating cells. Anal Biochem 417(1):112–121
29. Legant WR, Shao L, Grimm JB et al (2016) High-density three-dimensional localization microscopy across large volumes. Nat Methods 13(4):359–365
30. Klar TA, Hell SW (1999) Subdiffraction resolution in far-field fluorescence microscopy. Opt Lett 24(14):954–956
31. Vicidomini G, Moneron G, Han KY et al (2011) Sharper low-power STED nanoscopy by time gating. Nat Methods 8(7):571–573
32. Mortensen KI, Churchman LS, Spudich JA et al (2010) Optimized localization analysis for single-molecule tracking and super-resolution microscopy. Nat Methods 7 (5):377–381
33. Endesfelder U, Malkusch S, Fricke F et al (2014) A simple method to estimate the average localization precision of a single-molecule localization microscopy experiment. Histochem Cell Biol 141(6):629–638
34. Ester M, Kriegel H-P, Sander J et al (1996) A density-based algorithm for discovering clusters in a density-based algorithm for discovering clusters in large spatial databases with noise. Data Min Knowl Discov Databases 34:226–231
35. Hein B, Willig KI, Hell SW (2008) Stimulated emission depletion (STED) nanoscopy of a fluorescent protein-labeled organelle inside a living cell. Proc Natl Acad Sci U S A 105 (38):14271–14276
36. Loschberger A, Niehorster T, Sauer M (2014) Click chemistry for the conservation of cellular structures and fluorescent proteins: ClickOx. Biotechnol J 9(5):693–697
37. Subach FV, Malashkevich VN, Zencheck WD et al (2009) Photoactivation mechanism of PAmCherry based on crystal structures of the protein in the dark and fluorescent states. Proc Natl Acad Sci U S A 106(50):21097–21102
38. Tokunaga M, Imamoto N, Sakata-Sogawa K (2008) Highly inclined thin illumination enables clear single-molecule imaging in cells. Nat Methods 5(2):159–161

Chapter 21

Procedures for Model-Guided Data Analysis of Chromosomal Loci Dynamics at Short Time Scales

Marco Gherardi and Marco Cosentino Lagomarsino

Abstract

This chapter provides theoretical background and practical procedures for model-guided analysis of mobility of chromosomal loci from movies of many single trajectories. We guide the reader through existing physical models and measurable quantities, illustrating how this knowledge is useful for the interpretation of the measurements.

Key words Chromosomal loci dynamics, Theoretical models, Data analysis, Subdiffusion, Viscoelasticity

1 Introduction

This chapter provides practical procedures to analyze mobility of chromosomal loci from movies of many single trajectories [1, 2]. The ideal target is a biologist oriented to quantitative work, and able to produce data of this kind. We have in mind experiments following single chromosomal loci tagged with FROS or ParB-GFP in bacteria, at time scales that are shorter than the "macroscopic" movements due to segregation (which become evident observing for minutes). At such time scales, fluctuations of single loci report on features of the chromosomal organization, biochemical noise, and active forces/transport. We will mainly refer to the bacterial literature and the related scientific questions [3–6]. However, some of the procedures are valid for other systems such as tagged chromosome loci in eukaryotes or other labeled protein foci/organelles/cytoplasmic particles [7, 8]. Future endeavors may combine such model-guided techniques systematic genetic and physiological perturbations and have the ultimate goal of specifying how the physical nature of chromosome organization and segregation affects key biological processes such as transcription and cell-cycle progression.

Olivier Espéli (ed.), *The Bacterial Nucleoid: Methods and Protocols*, Methods in Molecular Biology, vol. 1624, DOI 10.1007/978-1-4939-7098-8_21,

We will assume here that the analysis starts with a large set of already tracked trajectories. A detailed description of how to carry out the tracking experiments and how to treat the raw data has been provided elsewhere [9]. The goal of this chapter is to give access to more sophisticated ways to analyze the data, linking to existing theoretical models, and advocate the use of models to guide and enhance the physical interpretation of such data. We will link to cases where such methodology has been effective for answering key questions. A secondary goal is to give the reader a compass to be able to explore the jungle of existing modeling approaches.

Physical vs Phenomenological Models The measurable quantities, framed into the appropriate mathematical model, allow to investigate the physical properties of the medium (dynamical, rheological, elastic, etc.). Modeling in physics usually follows either of two strategies: physical and phenomenological. "Physical model" refers to a mechanistic description of the processes acting in the system, derived from an established physical theory. It needs not incorporate all known details underlying a process, but only those assumed to be relevant to the dynamics of the quantities under study. A "phenomenological" model, instead, is less aware of the physics, and aims at capturing the behavior of the system by a simple mathematical description that is not derived from a physical theory, with the least possible number of free parameters. In practice, the difference between physical and phenomenological models may be subtle in some cases. Importantly, as we will see, both approaches can lead to physical interpretation of the parameters, the main difference between the two kinds of models being the derivation from an underlying theoretical framework or lack thereof.

2 Materials

This section reviews the basic observables and the main models (definitions and meaning) that are commonly used in the context of tracking chromosomal loci (Fig. 1). It contains the formulas that define the main observables and model equations, but all definitions are also explained in words.

2.1 Anomalous Diffusion

The mean-squared displacement (MSD) of the particle is a simple informative quantity, and the most commonly used observable. For an ensemble of N trajectories $x_i(t)$, it is defined as the average of the square displacement between the position at the initial time and any subsequent position, and it is thus a function of time,

$$\mathrm{MSD}(t) = \left\langle [x(t) - x(0)]^2 \right\rangle = \frac{1}{N}\sum_{i=1}^{N}[x_i(t) - x_i(0)]^2. \quad (1)$$

In normal (Brownian) diffusion, the MSD depends linearly on time, at least for long times. When this law is violated, diffusion is

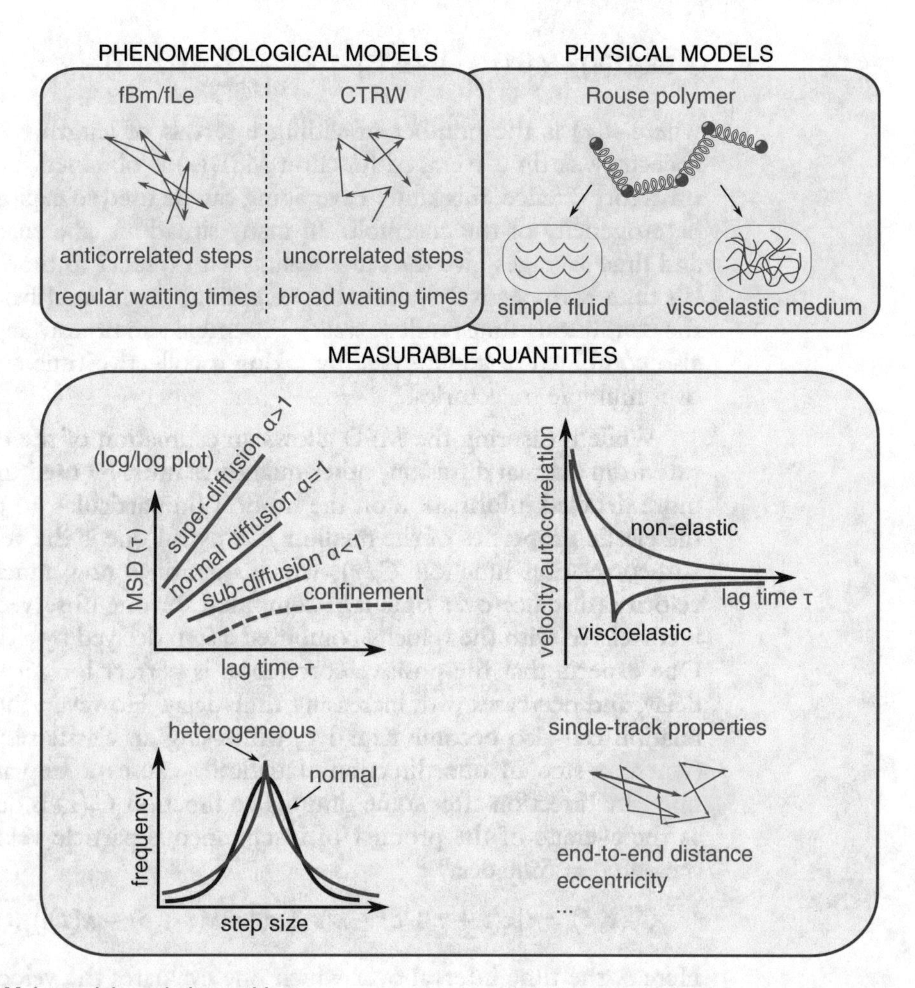

Fig. 1 Main models and observables

termed *anomalous.* The terms *subdiffusion* and *superdiffusion* are used to refer to asymptotically sub-linear and super-linear cases, respectively. In many cases, the MSD is found to grow as a power of time, for non-confined motion:

$$\mathrm{MSD}(t) \sim t^{\alpha}. \tag{2}$$

This trend is usually a signature of interesting physical features (see below) and defines the exponent α, which quantifies the deviation from Brownian behavior.

The average in Eq. 1 is an *ensemble average*, meaning that it is calculated on the whole set of trajectories, at fixed time t. Another commonly used procedure is to average over lag times τ without averaging on the ensemble, i.e., considering a trajectory, taking the mean of the steps of a given time difference ("lag time") starting from all the measured time frames. This gives the *time-averaged* MSD for each trajectory:

$$\mathrm{MSD}(\tau) = \left\langle [x(t+\tau) - x(t)]^2 \right\rangle_t = \frac{1}{n(\tau)} \sum_t [x(t+\tau) - x(t)]^2, \quad (3)$$

where $n(\tau)$ is the number of sliding intervals of length τ in the trajectory. A time-averaged function $\mathrm{MSD}(\tau)$ is obtained for each trajectory. Hence, this kind of averaging can be used to explore the heterogeneity of the ensemble. In many situations, the ensemble and time averages give the same scaling with respect to time t and lag time τ, respectively (a notable exception, mentioned below, is the continuous-time random walk). Ensemble and time average are also combined in some cases, by taking a collective time average over multiple trajectories.

While measuring the MSD allows an estimation of the departure from normal diffusion, other quantities must be used to gain more detailed information on the motion (in particular, to probe the elastic properties of the medium). A useful one is the velocity autocorrelation function $C_v(\tau)$, which quantifies how much the velocity (distance over time lag) computed for one observed step is correlated with the velocity computed a step delayed by a time τ. One expects that the positive correlation is perfect for zero time delay, and decreases with increasing time delay. However, the correlation can also become negative, witnessing an elastic response (where a step in one direction statistically causes a step in the opposite direction after some time). The function $C_v(\tau)$ is defined as the average of the product of instantaneous particle velocities measured at time delay τ,

$$C_v(\tau, \delta) = \langle [x(t+\tau+\delta) - x(t+\tau)] \cdot [x(t+\delta) - x(t)] \rangle_t. \quad (4)$$

Here δ, the time interval over which one evaluates the velocity, is usually taken to be a single frame (so that the velocity is computed by differences of consecutive frames). However, it is sometimes useful to compute $C_v(\tau, \delta)$ for several values of δ in a range [10]. *See* Subheading 3.4 for a use case.

2.2 Phenomenological Models of Subdiffusion

We describe here the main theoretical models that are relevant to understand the dynamics of tracked loci (Fig. 1).

2.2.1 Fractional Brownian Motion

Fractional Brownian motion (fBm) was introduced and described in the mid-twentieth century [11–13] motivated by empirical observations of anomalously slow-decaying autocorrelations in river levels and economic time series. It is a stationary process $B_H(t)$ (i.e., it is invariant under translations of time), with Gaussian distribution and variance (i.e., MSD), that increases as a power law with time $B_H(t)^2 - B_H(t)^2 = t^{2H}$. This relation is the definition of anomalous diffusion, Eq. 2. The exponent $H = \alpha/2$ is called the *Hurst exponent*, and measures the deviations from normal diffusion, which is recovered for $H = 1/2$. The important ingredient of this process is that *steps are correlated*, positively or negatively, depending on the value of H. For $H < 1/2$ the process is subdiffusive and the increments $X(k)$ are anti-correlated, while for $H > 1/2$ the process is super-diffusive and the increments are positively correlated.

The fBm, like ordinary Brownian motion, has the notable property of being self-similar (fractal), meaning that a rescaling of time $t \mapsto at$ is equivalent to a rescaling of space $B_H(t) \mapsto |a|^H B_H(t)$. In fact, it is the most general class of Gaussian processes with this property.

The covariance (autocorrelation) function of fBm contains power-law contributions, and is often used to define the process.

$$\langle B_H(t) B_H(s) \rangle = \frac{1}{2}\left(t^{2H} + s^{2H} - |t - s|^{2H}\right). \quad (5)$$

Note that for $H = 1/2$, this expression reduces to the case of the random walk, i.e., $\max(t, s)$.

The increment process $X(k) = B_H(k + 1) - B_H(k)$, $k = 0, 1, \ldots$, defined by the "steps" of fBm is called fractional Gaussian noise, and is the analogue of white noise for this self-similar correlated process.

2.2.2 Viscoelasticity and the Fractional Langevin Equation

A viscoelastic material is one displaying both viscous (i.e., it flows like a fluid) and elastic response properties. Contrary to purely elastic materials, a viscoelastic substance has notable physical properties, it dissipates energy when subject to stress, and its stress--strain curves display hysteresis. There are indications that the cytoplasm behaves as a viscoelastic medium, owing to molecular crowding and to the presence of elastic elements [1, 14, 15]. As a particle moves in a viscoelastic medium, the medium reacts elastically to the motion of the tracer, thus generating anti-correlations in the trajectory. This phenomenon can give rise to subdiffusion.

The fBm is a reasonably good phenomenological model of tracer subdiffusion in a viscoelastic medium, because its anti-correlated steps represent well the transient elastic response. However, the fBm has a property that (from the physics viewpoint) makes it not adapted to describe fluctuating systems in thermodynamic equilibrium: it does not satisfy the so-called fluctuation-dissipation relation. This relation is an expected property of a physical fluctuating system at thermal equilibrium [16]. Incidentally, note that the fluctuation-dissipation relation is not guaranteed to apply in biological systems, thinking, for example, that the system might be kinetically constrained (i.e., subject to constraints preventing its equilibrium relaxation) or driven far from equilibrium by ATP-dependent active forces. However, physicists generally agree that the basal equilibrium behavior should be first reproduced in a model, as a control, and more complex effects should be considered as additional ingredients on top of the familiar equilibrium behavior.

Simply put, the fluctuation-dissipation relation states that the relaxation of the system from a thermal fluctuation is the same as that from an applied perturbation. Such perturbation can be, for example, an external force exerted by a cellular element or, in a more controlled way, by the experimental setup. The classical example is a microscopic object in a fluid following a simple random walk. Here, as in our case, the fluctuating quantity is the position $x(t)$, and the mobility μ (i.e., the inverse of the viscous drag) quantifies the system response (net movement) in presence of an external force. In this case, the fluctuation-dissipation theorem can be formulated as the classic "Stokes–Einstein" relation, between the diffusion constant (characterizing the fluctuating quantity) and the mobility (characterizing the behavior under perturbation), $D = \mu k_B T$.

Since the diffusion constant D is related to the mean-square displacement, this expression can be also written as

$$\frac{1}{2}\langle x^2(t)\rangle_{F=0} = k_B T \frac{1}{F}\frac{d}{dt}\langle x(t)\rangle_F,$$

stating that a Brownian particle under a constant force F moves on average with constant speed, and the same law is obeyed by the mean-square displacement. Now, this property is not verified for a particle following fractional Brownian motion (unless $H = 1/2$, where this model describes the conventional Brownian motion). Such model particle moves at constant speed under an applied external force, but its fluctuations may be characterized by a different power law.

The fluctuation-dissipation relation is restored by another phenomenological model, called the fractional Langevin equation (fLe) [17–20]. In order to understand this process we need to take a few steps back. The Langevin equation is a force-balance equation in an overdamped medium, i.e., when inertia is negligible as it is at the microscale, explicitly taking into account sources of molecular noise (e.g., thermal noise). Going back to the random walk, the process results from the Langevin equation of a free particle under thermal white noise. Equally, fBm can be expressed as an overdamped Langevin equation, where the forcing term is fractional Gaussian noise.

The fLe is a generalized Langevin equation where the force is fractional Gaussian noise (with Hurst exponent H) and the particle feels the effects of its past trajectory as an additional forcing term. Thanks to this additional memory term, absent for the fBm, the behavior under external force can correspond again to the scaling law of the mean-square displacement. Hence, the fLe allows the definition of subdiffusive motions satisfying the fluctuation-dissipation relation. The MSD in this model presents a crossover from a short-time ballistic motion (where the MSD is proportional to time) to a long-time subdiffusion with exponent $2 - 2H$. Note that here subdiffusion is generated by positively correlated noise (which is often not easy to interpret), as opposed to anti-correlated noise as in the fBm.

While the processes described above are defined in one dimension, an isotropic d-dimensional fBm (or fLe) is readily defined by assuming that each coordinate follows a one-dimensional fBm (or fLe) independently and with the same parameters. When these processes are confined in a finite volume, the understanding of their dynamics is less systematic, but signatures of confinement can be obtained from empirical data. The time-averaged MSD saturates to a constant (volume dependent) value for large times. Both fBm and fLe follow similar behavior under confinement [21].

2.2.3 Glassiness and the Continuous-Time Random Walk

The continuous-time random walk (CTRW), introduced in [22], is a classic phenomenological model of "cage dynamics." It separates the microscopic motion into two components: confined diffusion in a cage and hopping dynamics between cages. The CTRW differs from the fractional Brownian–Langevin description in that tracer motion is realized by periods of immobility followed by instantaneous jumps. The jumps are uncorrelated (they are independent and identically distributed random variables), and the phenomenology of the system modeled is encoded in the distribution $\psi(\tau)$ of the waiting times τ between jumps. In many cases, the energy landscape the tracer is subject to is heterogeneous, and can transiently trap the particle for long times, possibly with infinite mean waiting time. In such cases, τ is distributed as a power law $\psi(\tau) \approx \tau^{-(1+\alpha)}$ with $0 < \alpha < 1$, meaning that there is a non-negligible probability that the particle remains trapped for very long times, resulting in macroscopically intermittent motion. The anomalous diffusion in this case is determined by the divergence of typical microscopic time scales [23–25], and the MSD scales as t^{α}.

When geometrical inhomogeneities, rather than transient binding, are believed to be the main responsible for anomalous transport, the motion is often modeled as (normal) diffusion confined in a fractal. Diffusion in a fractal leads to the same anomalous scaling laws for the MSD as above. The phenomenology is similar to that of the CTRW, but the two models can be discriminated by analyzing first-passage properties (*see* [26] for details). It is important to note that ensemble averages and time averages are not equivalent in the CTRW (a phenomenon related to the "breaking of ergodicity") [27]. In particular, while the ensemble-averaged MSD bears the hallmark scaling of anomalous diffusion, the time-averaged MSD has a simple diffusive behavior. The doubly averaged MSD scales as $T^{\alpha-1}\tau$, where τ is the lag time and T (assumed much larger than τ) is the time span over which the time average is computed [28]. (A sub-linear scaling both in trajectory time and in lag time has been interpreted as evidence for a CTRW restricted on a fractal substrate [25].)

In contrast to the CTRW, ergodicity holds for the fBm and fLe in free space, meaning that time averages are equivalent to ensemble averages, although with slow convergence [29] (weak ergodicity breaking is observed in confined space [21]). In most situations, it is therefore advisable to perform both time and ensemble averages on the MSD, in order to improve the statistics.

2.3 Physical Models of Polymer Dynamics

2.3.1 The Rouse Model

The simplest physical model that can be used to evaluate the dynamical properties of a polymer is the Rouse model. The coarse grained description is that of a linear polymer, whose monomer segments are subject to hydrodynamic drag and random forces as in Brownian motion. In the Rouse model, consecutive monomers are connected by entropic springs, i.e., by Gaussian chains that follow Hooke's law, with elastic constant $k = 3k_B T b^{-2}$. Here T is the temperature, k_B is Boltzmann's constant, and b is the Kuhn length, i.e., the typical length of a segment. The motion of each segment is considered overdamped with the same friction ζ, and different segments do not interact with each other ("ghost" chain). The Rouse polymer model is simple enough to be solved analytically [30]. The dynamics of a monomer in the chain is diffusive up to the time $t_0 \approx \zeta b^2/3k_B T$ when it begins to be influenced by the surrounding polymer. Then it is subdiffusive with exponent $\alpha = 1/2$ up to the crossover time $t_1 \approx \zeta b^2(N+1)^2/3\pi^2 k_B T$ (known as Rouse time). After the Rouse time, segmental motion is dominated by the center of mass of the whole polymer, and is again diffusive ($\alpha = 1$). Note that t_1 increases quadratically with the number of monomers N.

Including hydrodynamic interactions in the Rouse description yields the Zimm model, which predicts a scaling exponent $\alpha \approx 0.67$. However, it is commonly believed that crowding in the cytosol screens these interactions, thus making them irrelevant in vivo. When inter-chain interactions are important, such as at high concentrations in a random "melt," where entanglement between different chains strongly constrains their dynamics, the *reptation* dynamics is often assumed [31]. In such a description, the polymers are constrained to move inside "tubes" by movements resembling those of snakes. Compared to the Rouse model, the MSD has an additional regime, where the subdiffusion has exponent $\alpha = 1/4$, corresponding to longitudinal diffusion in a tube shaped as a random walk.

2.3.2 Non-Newtonian and Active Media

The extreme long-time and short-time regimes in these models are diffusive, due to the polymer being immersed in a Newtonian fluid (i.e., a simple viscous fluid). How the foregoing picture changes for polymers moving in a viscoelastic medium has been addressed in [32]. As described above, the viscoelasticity of the medium makes a free tracer subdiffuse; let us call its exponent α_{free}. As in the Rouse model, the intra-polymer forces give rise to an intermediate-time regime where monomers diffuse with a halved exponent with respect to the free motion. Namely, a crossover is observed between subdiffusion with $\alpha = \alpha_{\text{free}}/2$ before the Rouse time, to free subdiffusion of the center of mass with $\alpha = \alpha_{\text{free}}$ for large times.

The foregoing models rely on the assumption of a passive medium. However, particles in the cytoplasm and chromosomal loci are subject to additional active forces, notably during segregation. Consider a Rouse model in a subdiffusive medium, characterized by the exponent α_{free}. A constant force applied to a monomer in such a model gives a mean displacement proportional to $t^{\alpha_{\text{free}}/2}$ [33, 34] (a more general elastic model was treated in ref. [35]). Fluctuations around this drift can be measured by the de-drifted MSD, as $\text{VarD}(t) = \langle |x(t) - x(0) - \text{MD}(t)| \rangle$, where $\text{MD}(t)$ is the mean displacement at time t, measured in the direction of the force. These quantities satisfy the Einstein relation $\text{MD}(t)/\text{VarD}(t) = F/(2k_\text{B}T)$, where F is the force and T the temperature, thus allowing to estimate the force. Note that the behavior under constant force can be used to discriminate fBm and fLe, since, as we discussed, the latter process satisfies the Einstein relation, while $\text{MD}(t) \propto t$ for the fBm under constant forcing.

2.3.3 Polymer Networks

Descriptions of bacterial chromosomes as simple linear polymers are a gross simplification, as the real topologies are complicated by the presence of nucleoid-associated and other DNA-binding proteins [5]. These proteins can link distant regions of the chromosome, thus mediating long-range interactions along the chain. Classic models of this situation are Gaussian networks, i.e., generalizations of the Rouse model where sets of ghost springs are interconnected in network structures of chosen topology, which can range from a main backbone with loops, to random or fractal-like networks [5]. Such models are quite developed and largely applied to proteins [36–38] but have never (to the best of our knowledge) been used for comparisons with dynamics of chromosomal loci. Thus, they make an interesting territory for exploration. A recently developed hybrid approach bypasses the specification of a precise topology for the network of springs, by inserting its long-range features directly in the model [39].

3 Methods

We describe here some of the main analyses that can be performed on the data in order to answer specific questions about the physical processes at play. Chromosomal loci in bacteria follow a basal subdiffusive dynamics (see below) and are subject to active noise and active drives.

Table 1
Resources for simulation code and packages available on the web

http://www.mathworks.com/matlabcentral/fileexchange/38935-fractional-brownian-motion-generator	MATLAB	fBm
http://www.columbia.edu/~ad3217/fbm.html	C and R	fBm
http://github.com/yikelu/fbm	Python	fBm
http://demonstrations.wolfram.com/OneDimensionalFractionalBrownianMotion	Mathematica	fBm
http://technion.ac.il/textasciitildepavel/comphy/code.htm	MATLAB	fBm
http://www.weizmann.ac.il/EPS/People/Brian/CTRW/software	MATLAB	CTRW

3.1 Using Physical Models as Null Models

Once a candidate process is selected to describe the data, observables generated from numerical simulations of the process can be directly compared to the data. This allows to fix (at least some of) the parameters in the model by fits. Several examples of these comparisons are presented below, mostly based on fractional Brownian motion and continuous-time random walk. More detailed physical/phenomenological models may need the development and analysis of user-adapted variants.

Unfortunately, no simply packaged tool to run such simulations is readily available today. However, working code for the basic processes of continuous-time random walk and fractional Brownian motion is available as MATLAB or mathematica packages, or C/C++/python sources (Table 1), which can be used for simple applications by an end user without having to cross the barrier of coding them from scratch, and making them accessible to a wider audience of researchers, including the more tech savvy biologists.

3.2 Estimating the Subdiffusive Exponent

The simplest way to estimate the subdiffusion exponent α is by looking at a log–log scale plot of the ensemble-averaged MSD. This can be compared with time averages of the MSD. Many alternative ways of evaluating the scaling exponent are available including autocorrelation functions (see below), considering the maximum displacement of a test particle in a fixed time [40], and p-variations [41].

3.3 Testing Ergodicity and Other CTRW Features

CTRW behavior can also be detected by looking at signatures of ergodicity breaking in the averages. As mentioned in Subheading 2.2.3, if the system is non-ergodic, the time averages do not converge to the ensemble averages for long lag times. Therefore, in principle one way to test ergodicity is to consider the trend of the distribution of apparent diffusion constants from time averages of single tracks at longer and longer times. In practice, this is not always easy, but many tests of ergodicity based on simplified observables are available in the literature and can be used in this context (e.g., *see* [28, 42]).

Note that a variability between time averages of single tracks is expected also in ergodic systems, caused by fluctuations due to finite track size. For track sampling on the order of tens to hundreds of frames, as the experimentally accessible ones, such variability can be considerable. A simple test to perform is to verify whether the distribution of apparent diffusion constants from time-averaged tracks is consistent with a parameter-matched fractional Brownian motion [1, 2]. In such case, CTRW behavior can be dismissed as less likely (or cannot be easily verified in the given conditions). Parameter-matched simulations of CTRW are also possible, but they need an hypothesis on the distribution $\psi(\tau)$ of waiting times.

3.4 Testing the Viscoelastic Nature of the Medium

An efficient way to test whether the motion exhibits viscoelasticity, linked to direct physical interpretation is to consider the velocity autocorrelation function $C_v(\tau, \delta)$, Eq. 4. This function has a negative dip at a characteristic delay time which is a signature of viscoelasticity, built in as step anti-correlation in models like fractional Brownian motion and fractional Langevin equations. Additionally, the long-time decay of the velocity autocorrelation function is a power law with exponent $\alpha - 2$, and hence it can be used as a test for the subdiffusive scaling exponent. For a CTRW, the velocity autocorrelation function C_v is non-negative, due to the lack of directional correlations. Note, however, that in presence of strong confinement a CTRW can also show negatively correlated velocities as a signature of the confinement size on which steps become anti-correlated. Hence, in presence of evidence of confinement (see below) one has to be careful.

One important caveat is that in empirical data, negative auto-correlation may emerge from localization errors [10]. It is simple to rationalize why this should be the case. Consider, for example, the positions x_1, x_2, x_3 of a focus at three consecutive times. Each position is subject to an error. In particular the error on the central position x_2 has opposite effects on the two consecutive one-frame velocities $x_2 - x_1$ and $x_3 - x_2$. Hence, a negative correlation dip is expected (and found) at a time scale corresponding exactly to the time difference between consecutive frames. To avoid this problem, one simply needs to consider the correlation $C_v(\delta, \tau)$ computed for all steps of time span δ at time lag τ, and verify the presence of negative correlations for non-consecutive time-steps [10]. Once carried out, this procedure has the advantage of giving an estimate of the localization error.

3.5 Testing Confinement, Glass-Like Properties

The simplest way to test confinement is looking at the appearance of plateaus in differently averaged mean-square displacements. Plateaus may emerge from ensemble averages as a global characteristic scale. In single tracks time averages, they should be compared to a suitable null model (see below) in order to verify that they are not an effect of track-to-track fluctuations. More in general, all features that are consistent with a continuous- time random walk can usually be understood as signatures of glassy behavior, including ergodicity breaking, non-negative velocity autocorrelation, etc. First-passage time observables also distinguish the different kinds of subdiffusion [26], but they are rarely used in practice. Detection of temporary (or local) lateral confinement in single tracks is also possible, but has not, to date, been systematically applied to chromosomal loci. The available tools detect sub-tracks where the diffusivity is inconsistent with normal diffusion [43, 44]. The generalization of these techniques to anomalous diffusion is straightforward. Finally, Rouse-like models have been recently used to infer elastic tethering forces from dynamics of chromosomal loci [45].

3.6 Step-Size Distributions and Stationarity

The probability distribution for the tracer to be at position x at fixed time t is Gaussian if the anomalous diffusion is modeled by the fractional Brownian motion (or the fLe). However, this is not necessarily the case in complex media and glassy liquids. In particular, step-size distributions of objects in intracellular media are commonly found to have deviations from Gaussian behavior in the tails. Step-size distributions are simple to obtain from data as histograms of signed steps in an arbitrary direction, or as radial distribution functions of vector steps. Deviations from Gaussianity of the distribution of displacements [46] can be quantified by the so-called non-Gaussian parameter, which is calculated via the averages of the second and fourth powers of the absolute displacements $r = |x(t) - x(0)|$ as

$$\alpha_2 = 3/5\langle r^4\rangle/\langle r^2\rangle^2 - 1$$

in three dimensions [47]. Values of α_2 different from zero can be indications of highly crowded environments and glass-forming liquids close to the glass transition [7, 48, 49].

Non-Gaussian processes are also common in spatially heterogeneous media, where particles may get different diffusion constants depending on where they are or on time [46]. In such case, the step-size distribution becomes a convolution of different Gaussians, and hence deviate from being Gaussian. Note that, as mentioned above, the CTRW with broad waiting-time distributions is typically non-Gaussian, and glassy behavior can be seen as a special kind of heterogeneity.

Stationarity (i.e., invariance for time translation) of the process can also be tested, for example, using the time evolution of the increments (plotted as a function of time) of the quantiles of the probability distributions of displacements at fixed times [41]. Constant increments correspond to a stationary distribution. The quantiles can also be used for a further complementary estimate of the scaling exponent.

3.7 Testing Specific Behavior with Conditional Observables

All the above analyses can generate significant insight when restricted to a subset of tracks with specific properties. Examples of relevant physical and biological properties are locus position on the chromosome, cell-cycle stage, cell size, foci intracellular position. For example, conditional mean-square displacements may show that when a focus is close to a specific cell area or in a specific time into the cell cycle, its diffusive properties change, enabling detection of conformational transitions, cellular substructures, etc. Gated averages can also be performed based on track properties [43, 50]. For example, one can compute velocity autocorrelation functions for trajectories with fixed eccentricity, or end-to-end distance.

3.8 Testing the Presence and Effects of Active Processes

An additional feature of the chromosome is that it is driven far from equilibrium by active forces from cellular processes. Such features can be detected from loci displacement. For example, scaling of MSD with temperature can be used to find non-thermal contributions to the noise felt by the tracked locus [51].

Single track properties such as elongation and end-to-end distance are also useful to detect drifts and active components of the motion, for example, by evaluating mean-square displacements restricted to elongated tracks or to tracks where the positions tend to be more time-ordered in space. In this application, using fractional Brownian motion as a null model is extremely important [50], as it allows to determine whether specific properties (for example, the end-to-end distance) of empirical trajectories are significantly different compared to simulated tracks (of, e.g., an fBm or fLe) giving the same overall ensemble-averaged MSDs. Detection of directional drifts and connection to physical models [33] has also lead to significant conclusions on the segregation process at larger time scales.

3.9 Needed Statistics

Given the finite-size stochasticity of all subdiffusive processes, and the variability of data due to biologically relevant variables, a wide statistics is highly desirable. Ideally, data should be taken with as much as possible control on biological variability, i.e., with cell-cycle resolution (as many movies where the time since cell birth is tracked), segmenting cell shape, and considering foci positions in the cell. However, such experiments are not always possible, and gating on all the biological control variables may severely restrict the size of the sample of recorder tracks available for a specific average, compromising its quality.

The amount of statistics needed can be computed precisely by simulating an underlying process with matched parameters or with different ones and evaluating how much they can be distinguished. As a rule of thumb, since the number of time points per track may range in most systems from a few tens to a few hundreds because of photobleaching, one can say that at least 500 tracks are desirable to obtain reliable and robust averages. In practice, while some recent studies have scaled up to tens of thousands in a given condition, many experiments may be difficult and may allow very low track numbers. In such cases, comparison with simulations is highly desirable, and one can also test the robustness of the results by subsampling the averages.

Glossary

Anomalous diffusion. A diffusion process where the mean-square displacement of particles is a non-linear function of time. Sub-linear and super-linear functions, in particular, identify "sub-diffusion" and "superdiffusion," respectively.

Autocorrelation. The correlation of a time series or signal with itself at a delayed time. Autocorrelation can reveal oscillatory behavior and characteristic decay times of a process.

Crowding. Complex set of (only partially understood) processes due to the high concentration of macromolecules within a cell.

Eccentricity of a track. Measurable quantity used to quantify deviation from isotropy of a track. It is defined as the eccentricity of the "ellipse of inertia" (or ellipsoid in more than two dimensions), whose semiaxes are numerically equal to the principal radii of gyration of the track, seen as a cloud of points of equal weight.

Ensemble average. The mean of a single-track observable computed on all (or a subset of) tracks (usually denoted $\langle \cdot \rangle$).

Gated averages (conditional averages). Averages of a parameter computed with constraints on another measurable quantity (e.g., average displacement over trajectories with given range of end-to-end distance).

Newtonian fluid. A simple fluid where the viscous stress is always linearly proportional to the strain. Non-Newtonian fluids exhibit more complex behavior, such as viscoelasticity.

Time average. A mean computed on a single track by averaging on all (or a subset of) sub-tracks spanning a given time window (usually denoted $\langle \cdot \rangle_t$).

Viscoelastic medium. Material (e.g., polymer solution) exhibiting both viscous and elastic response to deformation.

White noise. A white noise contains many frequencies with equal intensities. Thermal fluctuations at thermodynamic equilibrium are best approximated by white uncorrelated noise. Brownian motion can be defined as the process whose derivative ("velocity") is white noise.

Acknowledgements

This work was supported by the International Human Frontier Science Program Organization, grant RGY0070/2014. We are grateful to P. Cicuta, A. Javer, and A. Taloni for useful discussions.

References

1. Weber SC, Spakowitz AJ, Theriot JA (2010) Bacterial chromosomal loci move subdiffusively through a viscoelastic cytoplasm. Phys Rev Lett 104:238102
2. Javer A, Long Z, Nugent E, Grisi M, Siriwatwetchakul K, Dorfman KD, Cicuta P, Cosentino Lagomarsino M (2013) Short-time movement of E. coli chromosomal loci depends on coordinate and subcellular localization. Nat Commun 4:3003
3. Cosentino Lagomarsino M, Espéli O, Junier I (2015) From structure to function of bacterial chromosomes: evolutionary perspectives and ideas for new experiments. FEBS Lett 589(20 Pt A):2996–3004
4. Kleckner N, Fisher JK, Stouf M, White MA, Bates D, Witz G (2014) The bacterial nucleoid: nature, dynamics and sister segregation. Curr Opin Microbiol 22:127–137
5. Benza VG, Bassetti B, Dorfman KD, Scolari VF, Bromek K, Cicuta P, Cosentino Lagomarsino M (2012) Physical descriptions of the bacterial nucleoid at large scales, and their biological implications. Rep Prog Phys 75:076602
6. Espeli O, Mercier R, Boccard F (2008) DNA dynamics vary according to macrodomain topography in the E. coli chromosome. Mol Microbiol 68(6):1418–1427
7. Parry BR, Surovtsev IV, Cabeen MT, O'Hern CS, Dufresne ER, Jacobs-Wagner C (2014) The bacterial cytoplasm has glass-like properties and is fluidized by metabolic activity. Cell 156(1):183–194
8. Bronstein I, Israel Y, Kepten E, Mai S, Shav-Tal Y, Barkai E, Garini Y (2009) Transient anomalous diffusion of telomeres in the nucleus of mammalian cells. Phys Rev Lett 103(1):018102
9. Javer A, Cosentino Lagomarsino M, Cicuta P (2016) Bacterial chromosome dynamics by locus tracking in fluorescence microscopy. Springer, New York, pp 161–173
10. Weber SC, Thompson MA, Moerner WE, Spakowitz AJ, Theriot JA (2012) Analytical tools to distinguish the effects of localization error, confinement, and medium elasticity on the velocity autocorrelation function. Biophys J 102:2443–2450
11. Kolmogorov AN (1940) Wiener spirals and some other interesting curves in Hilbert space. Dokl Akad Nauk SSSR 26:115
12. Mandelbrot BB (1965) Une classe de processus stochastiques homothétiques à soi. application à la loi climatologique de h. e. hurst. Compt Rendus 260:3274–3277
13. Mandelbrot BB, Van Ness JW (1968) Fractional Brownian motions, fractional noises and applications. SIAM Rev 10(4):422–437
14. Pan W, Filobelo L, Pham NDQ, Galkin O, Uzunova VV, Vekilov PG (2009) Viscoelasticity in homogeneous protein solutions. Phys Rev Lett 102:058101
15. Wirtz D (2009) Particle-tracking microrheology of living cells: principles and applications. Annu Rev Biophys 38(1):301–326. PMID: 19416071
16. Kubo R (1966) The fluctuation-dissipation theorem. Rep Prog Phys 29(1):255
17. Granek R, Klafter J (2001) Anomalous motion of membranes under a localized external potential. Europhys Lett 56(1):15
18. Kou SC, Xie XS (2004) Generalized Langevin equation with fractional Gaussian noise: subdiffusion within a single protein molecule. Phys Rev Lett 93:180603

19. Taloni A, Lomholt MA (2008) Langevin formulation for single-file diffusion. Phys Rev E 78:051116
20. Lutz E (2001) Fractional Langevin equation. Phys Rev E 64:051106
21. Jeon J-H, Metzler R (2010) Fractional Brownian motion and motion governed by the fractional Langevin equation in confined geometries. Phys Rev E 81:021103
22. Montroll EW, Weiss GH (1965) Random walks on lattices. II. J Math Phys 6(2):167
23. Scher H, Montroll EW (1975) Anomalous transit-time dispersion in amorphous solids. Phys Rev B 12:2455–2477
24. Wong IY, Gardel ML, Reichman DR, Weeks ER, Valentine MT, Bausch AR, Weitz DA (2004) Anomalous diffusion probes microstructure dynamics of entangled f-actin networks. Phys Rev Lett 92:178101
25. Weigel AV, Simon B, Tamkun MM, Krapf D (2011) Ergodic and nonergodic processes coexist in the plasma membrane as observed by single-molecule tracking. Proc Natl Acad Sci 108(16):6438–6443
26. Condamin S, Tejedor V, Voituriez R, Bénichou O, Klafter J (2008) Probing microscopic origins of confined subdiffusion by first-passage observables. Proc Natl Acad Sci USA 105 (15):5675–5680
27. He Y, Burov S, Metzler R, Barkai E (2008) Random time-scale invariant diffusion and transport coefficients. Phys Rev Lett 101:058101
28. Lubelski A, Sokolov IM, Klafter J (2008) Nonergodicity mimics inhomogeneity in single particle tracking. Phys Rev Lett 100:250602
29. Deng W, Barkai E (2009) Ergodic properties of fractional Brownian-Langevin motion. Phys Rev E 79:011112
30. Doi M, Edwards SF (1986) The theory of polymer dynamics. Oxford University Press, Oxford
31. de Gennes PG (1971) Reptation of a polymer chain in the presence of fixed obstacles. J Chem Phys 55(2):572–579
32. Weber SC, Theriot JA, Spakowitz AJ (2010) Subdiffusive motion of a polymer composed of subdiffusive monomers. Phys Rev E 82:011913
33. Lampo TJ, Kuwada NJ, Wiggins PA, Spakowitz AJ (2015) Physical modeling of chromosome segregation in *Escherichia coli* reveals impact of force and DNA relaxation. Biophys J 108(1):146–153
34. Vandebroek H, Vanderzande C (2014) Transient behaviour of a polymer dragged through a viscoelastic medium. J Chem Phys 141 (11):114910
35. Taloni A, Chechkin A, Klafter J (2010) Generalized elastic model yields a fractional Langevin equation description. Phys Rev Lett 104:160602
36. Jasch F, von Ferber Ch, Blumen A (2003) Dynamics of randomly branched polymers: configuration averages and solvable models. Phys Rev E Stat Nonlin Soft Matter Phys 68 (5 Pt 1):051106
37. Burioni R, Cassi D, Cecconi F, Vulpiani A (2004) Topological thermal instability and length of proteins. Proteins 55(3):529–535
38. Reuveni S, Klafter J, Granek R (2012) Dynamic structure factor of vibrating fractals: proteins as a case study. Phys Rev E Stat Nonlin Soft Matter Phys 85(1 Pt 1):011906
39. Amitai A, Holcman D (2013) Polymer model with long-range interactions: analysis and applications to the chromatin structure. Phys Rev E Stat Nonlin Soft Matter Phys 88 (5):052604
40. Tejedor V, Bénichou O, Voituriez R, Jungmann R, Simmel F, Selhuber-Unkel C, Oddershede LB, Metzler R (2010) Quantitative analysis of single particle trajectories: mean maximal excursion method. Biophys J 98 (7):1364–1372
41. Burnecki K, Kepten E, Janczura J, Bronshtein I, Garini Y, Weron A (2012) Universal algorithm for identification of fractional Brownian motion. a case of telomere subdiffusion. Biophys J 103(9):1839–1847
42. Kepten E, Bronshtein I, Garini Y (2011) Ergodicity convergence test suggests telomere motion obeys fractional dynamics. Phys Rev E Stat Nonlin Soft Matter Phys 83(4 Pt 1):041919
43. Saxton MJ (1993) Lateral diffusion in an archipelago. single-particle diffusion. Biophys J 64 (6):1766–1780
44. Simson R, Sheets ED, Jacobson K (1995) Detection of temporary lateral confinement of membrane proteins using single-particle tracking analysis. Biophys J 69(3):989–993
45. Amitai A, Toulouze M, Dubrana K, Holcman D (2015) Analysis of single locus trajectories for extracting in vivo chromatin tethering interactions. PLoS Comput Biol 11(8): e1004433
46. Wang B, Kuo J, Bae SC, Granick S (2012) When Brownian diffusion is not Gaussian. Nat Mater 11(6):481–485

47. Höfling F, Franosch T (2013) Anomalous transport in the crowded world of biological cells. Rep Prog Phys 76(4):046602
48. Ghosh SK, Cherstvy AG, Grebenkov DS, Metzler R (2016) Anomalous, non-Gaussian tracer diffusion in crowded two-dimensional environments. New J Phys 18(1):013027
49. Shell MS, Debenedetti PG, Stillinger FH (2005) Dynamic heterogeneity and non-Gaussian behaviour in a model supercooled liquid. J Phys Condens Matter 17(49):S4035
50. Javer A, Kuwada NJ, Long Z, Benza VG, Dorfman KD, Wiggins PA, Cicuta P, Cosentino Lagomarsino M (2014) Persistent super-diffusive motion of Escherichia coli chromosomal loci. Nat Commun 5:3854
51. Weber SC, Spakowitz AJ, Theriot JA (2012) Nonthermal ATP-dependent fluctuations contribute to the in vivo motion of chromosomal loci. Proc Natl Acad Sci USA 109:7338–7343

Part V

Biophysics of the Bacterial Nucleoid

Chapter 22

Isolation and Characterization of Bacterial Nucleoids in Microfluidic Devices

James Pelletier and Suckjoon Jun

Abstract

We report methods for isolation of *Escherichia coli* nucleoids in microfluidic devices, allowing characterization of nucleoids during a controlled in vivo to in vitro transition. Biochemically, nucleoids are isolated by gentle osmotic lysis, which minimally perturbs nucleoid-associated proteins (NAPs). Biophysically, nucleoids are isolated in microfluidic chambers, which mimic confinement within the cell, as well as facilitate diffusive buffer exchange around nucleoids without subjecting them to flow. These methods can be used to characterize interactions between NAPs and whole nucleoids, and to investigate nucleoid structure and dynamics in confinement. We present protocols for isolation, quantification, and perturbation of nucleoids in microfluidic confinement.

Key words Nucleoid isolation, Microfluidics, Nucleoid-associated proteins, Molecular crowding

1 Introduction

During the *Escherichia coli* cell cycle, physiological processes such as regulation of gene expression, and genome replication and segregation, depend on nucleoid structure and dynamics. How do nucleoid-associated proteins (NAPs) and confinement within the cell facilitate these complex functions? About ten known NAPs contribute to nucleoid organization and have a wide range of functions [1, 2]. Their abundances depend on the growth physiology and change from exponential to stationary phase [3]. Diverse experiments have elucidated how the nucleoid is influenced by NAPs and thermodynamic driving forces such as macromolecular crowding and conformational entropy.

In vitro approaches complement and facilitate measurements and perturbations not possible in vitro, but they often involve simplified DNA substrates and NAP mixtures. We developed a microfluidic device to isolate whole nucleoids in chambers, offering a bridge between in vivo and in vitro regimes [4]. The devices offer

Olivier Espéli (ed.), *The Bacterial Nucleoid: Methods and Protocols*, Methods in Molecular Biology, vol. 1624,
DOI 10.1007/978-1-4939-7098-8_22,

control over the timing of lysis, allowing us to image the lysis process.

We adapted a gentle osmotic lysis method that minimally perturbs DNA–protein interactions and yields nucleoids free of cell envelope fragments [5]. Thus, the nucleoids begin with their full complement of NAPs at in vivo concentrations, at least an order of magnitude greater than typical in vitro concentrations.

After nucleoid isolation, the microfluidic devices offer biochemical and biophysical control over the nucleoids. The chambers enable rapid ambient buffer exchange via diffusion, without subjecting the nucleoids to flow. Furthermore, different micron-scale chamber geometries impose different degrees of confinement. We anticipate microfluidic nucleoid isolation will enable biochemical characterization of diverse NAPs as well as biophysical characterization of nucleoid structure and dynamics in confinement.

2 Materials

2.1 Microfluidics

- Polydimethylsiloxane (PDMS).
- Coring tool, such as 1.5 mm inner diameter, which is compatible with 1/16″ outer diameter tubing.
- Glass coverslips.
- Plasma system, to bond PDMS to glass.
- Poly-L-lysine (20 kDa) grafted with polyethylene glycol (5 kDa) (SuSoS AG, PLL(20)-g[3.5]-PEG(5)).
- Syringe, such as 250 μL Hamilton Gastight Luer tip.
- Luer to tubing adapter and fitting.
- Soft tubing, such as 0.01″ ID, 1/16″ OD HPFA+.
- Stiff tubing, such as 0.005″ ID, 1/16″ OD PEEK tubing.[1]
- Syringe pump.[2]
- Multiport selection valve.[3]

2.2 Buffers and Reagents

- Growth medium, such as Luria–Bertani (LB), stored at room temperature.[4]
- Sucrose buffer, stored at room temperature.

  20% sucrose (w/v).

[1] Stiffer tubing gives improved control over flows, for experiments in which the timing of buffer exchange is critical, such as nucleoid expansion after cell lysis.

[2] For experiments that require precise control of flow, we recommend a pressure-driven pump or hydrostatic pressure system [6].

[3] The valve facilitates buffer changes after nucleoid isolation.

[4] When physiology is important, we recommend optimized synthetic media.

100 mM NaCl.
10 mM NaPi, pH 7.3.
10 mM EDTA.

- Lysis buffer, stored at 4 °C.
  40–200 mM NaCl.
  20 mM Na-HEPES, pH 7.5.
  0.5 mg/mL bovine serum albumin (BSA).
- Lysozyme, from chicken egg white, stored at −20 °C.
- 0.1 mg/mL PLL-g-PEG in 10 mM Na-HEPES, pH 7.4. Store solution at 4 °C for up to 2 weeks.[5]
- Hoechst 33342, trihydrochloride, trihydrate, 100 mg (Thermo Fisher H1399), store powder at room temperature in dark. Prepare 10 mg/mL stock solution in water and store at 4 °C in dark.

3 Methods

3.1 Microfluidic Device

The microfluidic device (Fig. 1) facilitates gentle nucleoid isolation by osmotic shock then subsequent biochemical perturbation of the nucleoids, without subjecting the nucleoids to flow. In particular, nucleoids are isolated in microfluidic chambers. Like a cul-de-sac, each chamber opens on one end to a channel, whereas the other end is closed, so there is no flow through the chamber. Rather, buffer in the chambers exchanges with buffer in the channel via diffusion.

To investigate nucleoids in channel-like confinement, we used chambers shaped as rectangular prisms, 1.7 μm wide × 1.5 μm deep × 25 μm long, just wide and deep enough to accommodate a single row of cells. Over that chamber length, diffusive buffer exchange happens in several seconds. Spaced 5 μm apart, to decrease scattering of fluorescence between them, chambers are arranged perpendicular to a channel, 100 μm wide × 30 μm deep × several cm long. The channel connects on one end to an inlet and on the other to an outlet. The devices are made of polydimethylsiloxane (PDMS) and a glass coverslip.

3.1.1 Microfluidic Master Fabrication

Detailed microfluidic master fabrication protocols are available on the Jun lab website (https://jun.ucsd.edu). In contrast to the cell growth devices, which are treated with pentane then acetone to remove uncured PDMS toxic to cells, nucleoid devices are not treated, as any treatment can distort the chamber geometries.

[5] Static electricity causes the highly charged powder to fly unpredictably, making it challenging to dispense small amounts. A long, beveled metallic needle worked well to transfer PLL-g-PEG. Store PLL-g-PEG powder at −20 °C.

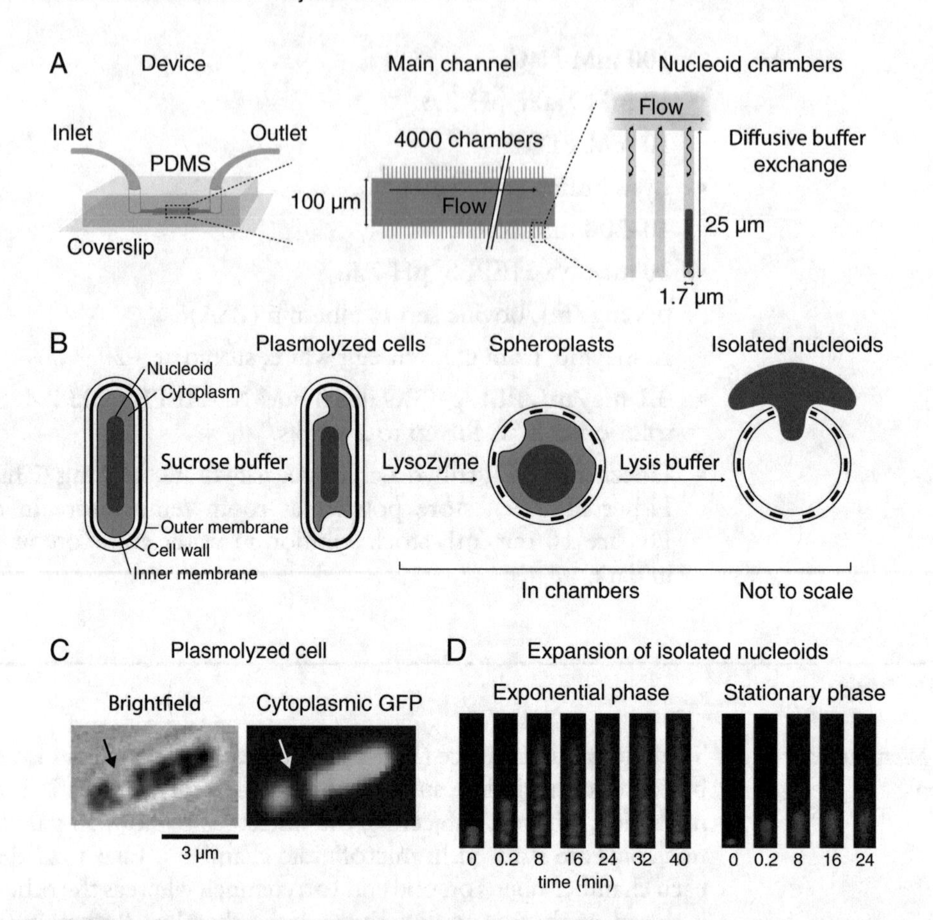

Fig. 1 Microfluidic lysis procedure. (**a**) The microfluidic device facilitates sequential buffer exchange around cells and isolated nucleoids. A PDMS device bonded to a glass coverslip contains a main channel with thousands of side chambers in which nucleoids are isolated. Diffusion between the channel and chambers enables rapid buffer exchange without subjecting the nucleoids to flow. (**b**) To isolate nucleoids, cells are first plasmolyzed in hypertonic sucrose buffer, relieving turgor pressure on the cell wall. Next, cells are incubated with lysozyme which digests the cell wall, causing cells to lose their rod shape and become more spherical. Last, rapid switch to a hypotonic lysis buffer restores turgor pressure and lyses the vulnerable cell, releasing the nucleoid into the chamber as the cytoplasm diffuses away. (**c**) In plasmolyzed cells, the inner membrane recedes from the cell wall, visible as a region of decreased contrast in brightfield images, and as a region without cytoplasmic GFP fluorescence. (**d**) After lysis, the nucleoid expands rapidly in seconds, then more slowly over tens of minutes. Nucleoid structure varies dramatically with cell physiology, with much larger and more heterogeneous nucleoids from exponential phase cells than stationary phase cells

3.1.2 Device Assembly and Surface Passivation

To prevent nonspecific adsorption of nucleoids, microfluidic surfaces were passivated with 20 kDa poly-L-lysine covalently grafted (grafting ratio 3.5) to 5 kDa polyethylene glycol, PLL(20)-g[3.5]-PEG(5) (SuSoS AG) [7]. We attempted other passivation methods, including bovine serum albumin (BSA), sheared salmon sperm DNA, and *E. coli* lipid extract. In our experience, PLL-g-PEG worked best to reduce nonspecific interactions between nucleoids and microfluidic surfaces. If surface passivation is effective, then nucleoids slide along the chambers in response to osmotic buffer changes, and nucleoids near the chamber exits may get pulled out of the chambers altogether.

1. Clean glass coverslips.[6]
2. Using a coring tool, core holes through the PDMS device for the inlet and outlet. Core the holes from the surface with channels to the opposite smooth surface, so the holes align with the channels, as shown in Fig. 1.
3. Bond PDMS device to glass coverslip.
4. After bonding, bake device for at least 10 min at 65 °C.
5. Let the device cool to room temperature, then immediately fill the device with 0.1 mg/mL PLL-g-PEG solution in 10 mM Na-HEPES, pH 7.4 (Subheading 2.2).[7]
6. Incubate the device with PLL-g-PEG solution for 30 min at room temperature.
7. Wash the device with sterile water.[8]

3.2 Nucleoid Isolation

3.2.1 Preparation of Cell Cultures

Nucleoid structure depends on cell physiology, so culture conditions influence nucleoid structure. In any growing culture, cells in different cell cycle stages contain different numbers of chromosome equivalents. In steady state growth, if the chromosome replication time exceeds the doubling time, *E. coli* perform multifork replication. In non-steady state growth, the number of chromosome equivalents per cell may change over time. For example, in LB medium the number of chromosome equivalents decreases from exponential to stationary phase [8]. The distribution of nucleoid structures in the culture is influenced by its history, not just its current optical density. While this protocol uses LB, we also recommend synthetic media and further optimization when physiology is important.

1. The day before the experiment, inoculate cells from an −80 °C glycerol stock into growth medium. We grew cells in 3 mL LB medium at 37 °C in 15 mL polypropylene round bottom tubes, shaken for aeration.
2. Grow the cells to early exponential phase, then dilute the culture at least 1000-fold to increase the number of generations after inoculation.
3. Grow cells to final optical density.

---

[6] We cleaned coverslips with ethanol then deionized water before the ethanol dried. If necessary, use a more elaborate cleaning procedure.

[7] The polycationic PLL associates with hydroxyl moieties generated by plasma treatment of the PDMS and glass surfaces.

[8] When replacing one buffer with another, infuse several times the total volume of the device. It is preferable to use a syringe pump to flow 50 μL at 10 μL/min, though it is possible to gently infuse at a similar rate by hand. We use HPFA+ tubing to infuse all buffers, with the exception of the lysis buffer for which we use the stiffer PEEK tubing (Subheading 2.1).

For exponential phase cells, we grow cultures to OD 0.3.

For stationary phase cells, track the number of hours after an exponential phase OD reference point.

3.2.2 Spheroplast Formation

Nucleoid isolation requires effective spheroplast formation, which in turn requires effective plasmolysis. The hypertonic sucrose buffer induces plasmolysis—recession of the inner membrane from the cell wall [9]. At 20% (w/v) (0.58 M) sucrose, cells remain viable [10]. Cells are plasmolyzed outside the device then loaded into microfluidic chambers. Spheroplasts are more delicate than plasmolyzed cells, so they are formed in the chambers rather than flowed into the device. Lysozyme treatment digests the peptidoglycan cell wall, making the cells susceptible to osmotic pressure changes, causing cells to rupture in hypotonic solution.

1. Centrifuge 750 μL cell culture in a 1.5 mL tube for 1 min at 16×10^3 rcf.[9]
2. Remove as much supernatant as possible, using a 200 μL pipette tip.
3. Resuspend cells in 750 μL sucrose buffer (Subheading 2.2), pipetting gently to disperse the pellet. Do not vortex.
4. Leave plasmolyzed cells gently rotating on a rotisserie at room temperature until loading them to a passivated microfluidic device, no more than 1 h after resuspending them in sucrose buffer.
5. Flow sucrose buffer into the microfluidic device, to displace water after PLL-g-PEG surface treatment.
6. Prepare a 10 mg/mL lysozyme stock solution in sterile water. Keep on ice and use on the same day. Prepare 300 μg/mL lysozyme in sucrose buffer.
7. Centrifuge plasmolyzed cells for 1 min at 16×10^3 rcf, then resuspend in 100 μL sucrose buffer without lysozyme.
8. Flow plasmolyzed cells into the device.
9. Load cells into chambers. Depending on the experiment, we aimed for one or two cells per chamber.

   Option 1. Allow cells to drift into chambers, which works well for concentrated stationary phase cells. PDMS is permeable to water [11], and the permeation flux helps load cells into the chambers.

   Option 2. Spin the entire device using a small microcentrifuge, which helps load larger exponential phase cells into

[9] Depending on the optical density of the culture, the volume of the microfluidic device, and the strategy to load cells into chambers (**step 9**), scale the volume up or down.

chambers, by increasing the concentration of cells near the chamber entrances.

10. Wait for several minutes for cells to drift to the chamber ends.
11. Flow lysozyme in sucrose buffer into the device.
12. Incubate cells in lysozyme for 30 min at 30 °C.

3.2.3 Cell Lysis and Nucleoid Isolation

After lysozyme treatment, hypotonic buffer lyses vulnerable spheroplasts, releasing nucleoids into the chamber. After lysis, the cytoplasm diffuses out of the chamber, but the cell envelope remains in the chamber. Lysis buffer should be infused as fast as possible, to thoroughly remove lysozyme at the moment of lysis, as lysozyme is cationic and can compact DNA.

As the salt concentration in the lysis buffer increased, the synchronization of lysis decreased. At 100 mM NaCl, almost all cells lysed at the moment lysis buffer arrived. At 150 mM NaCl, many cells lysed minutes after lysis buffer arrived. At 200 mM NaCl, the highest salt concentration studied, we observed little to no lysis. Salts increase the osmotic pressure and thus decrease the osmotic shock. Furthermore, monovalent ions increase membrane stability by screening electrostatic repulsion between lipids, whereas multivalent ions dramatically increase membrane stability [12].

1. Fill PEEK tubing with lysis buffer (Subheading 2.2).
2. Arrange the tubing so it does not torque the microfluidic device.[10]
3. Make sure the flow is stopped completely.[11]
4. Press tubing into inlet, leaving about 0.5 μL dead volume in the inlet.
5. Mount the device on the microscope.[12]
6. To image nucleoid expansion after lysis, start acquisition before starting the flow of lysis buffer.[13]
7. Start the flow of lysis buffer at 2000 μL/h.
8. After lysis, continue to flow lysis buffer at 50 μL/h, for the duration of the experiment.[14]

[10] Tape the tubing to the stage, so that the device and tubing move together as a firm unit when the stage moves. Add a short 90° bend just above the device inlet, so that the tubing is not pushed by the condenser lens.

[11] Otherwise, spheroplasts may lyse before the start of image acquisition. A pressure-driven pump or hydrostatic pressure system [6] offers precise control of flow.

[12] To prevent motion of the device during the experiment, adhering the device with double-sided tape to a stage insert works well.

[13] At 100 mM NaCl, lysis is nearly instantaneous upon arrival of lysis buffer. Since nucleoid expansion after lysis is rapid, per experiment we often follow just one field of view containing about ten nucleoids. At higher salt concentrations, such as 150 mM NaCl, many cells lyse minutes after infusion of lysis buffer.

[14] This helps to maintain a constant biochemical environment, for example to wash dissociated proteins from the device.

3.3 Imaging

3.3.1 Microscopy

We imaged nucleoids and cells on an inverted microscope (Nikon Eclipse Ti-E) with a high numerical aperture 100× objective lens (Nikon CFI Apo TIRF 100× oil). To minimize photodamage and photobleaching, we imaged with an EM-CCD (Hamamatsu ImagEM C9100-13) with minimal illumination, using the ND32 neutral density filter on the lamp (Nikon Intensilight C-HGFI). We used widefield epifluorescence to image nucleoids, and differential interference contrast (DIC) to image the cell envelope after lysis. To image dynamics at short and long time scales, we acquired images in two phases.

1. For the first minute, we image at 20 frames per second, with the shutter open continuously, to image rapid expansion of the nucleoid at the moment of lysis.
2. For the rest of the experiment, we image at 5 s per frame, with the shutter closed between frames, to minimize exposure while imaging slow continued expansion and structural relaxation of the nucleoid.

3.3.2 Fluorescent Probes

As a native marker of the whole nucleoid, we used the nucleoid-associated protein HU fused to a fluorescent protein [13]. HU is highly abundant, with on the order of 10^4 copies per cell, and it exhibits low sequence specificity [14, 15]. HU is a dimer, composed of α or β subunits, encoded by the genes *hupA* or *hupB*, respectively. HU influences nucleoid structure, so to express the HU fluorescent fusion at natural levels, we used a strain with *hupA* replaced by *hupA-mCherry* or *hupA-GFP* at its native chromosomal locus [16, 17]. The fluorescent fusion was functional, as deletion of *hupB* did not cause cellular filamentation, characteristic of double mutants lacking both *hupA* and *hupB* [18].

DNA dyes, such as Hoechst 33342 (Subheading 2.2), are another nucleoid labeling option. Hoechst permeates cells, so before resuspending cells in sucrose buffer, add 10 μg/mL Hoechst to the growth medium and incubate for 20 min at room temperature. Alternatively, Hoechst may be included in the lysis buffer to label nucleoids after isolation. Hoechst permeates the PDMS, increasing the background fluorescence, but the signal from nucleoids was often high enough to discern the nucleoids over the background.

3.4 Biochemical Characterization of Nucleoid-Associated Proteins

Microfluidic nucleoid isolation enables biochemical characterization of nucleoid-associated proteins (NAPs), such as measurement of their unbound fractions and off-rates. During the in vivo to in vitro transition, the whole nucleoids start with a full complement of NAPs at in vivo concentrations. After lysis, NAPs dissociate from nucleoids and escape from the chambers.

Isolate nucleoids as described in Subheading 3.2, then measure the total mCherry intensity to estimate the amount of

HU-mCherry remaining on the nucleoid. To estimate photobleaching, we measured the fluorescence decrease in unlysed cells. During the first acquisition phase, with the shutter open continuously, photobleaching was significant, with an exponential decay constant of about 2 min. During the second phase, with the shutter closed between frames, photobleaching was negligible.

3.4.1 Measurement of Unbound Fraction

After lysis, cytoplasmic proteins not associated with the nucleoid, including transiently unbound NAPs, diffuse from the microfluidic chambers. At the moment of lysis, the total HU-mCherry intensity abruptly decreased several percent (Fig. 2), which we identify as the fraction of HU not initially bound to the nucleoid.

The maximum fractional drop $I_{HU}(t + 50\text{ ms})/I_{HU}(t)$ between subsequent 50 ms time points estimates a lower limit on the fraction of unbound HU-mCherry. This metric underestimates the unbound fraction, as some HU-mCherry may not escape within the first 50 ms after lysis, and some HU-mCherry remains trapped in the cell envelope after lysis.

3.4.2 Measurement of Off-Rate

After nucleoid isolation, the device enables measurement of the off-rate versus time. Soon after lysis, trajectories exhibited nonexponential decay, perhaps due to a range of binding site affinities (Fig. 2). For example, HU has a higher affinity for kinked, cruciform, and nicked DNA structures [19–21]. At later times, the trajectories converged to exponential decay. We estimated a lower limit on the off-rate based on the exponential regime. The half-life

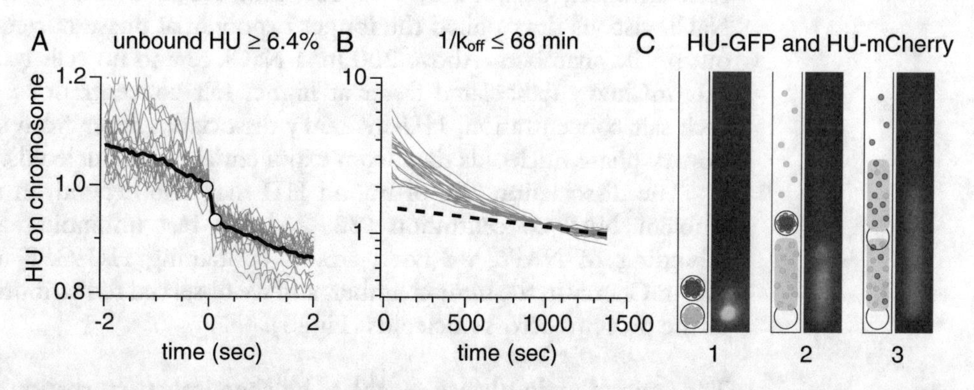

Fig. 2 Biochemical characterization of HU-mCherry by monitoring its dissociation from isolated nucleoids. (**a**) At the moment of lysis, the integrated HU-mCherry fluorescence abruptly drops a few percent, which we attribute to the rapid escape of cytoplasmic HU-mCherry not initially bound to the nucleoid. Cytoplasmic fluorescent proteins diffuse away abruptly after lysis (data unpublished). *Gray traces* represent individual trajectories, and the *black trace* represents the average trajectory. (**b**) HU-mCherry exhibits nonexponential decay at short times and approaches exponential decay at long times. Fitting the latter regime provides a lower bound on the off-rate. (**c**) HU redistributes across different nucleoids in the same chamber. The cell containing HU-GFP lyses first. After the cell containing HU-mCherry lyses, some HU-mCherry binds to the HU-GFP nucleoid and vice versa

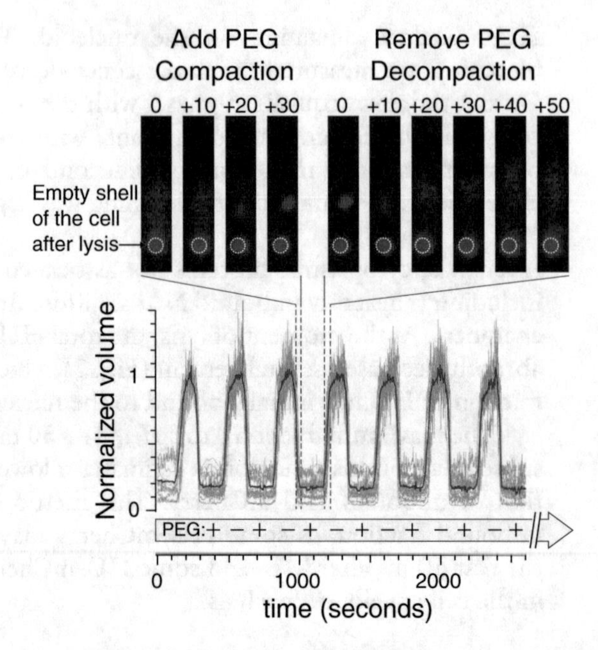

Fig. 3 Molecular crowding of isolated nucleoids. After nucleoid isolation, we oscillated the concentration of 20 kDa polyethylene glycol (PEG) in the ambient buffer. At 29% (w/v) PEG, nucleoids were abruptly compacted to in vivo size. At 3% (w/v) PEG, nucleoids expanded again, with dynamics comparable to expansion after lysis. Nucleoid compaction and decompaction was reversible and repeatable for many cycles

is a useful approximation for how long NAPs are bound to the nucleoid.

We measured dependence of the off-rate on the ambient salt concentration, between 40 and 200 mM NaCl. Below 40 mM NaCl, viscous drag pulled the longer exponential phase nucleoids out of the chambers. Above 200 mM NaCl, few to no cells lysed. HU-mCherry dissociated faster at higher salt concentrations. At each salt concentration, HU-mCherry dissociated faster from stationary phase nucleoids than from exponential phase nucleoids.

The dissociation rate of bound HU may also depend on the ambient NAP concentration [22, 23]. To test unbinding and rebinding of NAPs, we lysed strains containing HU-GFP and HU-mCherry in the same chamber, and we observed redistribution of the proteins across nucleoids (Fig. 3).

3.5 Osmotic Compression of Nucleoids

The microfluidic device enables biochemical perturbation of isolated nucleoids. For example, we exposed isolated nucleoids to different polyethylene glycol (PEG) solutions, to study the effects of molecular crowding on nucleoid structure.

For fast and precise control of the ambient buffer, we used a multiport valve (Subheading 2.1) to switch between solutions with 3% (w/v) or 29% (w/v) PEG, each controlled by its own syringe pump. We used LabVIEW software to coordinate switching the valve with changing the syringe pump infuse rates. Alternatively, use a microfluidic mixer, a module that merges and mixes streams from the different syringe pumps. In principle, a mixer can interpolate between the high and low concentrations, based on the relative infuse rates. In practice, make sure that the module sufficiently mixes the laminar streams, and characterize the transients after changing infuse rates. To decrease the delay after buffer change, minimize the length of tubing between the valve and device.

Acknowledgments

J.F.P. was supported by a Fannie and John Hertz Graduate Fellowship, and S.J. by Paul G. Allen Foundation, Pew Charitable Trusts, NSF CAREER, and NIH GM118565-01.

References

1. Dillon SC, Dorman CJ (2010) Bacterial nucleoid-associated proteins, nucleoid structure and gene expression. Nat Rev Microbiol 8(3):185–195
2. Stavans J, Oppenheim A (2006) DNA-protein interactions and bacterial chromosome architecture. Phys Biol 3:R1
3. Ali Azam T, Iwata A, Nishimura A, Ueda S, Ishihama A (1999) Growth phase-dependent variation in protein composition of the *Escherichia coli* nucleoid. J Bacteriol 20:6361
4. Pelletier J, Halvorsen K, Ha B-Y, Paparcone R, Sandler SJ, Woldringh CL, Wong WP, Jun S (2012) Physical manipulation of the *Escherichia coli* chromosome reveals its soft nature. Proc Nat Acad Sci U S A 109:E2649–E2656
5. Cunha S, Odijk T, Süleymanoglu E, Woldringh CL (2001) Isolation of the *Escherichia coli* nucleoid. Biochimie 83:149–154
6. Amir A, Babaeipour F, McIntosh DB, Nelson DR, Jun S (2014) Bending forces plastically deform growing bacterial cell walls. Proc Nat Acad Sci U S A 111:5778–5783
7. Lee S, Vörös J (2005) An aqueous-based surface modification of poly(dimethylsiloxane) with poly(ethylene glycol) to prevent biofouling. Langmuir 21:11957–11962
8. Akerlund T, Nordstrom K, Bernander R (1995) Analysis of cell size and DNA content in exponentially growing and stationary-phase batch cultures of *Escherichia coli*. J Bacteriol 177:6791–6797
9. Rojas E, Theriot JA, Huang KC (2014) Response of *Escherichia coli* growth rate to osmotic shock. Proc Natl Acad Sci U S A 111:7807–7812
10. Pilizota T, Shaevitz JW (2014) Origins of *Escherichia coli* growth rate and cell shape changes at high external osmolality. Biophys J 107:1962–1969
11. Randall GC, Doyle PS (2005) Permeation-driven flow in poly(dimethylsiloxane) microfluidic devices. Proc Natl Acad Sci U S A 102:10813–10818
12. Ha B-Y (2001) Stabilization and destabilization of cell membranes by multivalent ions. Phys Rev E 64:051902
13. Wery M, Woldringh CL, Rouviere-Yaniv J (2001) HU-GFP and DAPI colocalize on the *Escherichia coli* nucleoid. Biochimie 83:193–200
14. Ali Azam T, Hiraga S, Ishihama A (2000) Two types of localization of the DNA-binding proteins within the *Escherichia coli* nucleoid. Genes Cells 5:613–626
15. Krylov AS, Zasedateleva OA, Prokopenko DV, Rouviere-Yaniv J, Mirzabekov AD (2001) Massive parallel analysis of the binding specificity of histone-like protein HU to single- and double-stranded DNA with generic oligodeoxyribonucleotide microchips. Nucleic Acids Res 29(12):2654–2660
16. Centore RC, Lestini R, Sandler SJ (2008) XthA (Exonuclease III) regulates loading of RecA onto DNA substrates in log phase *Escherichia coli* cells. Mol Microbiol 67:88–101
17. Marceau AH, Bahng S, Massoni SC, George NP, Sandler SJ, Marians KJ, Keck JL (2011) Structure of the SSB-DNA polymerase III interface and its role in DNA replication. EMBO J 30:4236–4247
18. Dri AM, Rouviere-Yaniv J, Moreau PL (1991) Inhibition of cell division in hupA hupB mutant bacteria lacking HU protein. J Bacteriol 173:2852–2863

19. Pontiggia A, Negri A, Beltrame M, Bianchi ME (1993) Protein HU binds specifically to kinked DNA. Mol Microbiol 7:343–350
20. Kamashev D, Balandina A, Rouviere-Yaniv J (1999) The binding motif recognized by HU on both nicked and cruciform DNA. EMBO J 18:5434–5444
21. Kamashev D, Rouviere-Yaniv J (2000) The histone-like protein HU binds specifically to DNA recombination and repair intermediates. EMBO J 19:6527–6535
22. Graham JS, Johnson RC, Marko JF (2011) Concentration-dependent exchange accelerates turnover of proteins bound to double-stranded DNA. Nucleic Acids Res 39:2249–2259
23. Hadizadeha N, Johnson RC, Marko JF (2016) Facilitated dissociation of a nucleoid protein from the bacterial chromosome. J Bacteriol 198:1735–1742

Chapter 23

Modeling Bacterial DNA: Simulation of Self-Avoiding Supercoiled Worm-Like Chains Including Structural Transitions of the Helix

Thibaut Lepage and Ivan Junier

Abstract

Under supercoiling constraints, naked DNA, such as a large part of bacterial DNA, folds into braided structures called plectonemes. The double-helix can also undergo local structural transitions, leading to the formation of denaturation bubbles and other alternative structures. Various polymer models have been developed to capture these properties, with Monte-Carlo (MC) approaches dedicated to the inference of thermodynamic properties. In this chapter, we explain how to perform such Monte-Carlo simulations, following two objectives. On one hand, we present the self-avoiding supercoiled Worm-Like Chain (ssWLC) model, which is known to capture the folding properties of supercoiled DNA, and provide a detailed explanation of a standard MC simulation method. On the other hand, we explain how to extend this ssWLC model to include structural transitions of the helix.

Key words Monte-Carlo methods, DNA supercoiling, Worm-like chain, Plectonemes, DNA denaturation, Structural transitions, Multi-scale simulations

1 Introduction

In bacteria, chromosomes are partly structured by DNA supercoiling. Compared to its natural helicity, DNA is indeed often found in an underwound form in vivo, as a result of a yet-to-be-understood balance between transcription, replication and the action of topoisomerases and nucleoid associated proteins [1–3]. On one hand, DNA supercoiling leads to the formation of plectonemes. This can be explicitly shown using Worm-Like Chain (WLC) models of DNA that include supercoiling constraints and self-avoidance properties reflecting the impenetrable character of the DNA molecule [4–6]. Specifically, supercoiling constraints make molecules buckle so that they absorb, under the form of writhe, some of the excess or depletion of twist, while the short-range repulsion (self-avoidance) results in an effective entropic repulsive force that determines the radius of plectonemes [7]. Self-avoiding supercoiled

Olivier Espéli (ed.), *The Bacterial Nucleoid: Methods and Protocols*, Methods in Molecular Biology, vol. 1624,
DOI 10.1007/978-1-4939-7098-8_23, © Springer Science+Business Media LLC 2017

WLC (ssWLC) models are thus able to capture the folding properties of supercoiled DNA on the scale of several kilo base pairs (kbp), or tens of kbps [8], both for overwound (positive supercoiling) and underwound (negative supercoiling) DNA at low tension forces. On the other hand, negative supercoiling can induce structural transitions towards DNA forms different from the canonical B-DNA. Among a large number of possible alternative structures [9], supercoiled DNA can locally form denaturation bubbles [10, 11] or adopt left-handed DNA forms such as Z-DNA [12] or the so-called L-DNA [13, 14].

Since most bacterial chromosomes are negatively supercoiled, models dedicated to biological applications are destined to capture the balance between super-structuring (plectonemes) and local structural changes of the DNA-helix. Several recent approaches have thus been proposed to tackle this multi-scale problem. These include phenomenological models of the co-existence of the plectonemic and denatured states [12, 15, 16] as well as polymer models at the resolution of a single base [17] (*see* [18] for a recent review of single-base based models).

Here, we will explain how to design a discrete version of the ssWLC model in order to simulate the competition between plectoneme formation and structural transitions of the DNA helix, resulting in a 10-to-20-bps resolution model that can be used to simulate several kbps long molecules under negative supercoiling (Lepage and Junier, in prep). To this end, we first aim at recalling the definition of the discrete version of the ssWLC and at providing a detailed description of the methods used to simulate its equilibrium folding properties.

2 Methods

Methods are organised as follows. First, we recall the definition of the ssWLC and explain the discretization procedure to simulate it. We next explain how to parametrize the fundamental units of the model in order to solve the problem of the conservation of the linking number, which is at the root of the supercoiling constraints. We then recall principles of thermodynamics-oriented Monte-Carlo methods and discuss the problems of the detection of collisions and chain crossing, which are the most time-consuming steps of the simulations. Finally, we explain how to include structural transitions.

2.1 The Discrete Self-avoiding Supercoiled WLC Model

A WLC model provides a continuous description of a polymer chain. At the microscopic level, the simplest model is defined by a single bending modulus characterizing the cost for the chain to locally bend. The chain then has a certain persistence length (ℓ_p), below which it keeps memory of its orientation. Typically, $\ell_p \approx$ 50 nm for B-DNA at physiological salt concentration.

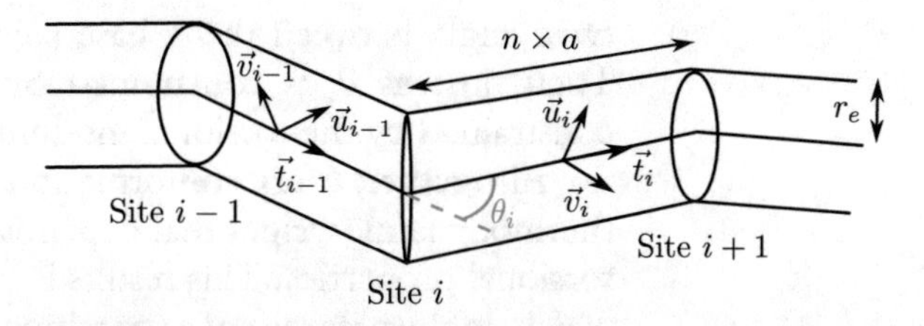

Fig. 1 A discrete model of the ssWLC. Here, two complete cylinders surrounding a site i are represented, along with their local frames ($\vec{t}$, $\vec{u}$, $\vec{v}$) used to compute the bending angle θ_i and the twist angle ϕ_i (not indicated, see text)

To investigate the spatial properties of a WLC, one generally resorts to a discretization procedure, which consists in dividing the chain into a succession of $N-1$ identical segments articulated by N sites (Fig. 1), counting the two sites located at the extremities of the chain—in the case of a circular molecule (e.g., a plasmid), there are as many segments as sites, and the first site joins the last and first segments together. For DNA molecules in physiological conditions of salt, an additional classical simplification consists in considering a hard-core description of the electrostatic repulsion of the negatively charged DNA backbone, such that segments of the discrete WLC are in fact impenetrable cylinders characterized by a radius r_e (Fig. 1), with r_e depending on the salt concentration [19] (for instance, $r_e = 2$ nm for [NaCl]=100 mM). Altogether, this framework defines the discrete self-avoiding WLC.

In this context, each site i is associated with a "discrete" bending modulus (K_d), which constrains the amplitude of the bending angle (θ_i, see next section for an operational definition) between the tangent vectors of the cylinders $i-1$ and i (Fig. 1), and whose value is adjusted depending on the level of discretization—note here that "the cylinder i" corresponds to the cylinder located between the sites i and $i+1$ (for simplicity, we drop the reference to the sites and cylinders in the indexing). Specifically, denoting k_B the Boltzmann constant, T the temperature ($T = 310$ K in physiological conditions), n the number of bps per cylinder, and a the average distance between any two consecutive bps ($a = 0.34$ nm for B-DNA) such that na is the length of a cylinder, the associated persistence length reads $\ell_p = na\frac{K_d}{k_B T}$. For a given value of ℓ_p, then, the smaller the discretization is, the larger the value of K_d. Note, here, that it is recommended to work with cylinders small enough to avoid discretization artifacts, a typical choice being five cylinders per ℓ_p [5] such that $n = 30$ bps for B-DNA (*see* **Note 1**).

In the case of a ssWLC model, a torsional modulus C must additionally be considered to account for the cost associated with the twist deformation of the chain (C is typically on the order of 100 nm for B-DNA [20, 21]). In this context, an average twist angle (ϕ_i, see next section for an operational definition) is associated with each site i of the chain [6, 22]. ϕ_i is equal to the average twist angle between the n base pairs located closest to the site i. Then, just as θ_i is constrained by the bending modulus, ϕ_i is constrained by the torsional modulus (*see* Eq. 1).

Altogether, every conformation $\mathcal{C}$ of the chain has an intrinsic thermodynamic weight that depends on the associated bending and torsional properties. This results in a conformational energy, $E(\mathcal{C})$, which, in the presence of a stretching force f (as in the case of single-molecule experiments), reads (*see* **Note 2**):

$$E(\mathcal{C}) = \frac{k_B T}{2} \sum_{i=1}^{N} \left[\frac{\ell_p}{na} \theta_i^2 + \frac{nC}{a} (\phi_i - \phi_0)^2 \right] - fz, \qquad (1)$$

with z the extension of the chain along the axis of the force, and ϕ_0 the average twist angle at rest, that is, the unconstrained helicity of DNA ($\phi_0 = 0.6$ for B-DNA). Note also that the bending angles for the extreme sites $i = 1$ and $i = N$ are defined with respect to the axis of the force.

Finally, to simulate the folding properties of such a ssWLC, topological constraints immanent to the double-stranded nature of DNA must be accounted for. Namely, for both circular DNA molecules and linear molecules whose ends cannot rotate, the linking number Lk($=$ Tw $+$ Wr) is an invariant quantity, meaning that the sum of the twist Tw $= (2\pi)^{-1}\sum_{i=1}^{N} \phi_i$ (the number of helices) plus the writhe Wr (the number of loops made by the axis of the molecule around itself, *see* **Note 3**) remains always the same, unless the molecule is cut by, e.g., an enzyme. In other words, should DNA be unwound (or overwound), the net change of the linking number will be distributed between the twist and the writhe. From a simulation viewpoint, this imposes strong constraints on the definition of the twist angle for the discrete ssWLC, as we now explain.

2.2 Bending and Twist Angles

Denoting $\vec{t}_i$ the tangent vector of the cylinder i, the bending angle θ_i between $i - 1$ and i is unambiguously given by (Fig. 1):

$$\cos\theta_i = \vec{t}_{i-1} \cdot \vec{t}_i \qquad (2)$$

Due to the fact that the ssWLC does not include any explicit representation of the DNA helix, the definition of the twist angle has been more ambiguous. Methods using Euler angles between local frames associated with the cylinders were first proposed, and shown to provide an excellent procedure to have a linking number that fluctuates around a fixed value [22]. More recent methods [8, 23, 24] based either on an explicit representation of the double-helix [8] or on a definition of the twist angle coming from the parametrization of the deformation of rigid bodies [23] allow to conserve the linking number exactly during the simulations. In particular, using the "parallel transport" approach developed in [23], it is possible to define a twist angle unambiguously such that after each block rotation of the cylinders (see below), the linking number remains constant. To this end, for every site i, define at the initial time of the simulation a vector orthogonal to $\vec{t}_i$, here called $\vec{u}_i$ (Fig. 1). Next, just as for the $\vec{t}_i$'s, continuously update these $\vec{u}_i$'s by applying the rotation matrix corresponding to the deformation of the chain (see below). In this context, using $\vec{v}_i = \vec{t}_i \times \vec{u}_i$, ϕ_i is given by [24]:

$$\cos \phi_i = \frac{\vec{u}_{i-1} \cdot \vec{u}_i + \vec{v}_{i-1} \cdot \vec{v}_i}{1 + \vec{t}_{i-1} \cdot \vec{t}_i} \tag{3}$$

$$\sin \phi_i = \frac{\vec{v}_{i-1} \cdot \vec{u}_i - \vec{u}_{i-1} \cdot \vec{v}_i}{1 + \vec{t}_{i-1} \cdot \vec{t}_i.} \tag{4}$$

The exact conservation of the linking number can then be verified explicitly by computing the twist and the writhe (*see* **Note 3**). In practice, the writhe involves a sum over all pairs of cylinders and, hence, should be computed only once in a while.

2.3 Monte-Carlo Method: Elementary Moves, Transition Probabilities, and Ergodicity Properties

Given the energy of the conformations (Eq. 1), the hard-core repulsion of the chain (self-avoidance) and the constraint of the conservation of Lk, the equilibrium sampling of the conformations is usually performed using a Monte-Carlo (MC) method. This consists in generating a large number of successive conformations (Fig. 2) such that the occurrence of conformations with energy E eventually

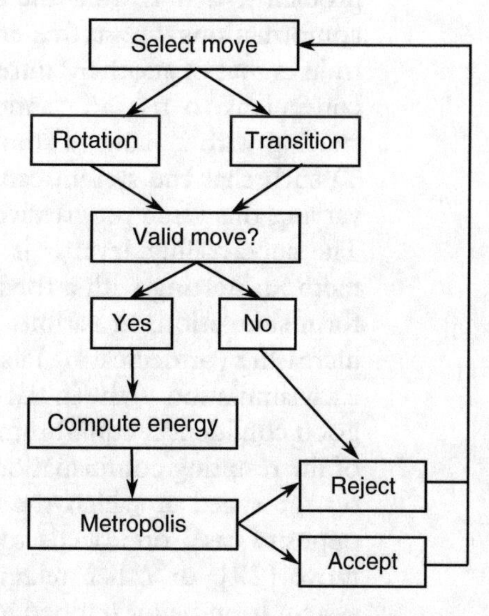

Fig. 2 Summary of the Monte-Carlo Metropolis algorithm. Note that the ssWLC requires only rotations, so that there is no need to select the type of elementary move to try in this case. The second type of move is used to take structural transitions into account. The elementary moves are described in Subheadings 2.4 and 2.6, and the validity check in Subheading 2.5

becomes proportional to $\mathcal{N}(E)\exp(-E/k_{\mathrm{B}}T)$ (Boltzmann's law), where $\mathcal{N}(E)$ is the total number of conformations with energy E. In practice, it is achieved by constructing each conformation from the previous one via trials of elementary moves, starting from an initial random conformation (*see* **Note 4**). To this end, the transformation of a conformation C into another conformation C' is accepted with a certain transition probability $W(C \to C')$ if no collision or crossing is generated during the applied elementary move. Different forms of transition probabilities can be used, the only constraint being that these must verify the detailed balance condition, $W(C \to C')/W(C' \to C) = e^{\frac{E(C)-E(C')}{k_{\mathrm{B}}T}}$, which ensures that at large enough time the chain will visit conformations according to their Boltzmann weight. In this regard, a classical choice is the Metropolis-Hastings transition rate [25, 26], such that:

$$W(C \to C') = \max\left\{1, \exp\left[\left(E(C) - E(C')\right)/k_{\mathrm{B}}T\right]\right\}. \quad (5)$$

Note that in this case any elementary move decreasing the energy is accepted.

A sufficiently high number of successive elementary moves must then be generated in order to get enough uncorrelated conformations such that the sampling of the conformations is representative of thermodynamic equilibrium. It is also important to check that the conformations visited during the simulation do not correspond to a metastable state only (the so-called ergodicity problem), that is, that the system is not trapped in a subset of conformations whose free energy is on the same order of magnitude as that of another "unreached" subset. To this end, it is often convenient to use an "annealing procedure," which consists in starting with a value of some parameter (usually the temperature T) such that the system can quickly reach equilibrium, and then varying this value progressively until the working value is reached. The supercoiling level σ is also a parameter well fitted for this method: starting with a torsionally relaxed molecule, one can perform simulations at various constant values of σ by continuously increasing (or decreasing) its value, using the last conformation of each simulation as the initial condition of the next one. Then, with good confidence, equilibrium is reached if the statistical properties of the resulting conformations at the working value do not depend on the speed at which the annealing has been realized. On the opposite case, one needs to resort to other types of elementary move [27], or other techniques of simulations to prevent the system from being trapped in specific states (see, e.g., [28]).

2.4 Rotations and Associated Change of Bending and Torsional Properties

Most commonly in the MC simulation of a WLC model, an elementary move consists in randomly rotating a block of contiguous cylinders, also called a crankshaft move (Fig. 3). To this end, first define once for the entire simulation a maximum number (M) of

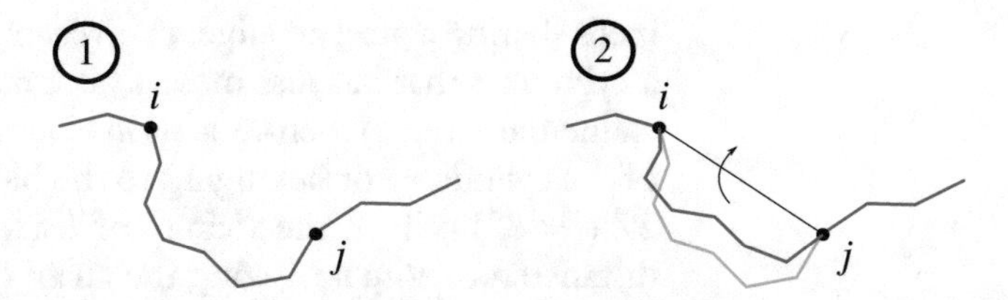

Fig. 3 Rotation: pick a random block [*i*, *j*] of cylinders (1) and rotate it around its axis according to a random angle (2)

cylinders allowed to rotate simultaneously and a maximum angle of rotation (α) (*see* **Note 5**). A rotation then consists in (1) picking a site i at random, (2) picking a number $m < M$ at random, (3) choosing a direction $s \in \{-1, 1\}$, (4) defining the axis of the rotation as the straight line connecting the two sites i and $j = i + s \times m + 1$ (modulo the number of sites), (5) choosing an angle of rotation randomly in $[-\alpha, \alpha]$, and (6) applying the corresponding rotation matrix to the vectors $\vec{t}$ and $\vec{u}$ associated with each site of the block $[i, j]$ (Fig. 3).

After having applied the rotation, test whether a collision or a crossing occurred (explained in Subheading 2.5). If this is the case, the rotation is rejected (*see* **Note 6**). In the opposite case, use Eqs. 2 and 4 to compute the new bending and twist angles at sites i and j (borders of the block). Then, compute the corresponding energies (*see* Eq. 1) to obtain the total energy variation generated by the rotation. Finally, accept the rotation according to the transition probability W (*see* Eq. 5).

For linear molecules, when the site i, the direction s, and the size m of the block are such that the extreme site j falls outside the chain ($j \leq 0$ or $j \geq N + 1$), the crankshaft rotation becomes ill-defined. In this case, apply a rotation to the block extending from i to the end of the molecule (site 1 or site N, depending on the direction s), around a random axis and according to an angle chosen at random in $[-\alpha, \alpha]$. Note that this type of elementary move is necessary to displace the end points of the linear chain; in such case, the extension (z) and, hence, the corresponding stretching energy ($-fz$) must also be updated. In the case of a linear molecule for which the linking number needs to be conserved, extra precaution must also be taken. In particular, it is necessary to define two walls bound to each end of the molecule and perpendicular to the stretching force [5, 8], and to prevent any cylinder from trespassing them (*see* **Note 7**). These walls can be viewed as a simple modelling of the fixed surface and the magnetic bead used in single-molecule experiments.

2.5 Detection of Collisions and Crossings

To detect both collisions and crossings, first build a mesh of the volume inside which the molecule is embedded, using a width of cells (w_{cell}) larger than $na + 2r_e$, so that collisions may only occur between cylinders whose center is in the same cell or in nearest neighbors (cells sharing a face, an edge, or a vertex). The collision detection for a cylinder i that has just moved (as a consequence of the trial of an elementary move) then consists in testing whether i collides with one of the cylinders not belonging to the block but located in one of the 27 ($= 3^3$) cells in the vicinity of i. Note, here, that cylinders at a distance too close to i along the chain (i.e., separated from i by less than $2r_e$ when measured along the axis of the chain) must be ignored since, by construction, they always overlap with i (*see* **Note 1**). Then, to test whether the new position $\vec{r}_i$ of i leads to a collision with a cylinder j at position $\vec{r}_j$, test whether $(\vec{r}_j - \vec{r}_i)\cdot \vec{n} > 2r_e$, where $\vec{n} = \vec{t}_i \times \vec{t}_j$ is the normal to the plane $(\vec{t}_i, \vec{t}_j)$. If the inequality holds, then there is no collision. In the opposite case, check first the distances between the four possible pairs of cylinder ends. If one of these distances is smaller than $2r_e$, then there is a collision. In the opposite case, there is still a possibility of collision in the middle of both cylinders. The intersection of the two straight lines (Site$_i$, $\vec{t}_i$) and (Site$_j$, $\vec{t}_j$) is located at the abscissa u along (Site$_i$, $\vec{t}_i$) and v along (Site$_j$, $\vec{t}_j$), where:

$$u = (\vec{r}_j - \vec{r}_i)\frac{\vec{t}_i - (\vec{t}_i \cdot \vec{t}_j)\vec{t}_j}{1 - (\vec{t}_i \cdot \vec{t}_j)^2} \tag{6}$$

$$v = -(\vec{r}_j - \vec{r}_i)\frac{\vec{t}_j - (\vec{t}_i \cdot \vec{t}_j)\vec{t}_i}{1 - (\vec{t}_i \cdot \vec{t}_j)^2}. \tag{7}$$

Then, the cylinders collide if and only if $0 < u < na$ and $0 < v < na$.

Next, in order to detect crossing events during the rotation of a cylinder i, first build a list of all the cells that may contain a cylinder crossed by i. This list corresponds to the cells whose center lies at a distance smaller than $a + \frac{\sqrt{3}}{2} w_{\text{cell}}$ from the arc formed by the rotation of the center of i (within these cells, just as in the case of collisions, only the non-moving cylinders have to be processed). Next, define a frame $(O', \vec{x}', \vec{y}', \vec{z}')$ associated with the rotation of i (Fig. 4), i.e. set the origin O' anywhere on the axis of the rotation and align $\vec{z}'$ with this axis. Then define $\vec{x}'$ and $\vec{y}'$ arbitrarily in order to complete an orthonormal base (for example, in Fig. 4, $\vec{x}'$ points toward $\vec{r}_i$). In that frame, the surface swept by the support segment of the cylinder i during the rotation is bound by z_0 and z_1, the coordinates of its ends along $\vec{z}'$ (Fig. 4). For any given z in this interval, the cylindrical coordinates of this surface are also easy to determine: $\rho(z)$ is constant and θ lies between $\theta_0(z)$ and $\theta_1(z)$, the initial and final angular coordinates at z. Thus, the coordinates (ρ_I, θ_I, z_I) of a crossing point (intersection between the support segment of a cylinder j and the surface swept by that of i) must verify:

$$z_0 < z_I < z_1 \tag{8}$$

$$\rho_I = \rho(z) \tag{9}$$

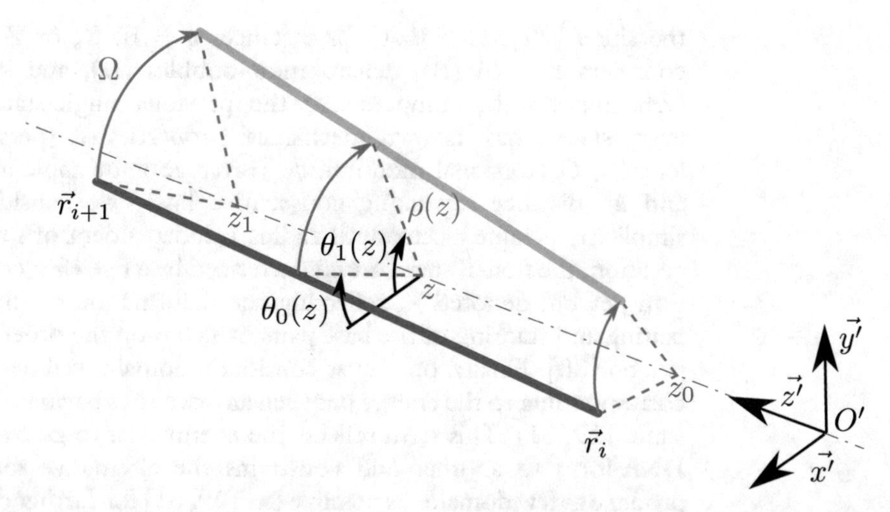

Fig. 4 Schematic representation of the rotation of the support segment of a cylinder (*thick lines*) from the *blue* position to the *green* one. Each point of the segment moves with constant coordinates z and $\rho(z)$ in the frame $(O', \vec{x'}, \vec{y'}, \vec{z'})$. The coordinate θ goes from $\theta_0(z)$ to $\theta_1(z) = \theta_0(z) + \Omega$, where Ω is the angle of the rotation. The *red dashed line* indicates the axis of rotation

$$\begin{cases} \theta_0(z_I) < \theta < \theta_1(z_I) & \text{if } \theta_0(z_I) < \theta_1(z_I) \\ \theta < \theta_0(z_I) \text{ or } \theta > \theta_1(z_I) & \text{otherwise} \end{cases} \quad (10)$$

In addition, this intersection must belong to the support segment of j, i.e. lie on its support line:

$$\rho_I \cos\theta_I = x_j + \frac{t_{jx}}{t_{jz}}(z_I - z_j) \quad (11)$$

$$\rho_I \sin\theta_I = y_j + \frac{t_{jy}}{t_{jz}}(z_I - z_j) \quad (12)$$

and be bound by the ends of j:

$$\begin{cases} z_j < z_I < z_{j+1} & \text{if } z_j < z_{j+1} \\ z_{j+1} < z_I < z_j & \text{otherwise} \end{cases} \quad (13)$$

Equations 9, 11, and 12 lead to a quadratic equation for z_I, the solutions of which (if any) yield ρ_I and θ_I. Then, if exactly one solution verifies the inequalities 8, 10 and 13, a crossing occurred and the trial must be rejected (*see* **Note 8**).

2.6 Including Structural Transitions

Structural transitions of the DNA-helix can be included and simulated in order to capture the multi-scale properties of negatively supercoiled molecules (Lepage and Junier, in prep). To this end, one first needs to modify the form of the conformational energy (Eq. 1) to account for the possibility that the sites can now be associated with different DNA-forms (called states hereafter). This is done by adding a new variable, s_i, reflecting the state of the site i [29], such that, for instance, s_i = B, D, or Z if one considers B-DNA (B), denaturation bubbles (D), and Z-DNA (Z). Importantly, compared to the previous single-state case, every state s has its own mechanical properties: ℓ_s (persistence length), C_s (torsional module), $\phi_{0,s}$ (average twist angle at rest), and a_s (distance separating consecutive bps)—we consider, for simplicity, a single electrostatic radius r_e independent of states. In addition, the non-B states are characterized by a free energy formation per bp, denoted γ_s, reflecting the deformation of the base-pairing and stacking of the base pairs, which is on the order of $k_B T$ per bp [30]. Finally, one must consider a domain wall penalty, J, corresponding to the energy between any two sites having different states [29, 31]. This term reflects the energy cost to go from one DNA-form to another and constrains the alternative forms to produce as few domains as possible (*see* [29, 31] for further details). Altogether, the new energy ($E'(N)$) of a conformation N with multiple possible forms along the chain reads:

$$E'(N) = \frac{k_B T}{2} \sum_{i=1}^{N} \left[n\gamma_{s_i} + \frac{\ell_{s_i}}{n a_{s_i}} \theta_i^2 + \frac{n C_{s_i}}{a_{s_i}} (\phi_i - \phi_{0,s_i})^2 \right] + J \sum_{i=1}^{N-1} \delta_{s_i, s_{i+1}} - f z. \quad (14)$$

where we considered $\gamma_B = 0$ (reference form) and where $\delta_{s_i, s_{i+1}} = 1$ if $s_i = s_{i+1}$, 0 otherwise.

Values of J, C_s, and ℓ_s have been estimated using single-molecule experiments [12–14, 31]. In addition, $\phi_{0,D} = 0$ (by definition), $\phi_{0,Z} = 0.52$ and $a_Z = 0.37$ nm [32] (from crystallographic measurements), while $a_D = 0.54$ nm has been previously used [13]. In this regard, because a_s differs from state to state, one must now consider different sizes of the cylinders. In practice, we use $n(a_i + a_{i+1})/2$ for the length of the cylinder defined by the sites i and $i+1$. As a consequence, each time the state of a site i is updated, the length of the two surrounding cylinders has to be updated, which is done in the following way.

First, pick a site i at random, whose state is going to be changed. Decide with probability 1/2 whether the first length adjustment will be performed towards decreasing indices or towards increasing indices; here, we consider, for example, the case of increasing indices (*see* Fig. 5). Second, stretch or shrink the cylinder $i-1$ to its new length $l_{i-1} = n(a_{i-1} + a_i)/2$. Now, the cylinder i has to be displaced in order to match the new position of the site i, and its length also has to be changed to $l_i = n(a_i + a_{i+1})/2$. To this end, choose a block with a random size between the sites $i+1$ and $j > i+1$ and find a rotation which, when applied to this block, will bring the site $i+1$ at the suitable distance l_i from the site i. More precisely, find first the intersection points between the three following objects: i) the sphere centered around the new site i with radius l_i (i.e., the possible positions of the site $i+1$ given the length of the cylinder i), ii) the sphere centered around the site j with radius the initial distance between $i+1$ and j

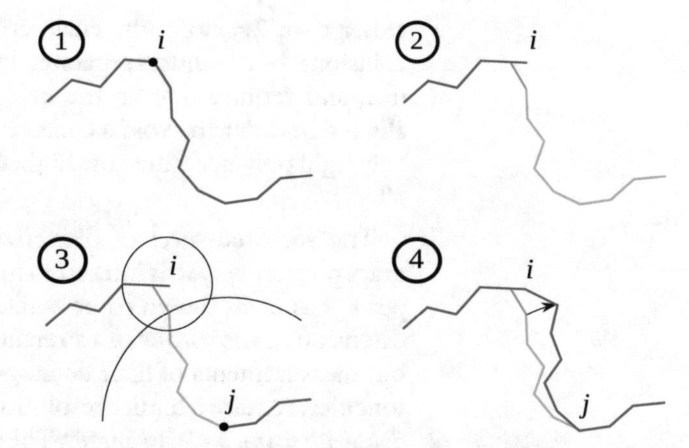

Fig. 5 Structural transition: pick a random site *i* (1). In this case, the transition is a denaturation (*red*), so the length increases (2). Pick another site *j* at random and find a suitable rotation for the block between *i* and *j* (3). Perform the rotation of the block and change the length of the cylinder on the right of *i* (4)

(i.e., the possible positions of the site $i + 1$ after a rotation of the block) and iii) the plane defined by the cylinders $i - 1$ and i (to ensure a minimal deformation of the conformation and thus minimize the number of rejections). If the intersection does not exist (which can happen when the cylinders are shrinked), reject the move. Otherwise, choose from the two possible points (intersections of the black circles in Fig. 5) the one which minimizes the variation in the bending angle between cylinders $i - 1$ and i (on the right in that example).

In this context, the final MC method is almost identical to that without structural transitions, the only difference being that at each step, one has to choose with some fixed probability the kind of elementary move to be performed (rotation or structural transition) (Fig. 2) and to consider the energy provided by Eq. 14. In this regard, depending on the parameters of the problem (the probability of occurrence of denaturation bubbles), it may prove more efficient to bias the probabilities in favor of one kind of move, as long as the resulting distribution of conformations does not depend on this implementation detail.

3 Notes

1. *Coarse-graining*. The level of coarse-graining should be as high as possible to speed up the simulations, but it is constrained by several factors. First, the effective radius r_e imposes a minimal length for the cylinders, which is typically equal to $2r_e$. In the absence of bending energy ($\ell_p = 0$), cylinders shorter than this distance indeed produce artifacts because all the cylinders

closer than $2r_e$ along the chain are ignored when looking for collisions, but cylinders separated by just a bit more than $2r_e$ are not, and require that all the cylinders in between be almost aligned in order to avoid a collision. This leads to an unexpectedly rigid polymer (thus, the higher ℓ_p is, the less important this effect).

The required level of discretization also depends on the exact properties one is interested in. For example, five cylinders per ℓ_p has been shown to be sufficiently accurate to study the extension or the torque of a stretched supercoiled molecule [5], but measurements of finer details, such as the number of plectonemes, require a higher resolution of ten cylinders per ℓ_p [8].

2. *Speed-up using a global energy.* The simulations can also be sped up by using a thermodynamically equivalent model featuring a global torsional energy [5], all the local fluctuations of the twist being integrated out [33]:

$$E(N) = \frac{k_B T}{2}\sum_{i=1}^{N}\frac{\ell_p}{na}\theta_i^2 + 2\pi^2 k_B T\frac{C}{aN}(\mathrm{Tw} - \mathrm{Tw}_0)^2 - fz \tag{15}$$

where Tw_0 is the total twist at rest for the molecule.

3. *Linking number.* At any time, it is helpful (if not necessary) to check a posteriori that the final conformation of the chain has the expected linking number and is unknotted (*see* **Note 9**). The conservation of Lk is verified by computing the sum Tw + Wr, where the writhe Wr is given by [34]:

$$\mathrm{Wr} = \frac{1}{2\pi}\sum_{i=2}^{N}\sum_{j<i}\Omega_{ij} \tag{16}$$

To obtain Ω_{ij}, define:

$$a_0 = \frac{1}{\sin^2\beta}(\vec{r}_j - \vec{r}_i)(\vec{t}_i \times \vec{t}_j) \tag{17}$$

$$a_1 = \frac{1}{\sin^2\beta}(\vec{r}_j - \vec{r}_i)(\cos\beta\,\vec{t}_j - \vec{t}_i) \tag{18}$$

$$a_2 = \frac{1}{\sin^2\beta}(\vec{r}_j - \vec{r}_i)(\vec{t}_j - \vec{t}_i\cos\beta) \tag{19}$$

$$F(x, y) = -\arctan\frac{xy + a_0^2\cos\beta}{a_0(x^2 + y^2 - 2xy\cos\beta + a_0^2\sin^2\beta)^{\frac{1}{2}}} \tag{20}$$

where β is the angle between $\vec{t}_i$ and $\vec{t}_j$. Then:

$$\Omega_{ij} = F(a_1 + l_i, a_2 + l_j) - F(a_1 + l_i, a_2) - F(a_1, a_2 + l_j) + F(a_1, a_2) \tag{21}$$

4. *Initial conditions.* For circular molecules, the initial conformation may be as simple as a planar, N-sided regular polygon (approximating a circle), since the rotations will quickly desorganize this structure. For linear chains, a straight line is however a bad idea: the axis of rotation of any inner block will pass exactly through the cylinders, and thus will have no effect. A solenoid is a better starting conformation, provided the pitch is high enough so that the cylinders have room to move without too many collisions occurring. Compute the writhe of this conformation (*see* **Note 3**) in order to know the initial twist.
5. *MC parameters and acceptance rates.* The maximal block size (M) and the maximal angle (α) for the rotations should be chosen such that the acceptance ratio during the simulation is neither close to 1, nor to 0. For a ratio close to 1, energies and conformations almost never vary; larger blocks and larger angles are then more efficient to generate uncorrelated conformations. For a ratio close to 0, most computation time is wasted into moves that are almost always rejected; smaller blocks and angles would then produce less collisions and crossings but also lower energy variations. Note also that the larger the surface swept by the block is, the longer the crossing-detection routine. Typically, in absence of chain confinement, a good compromise consists of blocks that span several persistence lengths (e.g., $Mna \approx 20\ell_p$), while $\alpha \approx 30°$.
6. *Pointers and better performances.* For better performances, avoid unnecessary copies of the data. In particular, it is not efficient to save a copy of the original chain before attempting a move and restore the copy if the move was rejected. Instead, create only two copies of each cylinder at the beginning of the simulation, and work only with pointers or references to these copies. One pointer designates the current state of the cylinder while the other one refers to a "draft." If the elementary move is accepted, swap both pointers so that the draft becomes the new reference for this cylinder. If the move is rejected, the reference stays unchanged.
7. *Walls for linear molecules.* During the simulation of a linear molecule, the walls may be crossed in different ways. First, a rotation may bring a cylinder beyond a wall, such that it is necessary to check that any moving site ends its rotation in between the walls. Second, the trespassing of a cylinder may happen during a rotation, even though both its initial and final positions are valid. To prevent this scenario, ensure that the extremal coordinates (along the axis of the stretching force) of any site during a rotation remain within the walls. Third, the

alternative type of rotation used with linear molecules allows the walls to move. If one of this wall movements goes toward the center of the chain, check that the non-moving cylinders remain between the walls.

8. *Crossing detection*. Due to the rounding errors inherently associated with floating-point computation, the detection algorithm may exceptionally fail to report a crossing, which results in an instantaneous variation of the linking number by ± 2, and the possibility that a knot is formed. In order to mitigate that risk, count every near-miss as a crossing (reject the move every time a cylinder is found too close (within some arbitrary threshold) to the surface swept by a moving cylinder). The few false-positive crossings will not affect the final results of the simulations. Other methods can be used to mitigate that risk, such as using higher-precision floating-point numbers or reducing the amplitude of the rotations, but all of them will come at a cost in performance.
9. *Knots.* No absolute invariant is known to discriminate between any two different knots, however a few partial invariants have proven to be useful in practice. In particular, the Alexander polynomial [35] has been used in this type of simulations because knots sharing the same Alexander polynomial as unknotted conformations are complex enough to be unlikely to appear during simulations. Moreover, the complete polynomial does not need to be computed: its value at $x = -1$ is sufficient to discriminate unknotted conformations from the simplest knots [36].

References

1. Thanbichler M, Viollier PH, Shapiro L (2005) The structure and function of the bacterial chromosome. Curr Opin Genet Dev 15:153–162
2. Blot N, Mavathur R, Geertz M, Travers A, Muskhelishvili G (2006) Homeostatic regulation of supercoiling sensitivity coordinates transcription of the bacterial genome. EMBO Rep 7:710–715
3. Lagomarsino MC, Espeli O, Junier I (2015) From structure to function of bacterial chromosomes: evolutionary perspectives and ideas for new experiments. FEBS Lett 589:2996–3004
4. Vologodskii AV, Cozzarelli NR (1994) Conformational and thermodynamic properties of supercoiled DNA. Annu Rev Biophys Biomol Struct 23(1):609–643
5. Vologodskii AV, Marko JF (1997) Extension of torsionally stressed DNA by external force. Biophys J 73:123–132
6. Klenin K, Merlitz H, Langowski J (1998) A Brownian dynamics program for the simulation of linear and circular DNA and other wormlike chain polyelectrolytes. Biophys J 74:780–788
7. Marko JF, Siggia ED (1995) Statistical mechanics of supercoiled DNA. Phys Rev E 52:2912–2938
8. Lepage T, Képès F, Junier I (2015) Thermodynamics of long supercoiled molecules: insights from highly efficient Monte Carlo simulations. Biophys J 109:135–143
9. Mirkin SM (2008) Discovery of alternative DNA structures: a heroic decade (1979-1989). Front Biosci 13:1064–1071
10. Benham CJ (1992) Energetics of the strand separation transition in superhelical DNA. J Mol Biol 225:835–847
11. Strick TR, Allemand J-F, Bensimon D, Croquette V (1998) Behavior of supercoiled DNA. Biophys J 74:2016–2028

12. Oberstrass FC, Fernandes LE, Bryant Z (2012) Torque measurements reveal sequence-specific cooperative transitions in supercoiled DNA. Proc Natl Acad Sci 109:6106–6111
13. Sheinin MY, Forth S, Marko JF, Wang MD (2011) Underwound DNA under tension: structure, elasticity, and sequence-dependent behaviors. Phys Rev Lett 107:108102
14. Vlijm R, Mashaghi A, Bernard S, Modesti M, Dekker C (2015) Experimental phase diagram of negatively supercoiled DNA measured by magnetic tweezers and fluorescence. Nanoscale 7:3205–3216
15. Marko JF (2007) Torque and dynamics of linking number relaxation in stretched supercoiled DNA. Phys Rev E 76:21926
16. Meng H, Bosman J, van der Heijden T, van Noort J (2014) Coexistence of twisted, plectonemic, and melted DNA in small topological domains. Biophys J 106:1174–1181
17. Matek C, Ouldridge TE, Doye JPK, Louis AA (2015) Plectoneme tip bubbles: coupled denaturation and writhing in supercoiled DNA. Sci Rep 5:7655
18. Manghi M, Destainville N (2016) Physics of base-pairing dynamics in DNA. Phys Rep 631:1–41
19. Rybenkov VV, Vologodskii AV, Cozzarelli NR (1997) The effect of ionic conditions on DNA helical repeat, effective diameter and free energy of supercoiling. Nucleic Acids Res 25 (7):1412–1418
20. Strick T, Bensimon D, Croquette V (1999) Micro-mechanical measurement of the torsional modulus of DNA. Genetica 106:57–62. doi:10.1023/A:1003772626927
21. Bryant Z, Stone MD, Gore J, Smith SB, Cozzarelli NR, Bustamante C (2003) Structural transitions and elasticity from torque measurements on DNA. Nature 424:338–341
22. Wu PG, Song L, Clendenning JB, Fujimoto BS, Benight AS, Schurr JM (1988) Interaction of chloroquine with linear and supercoiled DNAs. Effect on the torsional dynamics, rigidity, and twist energy parameter. Biochemistry 27:8128–8144
23. Bergou M, Wardetzky M, Robinson S, Audoly B, Grinspun E (2008) Discrete elastic rods. ACM Trans Graph 27:63:1–63:12
24. Carrivain P, Barbi M, Victor J-M (2014) In silico single-molecule manipulation of DNA with rigid body dynamics. PLoS Comput Biol 10:e1003456
25. Metropolis N, Rosenbluth AW, Rosenbluth MN, Teller AH, Teller E (1953) Equation of state calculations by fast computing machines. J Chem Phys 21(6):1087–1092
26. Newman M, Barkema G (1999) Monte Carlo methods in statistical physics. Clarendon Press, Oxford
27. Liu Z, Chan HS (2008) Efficient chain moves for Monte Carlo simulations of a wormlike DNA model: excluded volume, supercoils, site juxtapositions, knots, and comparisons with random-flight and lattice models. J Chem Phys 128:145104
28. Laio A, Gervasio FL (2008) Metadynamics: a method to simulate rare events and reconstruct the free energy in biophysics, chemistry and material science. Rep Prog Phys 71:126601
29. Manghi M, Palmeri J, Destainville N (2009) Coupling between denaturation and chain conformations in DNA: stretching, bending, torsion and finite size effects. J Phys Condens Matter 21:034104
30. SantaLucia J (1998) A unified view of polymer, dumbbell, and oligonucleotide DNA nearest-neighbor thermodynamics. Proc Natl Acad Sci USA 95:1460–1465
31. Oberstrass FC, Fernandes LE, Lebel P, Bryant Z (2013) Torque Spectroscopy of DNA:base-pair stability, boundary effects, backbending, and breathing dynamics. Phys Rev Lett 110 (17):178103
32. Rich A, Nordheim A, Wang AH (1984) The chemistry and biology of left-handed Z-DNA. Annu Rev Biochem 53(1):791–846
33. Gebe JA, Allison SA, Clendenning JB, Schurr JM (1995) Monte Carlo simulations of supercoiling free energies for unknotted and trefoil knotted DNAs. Biophys J 68:619–633
34. Klenin K, Langowski J (2000) Computation of writhe in modeling of supercoiled DNA. Biopolymers 54:307–317
35. Alexander JW (1928) Topological invariants of knots and links. Trans Am Math Soc 30:275–306
36. Vologodskii AV, Lukashin AV, Frank-Kamenetskii MD, Anshelevich VV (1974) The knot problem in statistical mechanics of polymer chains. Sov Phys JETP 39:1059

Chapter 24

Molecular Dynamics Simulation of Supercoiled, Knotted, and Catenated DNA Molecules, Including Modeling of Action of DNA Gyrase

Dusan Racko, Fabrizio Benedetti, Julien Dorier, Yannis Burnier, and Andrzej Stasiak

Abstract

A detailed protocol of molecular dynamics simulations of supercoiled DNA molecules that can be in addition knotted or catenated is described. We also describe how to model ongoing action of DNA gyrase that introduces negative supercoing into DNA molecules. The protocols provide detailed instructions about model parameters, equations of used potentials, simulation, and visualization. Implementation of the model into a frequently used molecular dynamics simulation environment, `ESPResSo`, is shown step by step.

Key words Molecular dynamics, Supercoiled DNA, DNA topoisomerases, DNA gyrase, DNA knots and catenanes

1 Introduction

1.1 Coarse-Grained Model

Molecular dynamics simulations of DNA molecules constitute complementary approach to the actual physical experiments. Results of the simulations provide illuminating insights that are frequently difficult to obtain in real experiments. All simulations have to be adjusted to the scale of details one is interested to study. If for example one is interested in the separation of base pairs resulting from excessive torsional stress, then the atomistic scale is appropriate. In atomistic simulations the properties of all modeled atoms in simulated DNA are based on quantum chemical calculations. The drawback of atomistic simulations is that with the current computer technology one is strongly limited with respect to the size (number of atoms) of the simulated system and also with respect to the corresponding physical time span over which simulated large molecules can be followed. Very recent molecular dynamics studies succeeded to simulate supercoiled DNA

Olivier Espéli (ed.), *The Bacterial Nucleoid: Methods and Protocols*, Methods in Molecular Biology, vol. 1624,
DOI 10.1007/978-1-4939-7098-8_24, © Springer Science+Business Media LLC 2017

minicircles with ca 300 bp, which is an impressive achievement [1]. However, studies of at least ten times larger plasmids are needed to reach the size of small bacterial plasmids that autonomously replicate and which are natural substrates for DNA topoisomerases that decatenate postreplicative catenanes arising during replication of circular DNA [2]. To model large DNA molecules it is still necessary to use the coarse-graining approach where groups of atoms are modeled as individual beads. Good coarse-graining should preserve many of the global properties of DNA molecule, such as excluded volume, mass, persistence length, or electrostatic charge.

The largest scale in the coarse grain polymer modeling is frequently limited by the diameter of studied polymers. Since the effective diameter of DNA under physiological conditions where electrostatic charges are nearly completely screened is of about 2.5 nm, this diameter is used here for bead building coarse-grained models of DNA [3]. We also show how to model torsional stiffness, and the action of DNA gyrase, which dynamically introduces supercoiling.

The outline of this chapter is as follows: In Subheading 2, we provide instructions for the basic setting of the dependencies in the operating system needed to run the simulation software. Furthermore, we show how to configure and compile `ESPResSo`. Next, we introduce topologically restrained molecules such as circular, knotted or catenated DNA molecules together with computer scripts in tcl/tk language that can be used to generate molecules with a given topology. In Subheading 3, we provide a step by step guide to set up the model and to run the molecular dynamics simulations of knotted and catenated molecules which are in addition supercoiled. The parts of the script provided here can be copied into a file and directly run with the `ESPResSo` package. In Subheading 4, further technical hints for improving the performance of your molecular dynamics simulations is provided.

2 Materials

2.1 Preparing the Operating System

Prior to installing the simulation package, one has to make sure that the Linux operating system contains preinstalled all necessary libraries and packages. In the following we show installation of the essential packages in Ubuntu Linux operating system. In order to make sure that all dependencies are installed type/paste into your terminal:

```
sudo apt-get install tcl8.6-dev tk8.4-dev cython fftw3 fftw3-dev pkg-
config povray python-numpy python-scipy openmpi-bin git
```

All scripts and codes necessary for running the simulations are explained along the chapter and the parts of code can be copied and pasted from the text of the electronic reprint of the chapter. The continuity of lines in the code is indicated by backslashes and by line numbers in the scripts. When doing copy and paste, one should not select over a page break.

Alternatively the scripts can be downloaded from the GitHub page by typing into terminal:

```
git clone https://github.com/dusanracko/mmb
```

2.2 The Initial Structure

Biologically relevant DNA knots, i.e., those that are observed in living cells belong mainly to torus and twist types of knots. Complex torus knots can arise as a consequence of site-specific recombination [4, 5] and complex twist knots can arise as a consequence of type II topoisomerase-mediated intramolecular passages occurring within supercoiled DNA molecules [6]. The initial structure describes the initial conditions in our model, and is part of parametrization. The coordinates of the initial structure can be in some cases set manually. Alternatively, starting configurations of knots and catenanes can be downloaded or generated by software KnotPlot [7] or by using parametric equations. To download the starting configuration of a given knot one needs to know its type and its topological notation. The Alexander–Briggs notation is most frequently used. For knots this notation is composed of two numbers where the first indicates the crossing number, i.e., the minimal number of crossings a given knot can have in a projection. The second number (written as subscript) indicates the tabular position of a given knot within knots with the same crossing number. So for example, the knot notation 9_2 indicates the knot type that in standard tables of knots such as those shown in KnotPlot [7] is represented by the second knot among the knots with nine crossings. For catenanes the notation is somewhat more complex and consists of three numbers. The first one, still indicates the crossing number, i.e., the minimal number of crossings a given type of catenane can have in a projection. The second number, written as a superscript, indicates the number of component a given catenane is composed of. In biological setting most important are catenanes composed of two DNA rings. The third number, written as subscript, indicates the tabular position a given catenane among catenanes with the same number of crossings and the same number of components. Again, the standard tables of catenanes can be found on such sites as KnotPlot [7].

In the following sections we address the cases of DNA molecules with above mentioned topologies and provide computer scripts to generate molecules with a given topology and size.

2.2.1 Torus Knots Including the Trivial Knot, Which Is a Simple Closed Ring

Torus knots are defined as closed trajectories that can be continuously placed without any self-crossings on the surface of a simple torus. Torus knots are conveniently characterized by two numbers p and q which indicate how many times a given trajectory encircles the torus along meridional and longitudinal direction, respectively [8]. The

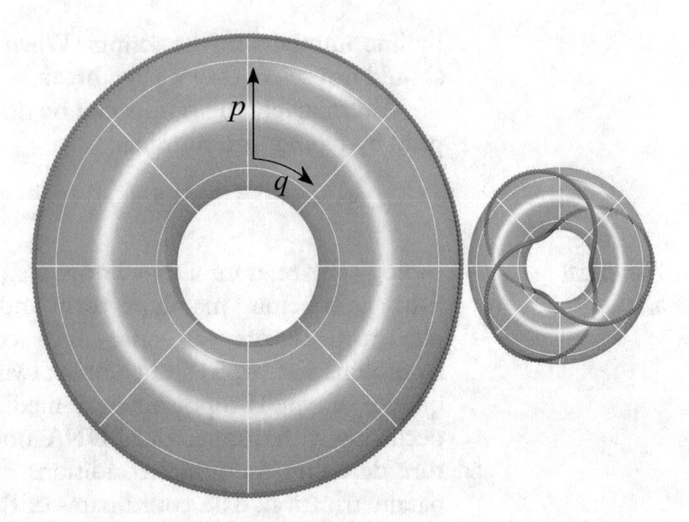

Fig. 1 A circle and a torus knot 3_1 (trefoil). A circle is a special case of a torus knot with no internal nodes—in this case $p = 1$ and q = 0; for 3_1 knot $p = 3$ and $q = 2$

trivial knot, which is unknotted circle can be placed on the surface of the torus without any intersections and can be described as $(1, 0)$ or $(0, 1)$ torus (*see* Fig. 1). The simplest nontrivial knot, which is the trefoil knot, is in fact a $(3, 2)$ torus knot [8]. The starting trajectories of all torus knots can be generated using parametric equations [8].

1. Download the code in a file `torus.tcl` (from https://github.com/dusanracko/mmb)
2. Choose your settings for *p*, *q* number.
3. Run the script from terminal by typing.

```
tclsh torus.tcl
```

4. You may view your generated structure from terminal.

```
vmd torus.vtf
```

2.2.2 Twist Knots and Other Structures

There is no simple parametric equation for twist knots. However, there is computer software for generating twist knots that can be downloaded from KnotPlot [7]. Another convenient option is downloading directly the coordinates of the knots from Knot Server [9] (Fig. 2).

1. Go to the URL address of the Knot Server and open page with coordinates of a particular knot of interest.
2. Select the coordinates and save them to a text file named `knot.txt`.
3. Download the code in a file `knofit.tcl` (from https://github.com/dusanracko/mmb)
4. Modify the size in terms of number of particles `NPART` you want to use for the given topology.

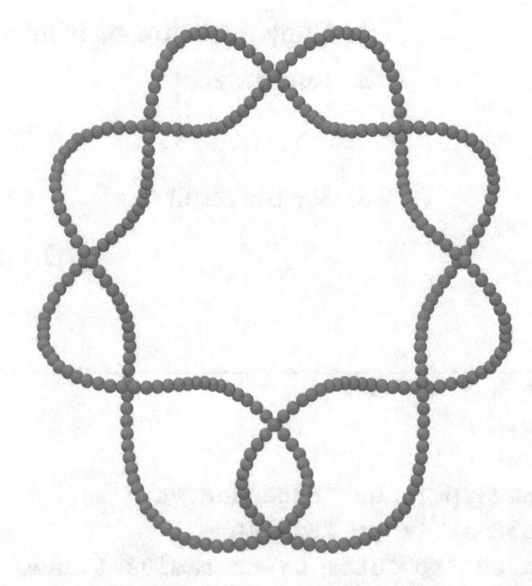

Fig. 2 Twist knot 9_2 obtained by interpolation of coordinates from Knot Server [9]

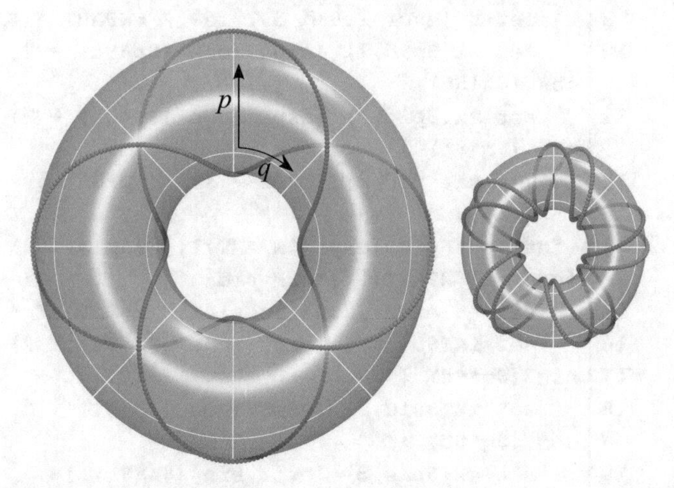

Fig. 3 Catenanes 4_1^2 and 10_1^2

5. Run the script.

```
tclsh knotfit.tcl
```

6. Check the resulting molecule with the required topology and interpolated particle positions.

```
vmd knot-300.vtf
```

2.2.3 Catenanes

Torus type, right-handed catenanes are obligatory intermediates during replication of circular DNA molecules [2]. Starting configurations of catenanes can be downloaded or generated using parametric equation. The easiest way to construct a torus type catenane is using two rings that wrap around the same torus, each running only once in the longitudinal direction and each encircling the torus the same number of times (at least once) in the meridional direction (Fig. 3).

1. Copy the below code into a file and save it into catenanes.tcl.
2. Run the script.

```
tclsh catenanes.tcl
```

3. See the result.

```
vmd catenane.vtf
```

```
 1  set NPART 300
 2  set p 2
 3  set q 1
 4  set fp [open "catenane.vtf" w]
 5  set at [expr 2*$NPART-1]
 6  puts $fp "atom 0:$at radius 1 name DNA"
 7  puts $fp "timestep ordered"
 8   for {set pid 1} {$pid <= $NPART} {incr pid} {
 9     set t [expr 2.0 * 3.14159 / $NPART * $pid ]
10     set xx($pid,1) [expr (cos ($p*$t) + 2) * \
    cos ($q*$t) ]
11     set xx($pid,2) [expr (cos ($p*$t) + 2) * \
    sin ($q*$t) ]
12     set xx($pid,3) [expr - sin ($p*$t) ]
13  }
14   for {set pid [expr $NPART+1]} {$pid <= \
    [expr 2*$NPART]} {incr pid} {
15     set t [expr (2.0) * 3.14159 / $NPART * $pid ]       Complementary
16     set xx($pid,1) [expr (cos ($p*$t) + 2) * \          (p,1)-torus ring
17  sin ($q*$t) ]
18     set xx($pid,2) [expr (cos ($p*$t) + 2) * \
    cos ($q*$t) ]
19     set xx($pid,3) [expr sin ($p*$t) ]
20  }
21    set ux(1) [expr $xx(2,1)-$xx(1,1)]
22    set ux(2) [expr $xx(2,2)-$xx(1,2)]
23    set ux(3) [expr $xx(2,3)-$xx(1,3)]
24    set nf [expr pow (($ux(1)*$ux(1)+$ux(2)*$ux(2) \  Normalize
    +$ux(3)*$ux(3)),0.5)]                               coordinates
25    for {set pid 1} {$pid <= [expr 2*$NPART]} {incr\
    pid} {
26     set xx($pid,1) [expr $xx($pid,1) / $nf]
27     set xx($pid,2) [expr $xx($pid,2) / $nf]
28     set xx($pid,3) [expr $xx($pid,3) / $nf]
29     puts $fp "$xx($pid,1) $xx($pid,2) $xx($pid,3)"
30   }
31  close $fp
```

2.3 Molecular Simulation Package

For the molecular simulations we are going to use the software package ESPResSo [10, 11]. ESPResSo stands for Extensible Simulation Package for Research on Soft matter. It is a highly versatile software for performing molecular dynamics simulations of many particle systems using coarse-grained or atomistic models, with a special focus on simulations of polymers. It is parallelized and can be employed on desktop computers, laboratory clusters as well as on supercomputers with hundreds of CPUs. The parallel code is controlled via the scripting language Tcl, which gives the software its great flexibility. In this section we are going to show how to get ESPResSo and how to configure the precompiler so that the package can be used for our particular purpose to simulate the supercoiled DNA.

1. Download ESPResSo (3.3.0) from the following address: http://espressomd.org/wordpress/download/.
2. Unpack.

```
tar -xvf espresso-3.3.0.tar.gz
```

3. Configure the precompiler by opening the file.

```
espresso-3.3.0/src/core/myconfig-default.hpp
```

4. Enable necessary features for our model of supercoiled DNA by editing the file myconfig-default.hpp in the following way.

```
1  /* global features */
2  #define PARTIAL_PERIODIC
3  #define ELECTROSTATICS
4  #define OLD_DIHEDRAL
5  #define MASS
6  #define EXCLUSIONS
7  #define LANGEVIN_PER_PARTICLE
8  /* potentials */
9  #define LENNARD_JONES
10 #define BOND_ANGLE_HARMONIC
11
12 #define MPI_CORE
13 #define FORCE_CORE
```

The routines that have to be compiled into ESPResSo

5. Later in our simulations it will be necessary to modify the source code of ESPResSo and implement new features—an active motor—as explained later (Subheading 3.4), and recompile the code. This modification however can be done already at this stage by replacing the file dihedral.hpp with the one provided as a downloadable material, or by editing the file as described in Subheading 3.4.
6. From the directory espresso-3.3.0/ of ESPResSo configure precompiler by running the script file and typing

```
./configure
```

7. Compile the executables

```
make -j 4
```

8. Test run ESPResSo by typing

```
./Espresso
```

9. Exit the program

```
exit
```

3 Methods

3.1 Setting Up Simulation Environment in ESPResSo

Modeling is a process of building the model and setting all the parameters. Then, the simulation is a process obtained by using the model. For the purpose of more efficient use of the computer power and reaching biologically interesting timescales, we have built a coarse grained model of the DNA molecule where sequential 8 bp segments are replaced, each by one bead with 2.5 nm diameter. For the commands reference see the ESPResSo User's Guide [12].

In the first step (1), test if all necessary features have been compiled into ESPResSo

```
1 require_feature LENNARD_JONES
2 require_feature ELECTROSTATICS
3 require_feature PARTIAL_PERIODIC
4 require_feature BOND_ANGLE_HARMONIC
5 require_feature LANGEVIN_PER_PARTICLE
6 require_feature MASS
7 require_feature EXCLUSIONS
8 require_feature OLD_DIHEDRAL
```

Routines which we need in our model

(2) Set global properties of the simulation environment, such as the time step, the simulation box size, the periodicity of the system. The box size `box_l` should be large enough if we do not want the periodic images of the molecules to interact. We may also change `periodicity` of the system, i.e., to define whether we want to have the box periodic in 3, 2, or 1 direction to simulate molecules in a bulk solvent or confined within a slit or channel, respectively. The standard setting 1 1 1 is for the bulk solvent situation.

```
9  setmd box_l 100. 100. 100.
10 setmd periodicity 1 1 1
```

Periodic in all directions

(3) Also set the skin depth to be used for the link cell/Verlet algorithm. This is the minimum of cell size or the maximal range of real space interactions. The larger the skin depth value, the less efficient is the parallelization. If the skin value is small, the bond may rip apart when beads are passed to another sub-cell computed on a different CPU. The value of the skin should be about the size

of the maximum interaction distance. In our model the maximum interaction distance is determined by the cutoff of the electrostatic interaction $r_{cut} = 8.0\,\sigma$. The time step should be set such that it is of the order of the reciprocal value of the largest force constant in the potentials used in our model.

```
11 setmd time_step 0.001                    Integration step
12 setmd skin 10.
```

3.2 Potential Model

Traditional approach in the coarse grained simulations of DNA, is to model the DNA as self-avoiding worm-like chain (WLC) [13]. WLC is well-suited to describe properties of semiflexible chains like DNA. In the WLC, the chain bends smoothly unlike the case of freely jointed model. The classical model for WLC consists of three terms: excluded volume, penalty for bending, such as Kratky–Porod interaction, and bonded interaction between the beads, such as the harmonic potential. In ESPResSo, interactions are created by using the inter command. The interactions are divided into two types: nonbonded and bonded interactions.

Nonbonded interactions only depend on the type of the two involved particles. This also applies to the electrostatic interactions, which require special care due to their long-ranged nature (*see* Subheading 3.2.3). The particle type and the charge are both defined using the `part type` and `charge` command. In the case of ***bonded*** interactions, particle identification numbers have to be used to create a particular bond by command `part bond`.

3.2.1 Nonbonded Interactions: Excluded Volume

(4) The traditional Lennard–Jones potential is usually used to describe particle–particle interactions in coarse-grained simulations. It is a simple model of the van der Waals interaction, and is attractive as soon as the interacting beads are closer than the cut-off distance, but is strongly repulsive at short distances to avoid beads interpenetration. A special case of the Lennard–Jones potential is the Weeks–Chandler–Andersen (WCA) potential [14], which one obtains by cutting the attractive part of the Lennard–Jones potential, i.e., choosing $r_{cut} = 2^{1/6}\,\sigma$ and shifting it by ε. The WCA potential is purely repulsive, and is often used to mimic hard sphere repulsion. WCA is also used in Kremer–Grest model of DNA. The bead size σ in our model represents the size of 8 bp, i.e., 2.5 nm. In ESPResSo, we use the following settings of the L-J potential to obtain WCA:

```
13 set ljsigma 1.0
14 set ljepsilon 1.0
15 set ljrcut [expr $ljsigma * pow (2., 1./6.)]       Cut attractive part at
16 set ljcshift [expr 0.25 * $ljepsilon]               the minimum
17 set ljroff 0.0
18 set ljrcap 3.0                                      Cut-off distance for
19 set ljrmin 0.0                                      force calculation
20 inter 0 0 lennard-jones $ljepsilon $ljsigma \
    $ljrcut $ljcshift $ljroff $ljrcap $ljrmin
```

3.2.2 Nonbonded Interactions: Electrostatic Interactions

(5) In our model of DNA the effect of charge located on phosphate groups and electrostatic interactions of charged parts of DNA molecule is included. In ESPResSo the electrostatic interaction is called by command `inter coulomb`. For a pair of particles at distance r with charges q_1 and q_2, the interaction is given by $U(r) = l_B k_B T q_1 q_2 / r$ where $l_B = \varepsilon_0^2/(4\pi k_B T)$ denotes the Bjerrum length, which measures the strength of the electrostatic interaction in screened solutions of salts [15].

```
21 set lB 0.28                                  Bjerrum length
22 set k 0.81
23 set relcut 8.0                               Potential cut-off
24 set bcharge 2                                charge on a bead
25 inter coulomb $lB dh $k $relcut
```

3.2.3 Bonded Interaction: Harmonic Potential

(6) For bonding beads in the beaded DNA chain we use harmonic potential $U(r) = 1/2K(r - R)^2$. The equilibrium length of the bond is set to oscillate around $R = 1.0$. The bond is kept very rigid by setting the force constant K to a very large value.

```
26 inter 1 harmonic 800. 1.0                    K = 800; R = 1.0
```

3.2.4 Bonded Interaction: Bending Stiffness

(7) Introducing bond angle interactions in the model makes modeled DNA molecules semiflexible with respect to bending. Semiflexible properties are set by a parameter of molecular stiffness known as the persistence length [16]. The persistence length of DNA is of about 50 nm which corresponds in our model to about 20 sigmas, $l_p = 20\,\sigma$ [17].

```
27 set stiffness 20.0                           Stiffness ~ P
28 inter 2 angle $stiffness
```

3.2.5 Creating Particles

(8) At this point the particles have to be created. Excluded volume and electrostatics are general properties, and can be set in advance. However, setting of individual bonded interactions requires identifying particles and their specific positions. Particles can be created in ESPResSo by command `part` followed by particle coordinates, particles' properties such as mass, friction coefficient with the solvent, particles' charges, or particle type identifier. Particle identification number is created automatically. As particle positions we use the coordinates created in Subheading 2.3. Particle type allows for creating groups of particles with different excluded volume interactions, which is discussed later (*see* Subheading 3.2.8).

```
29 set NBOND 300                                              Number of bonds and
30 set NPART $NBOND                                           beads
31 set mass 1.0                                               Mass, friction
32 set gammar 3.9                                             coefficient and
33 set bcharge 2.0                                            charge
34 \
35 set fp [open "knot.vtf" r]                                 Open .vtf file
36 gets $fp data                                              Read header
37 gets $fp data                                              Read coordinates
38 for {set pid 1} {$pid <= $NPART} {incr pid} {
39 gets $fp data
40 \
41 set xx($pid,1) [lindex $data 0]
42 set xx($pid,2) [lindex $data 1]
43 set xx($pid,3) [lindex $data 2]
44  part $pid pos $xx($pid,1) $xx($pid,2) \                   Real beads are of
   $xx($pid,3) mass $mass gamma $gammar \                     type 0
   type 0 q $bcharge}
```

3.2.6 Creating Harmonic Bonds

(9) The next step is bonding all the beads into a (circular) chain. At the end also the first and the last bead are joined in the case of circular DNAs, when the number of bonds equals the number of particles.

```
45 for {set pid 2 } {$pid <= $NPART} {incr pid} {             Bond between two
46 part [expr $pid-1] bond 1 $pid                             consecutive particles
47 }                                                          i and i-1
48
49 # close the ring                                           In circular DNA
50  if {$NPART==$NBOND} then {                                number of bonds
51    part $NPART bond 1 1                                    equals number of
52 }                                                          beads
```

3.2.7 Creating Bond Angle Interactions

(10) We also create all bond angle interactions. These bond angle interactions involve triplets of consecutive particles and are used to introduce bending stiffness into the model. In the case of circular DNA, two bonds are added to close the chain and provide uniform stiffness along the whole WLC model.

```
53 set rid 1                                                     Number of chains
54 set pid 3                                                     Number of beads
55 set bid 2                                                     Bond identification
56 set nickedpart 250
57 for {} {$pid <= $NPART} {incr pid} {                          Position of a nick
58 if {$pid<=[expr $nickedpart-0] || $pid>=[expr \
    $nickedpart+0]} then {
59 part [expr $pid-1+($rid-1)*$NPART] bond $bid [ \              Bond-angle inter-
   expr $pid-2+($rid-1)*$NPART] [expr $pid+( \                   action between three
   $rid-1)*$NPART]                                               consecutive particles
60  }
61 }
62 #join the ring
63  if {$NPART==$NBOND} then {                                   In circular DNA two
64   part [expr $NPART*$rid] bond $bid [ \                       additional angular
   expr $NPART*$rid-1] [expr $NPART*($rid-1)+1]                  bonds have to be
65   part [expr $NPART*($rid-1)+1] bond $bid [ \                 added
   expr $NPART*$rid] [expr $NPART*($rid-1)+2]
66 }
```

3.2.8 Adding Torsional Stiffness of DNA Helix

The setting presented above is sufficient for simulating wide range of interesting problems of DNA molecules, for instance *see* refs. 15, 18–22. However, the model described above is not torsionally constrained; hence, sequential beads can freely swivel with respect to each other, which is not the case of sequential 8 bp segments in the DNA double helix. The absence of torsional constraint makes the model unfit to simulate supercoiled DNA molecules. The bead spring model of DNA can be torsionally restrained for the price of introducing additional beads [23–25]. These beads are volumeless, i.e., without the excluded volume interaction, however they have some frictional coefficient governing hydrodynamic interactions with the solvent [23, 25]. Since these additional beads have no volume, we call them further as virtual beads. In order to keep the rotational and translational hydrodynamic drag of the molecule symmetric, for each real chain bead we add four of such virtual beads placed at the ends of four arms of orthogonal crosses placed perpendicularly with respect to the axis of modeled DNA molecule (*see* Fig. 4). Moreover, these four beads are displaced from the real bead onto a position in the middle of the bond. For this purpose attaching a fifth virtual bead is necessary. These five virtual beads, or vectors, attached to one real bead form a composite bead. The composite beads allow us also to introduce the torsional constraint by locking the dihedral angle (Subheading 3.4) between axially displaced beads belonging to consecutive composite particles.

(11) In the next step we create composite beads by using coordinates of the real beads. Positioning of the axial bead in the middle of the bond between real beads, away from the position of the real beads, brings more accuracy into torsional restraining under a given dihedral angle [23].

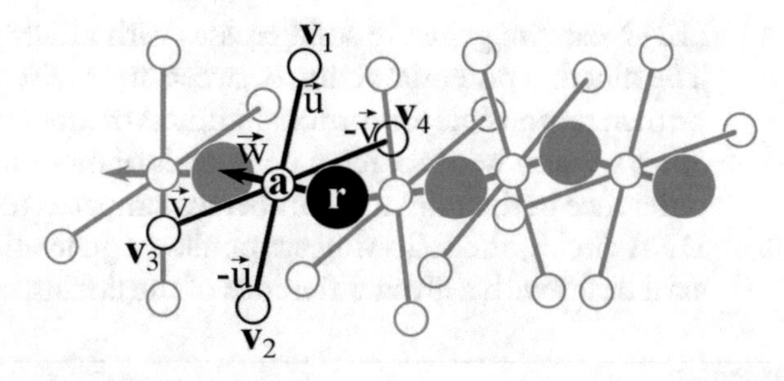

Fig. 4 Schematic construction of the composite beads consisting of one real bead **r** and five virtual beads (**a**, $\mathbf{v}_1$, $\mathbf{v}_2$, $\mathbf{v}_3$, $\mathbf{v}_4$) that are needed to model the effects of torsional stiffness and rotational drag

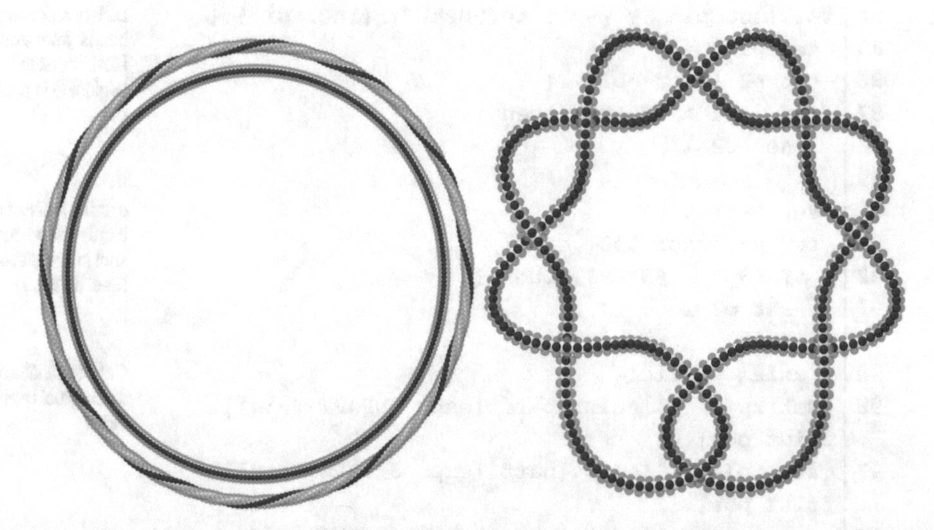

Fig. 5 The picture shows starting configurations of modeled circular/knotted DNA chains decorated with periaxial virtual particles by using the tcl/tk routine. *Left*: two circular DNAs have ΔLk $= 0$ and -12 (parameter on line 80 of the script in Subheading "Generate Periaxial Virtual Particles"), corresponding to supercoiling density of $\sigma_{\text{sup}} = 0$ and -0.04. The relaxed circular DNA in the inside was shrunk to 90% of size for the purpose of concentric placement in the picture. The configuration with ΔLk $= -12$ is not supercoiled as it is not equilibrated yet. In the *right*, a torsionally relaxed configuration of the 9_2 knot

```
67  set gammav 1.0                                              Mass and frictional
68  set massv 0.1                                               coefficient inducing
69  for {set pid 1 } {$pid <= $NPART} {incr pid} {              rotational drag
70  set p1 $pid
71
72  set p2 [expr $pid+1]
73  if {$p2 > $NPART} then {
74  set p2 1
75  }
76  set xp(1) [expr ($xx($p1,1)+$xx($p2,1))/2.0]                Center of the bond
77  set xp(2) [expr ($xx($p1,2)+$xx($p2,2))/2.0]                or position of a in Fig
78  set xp(3) [expr ($xx($p1,3)+$xx($p2,3))/2.0]                4
79  part [expr $NPART+$pid] pos $xp(1) $xp(2) $xp(3)            Virtual beads are of
    \ type 1 mass $massv gamma $gammav}                         type 1
```

Generate Periaxial Virtual Particles

(12) Next, we generate solid crosses with axially displaced particles. The dihedral potential acting between arms of sequential crosses will be then responsible for torsional stiffness of modeled DNA molecules. Since we may want to generate models of molecules with predefined difference in their linking number as compared to torsionally relaxed DNA circles, the following script allows generating positions of the axial beads with a given difference of the linking number ΔLk.

```
80  set dLk 0.0                                               See legend to Fig.5
81  set PI 3.14159
82  set arm 0.9
83
84  for {set pid 1} {$pid <= $NPART} {incr pid} {           Distance of periaxial
85   set p1 $pid                                             beads from axis
86   set p2 [expr $pid+1]                                    Move index to
87    if {$p2 > $NPART} then {                               beginning in circular
88     set p2 1                                              DNA
89   }
90   set extr 4                                              extr=Number of extra
91   set p3 [expr $p2+$extr]                                 beads added during
92    if {$p3 > $NPART} then {                               knot inter-polation
93     set p3 1                                              (Section 2.2.2)
94   }
95  # axial particle                                         Get coordinates of an
96   set xp(1) [lindex [part [expr $NPART+$pid]\             axial virtual bead a in
    print pos] 0]                                            Fig. 4
97   set xp(2) [lindex [part [expr $NPART+$pid]\
    print pos] 1]
98   set xp(3) [lindex [part [expr $NPART+$pid]\
    print pos] 2]
99
100 # axis of rotation                                       Get a directional
101 set ux(1) [expr ($xx($p2,1)-$xp(1))]                     vector of bond-angle
102 set ux(2) [expr ($xx($p2,2)-$xp(2))]                     w⃗
103 set ux(3) [expr ($xx($p2,3)-$xp(3))]
104 # real 1
105 set xr1(1) [lindex [part [expr $p1] print pos] 0]
106 set xr1(2) [lindex [part [expr $p1] print pos] 1]
107 set xr1(3) [lindex [part [expr $p1] print pos] 2]
108
109 # real 2
110 set xr2(1) [lindex [part [expr $p2] print pos] 0]
111 set xr2(2) [lindex [part [expr $p2] print pos] 1]
112 set xr2(3) [lindex [part [expr $p2] print pos] 2]
113
114 # real 3
115 set xr3(1) [lindex [part [expr $p3] print pos] 0]
116 set xr3(2) [lindex [part [expr $p3] print pos] 1]
117 set xr3(3) [lindex [part [expr $p3] print pos] 2]
118
119 # bond vectors
```

```
120 set b1(1) [expr $xr1(1)-$xr2(1)]
121 set b1(2) [expr $xr1(2)-$xr2(2)]
122 set b1(3) [expr $xr1(3)-$xr2(3)]
123
124 set b2(1) [expr $xr3(1)-$xr2(1)]
125 set b2(2) [expr $xr3(2)-$xr2(2)]
126 set b2(3) [expr $xr3(3)-$xr2(3)]
127
128 # cross product, normal to the bond angle
129 set px(1) [expr $b1(2)*$b2(3)-$b2(2)*$b1(3)]
130 set px(2) [expr $b1(3)*$b2(1)-$b2(3)*$b1(1)]
131 set px(3) [expr $b1(1)*$b2(2)-$b2(1)*$b1(2)]
132 # normalize
133 set px(1) [expr $arm*$px(1) / (pow
    ($px(1)*$px(1)\
    +$px(2)*$px(2)+$px(3)*$px(3),0.5)+0.00001)]
134 set px(2) [expr $arm*$px(2) / (pow
    ($px(1)*$px(1)\
    +$px(2)*$px(2)+$px(3)*$px(3),0.5)+0.00001)]
135 set px(3) [expr $arm*$px(3) / (pow
    ($px(1)*$px(1)\
    +$px(2)*$px(2)+$px(3)*$px(3),0.5)+0.00001)]
136
137 #set angle and pre-calculate trigonometric values
138 set theta [expr 2 * $PI / $NPART * $pid * $dLk]
139 set costheta [expr cos ($theta)]
140 set sintheta [expr sin ($theta)]
141
142 # pre-calculate
143 set u2 [expr $ux(1)*$ux(1)]
144 set v2 [expr $ux(2)*$ux(2)]
145 set w2 [expr $ux(3)*$ux(3)]
146 set L [expr ($u2 + $v2 + $w2)]
147
148 set M1(0,0) $px(1)
149 set M1(1,0) $px(2)
150 set M1(2,0) $px(3)
151 set M1(3,0) 1.0
152
153  set RM(0,0) [expr ($u2 + ($v2 + $w2) *
    $costheta) / $L]
154  set RM(0,1) [expr ($ux(1) * $ux(2) * (1 - \
     $costheta) - $ux(3) * sqrt($L) * $sintheta) /
    $L]
```

Directional vector of the periaxial bead $\overrightarrow{u_0}$ in Fig. 4

Initialization of quaternion rotational matrix

Rotate vector of the periaxial bead around the bond by angle theta given by linking number and position along the chain $\vec{u}=R_\theta\overrightarrow{u_0}$

```
155  set RM(0,2) [expr ($ux(1) * $ux(3) * (1 - \
      $costheta) + $ux(2) * sqrt($L) * $sintheta) /
     $L]
156  set RM(0,3) 0.0
157
158  set RM(1,0) [expr ($ux(1) * $ux(2) * (1 - \
      $costheta) + $ux(3) * sqrt($L) * $sintheta) /
     $L]
159  set RM(1,1) [expr ($v2 + ($u2 + $w2) * \
     $costheta) / $L]
160  set RM(1,2) [expr ($ux(2) * $ux(3) * (1 - \
     $costheta) - $ux(1) * sqrt($L) * $sintheta) / $L]
161  set RM(1,3) 0.0
162
163  set RM(2,0) [expr ($ux(1) * $ux(3) * (1 - \
     $costheta) - $ux(2) * sqrt($L) * $sintheta) / $L]
164  set RM(2,1) [expr ($ux(2) * $ux(3) * (1 - \
      $costheta) + $ux(1) * sqrt($L) * $sintheta) /
     $L]
165  set RM(2,2) [expr ($w2 + ($u2 + $v2) * \
     $costheta) / $L]
166  set RM(2,3) 0.0
167
168  set RM(3,0) 0.0
169  set RM(3,1) 0.0
170  set RM(3,2) 0.0
171  set RM(3,3) 1.0
172
173 for {set i 0} { $i < 4 } { incr i } {
174  for {set j 0} { $j < 1 } { incr j } {
175   set M2($i,$j) 0.0
176     for {set k 0} { $k < 4 } { incr k } {
177 set M2($i,$j) [expr $M2($i,$j) + \
      $RM($i,$k) * $M1($k,$j)]
178   }
179  }
180 }
181
182 #shift the rotated point to the particular bond
183 set pv1(1) [expr $M2(0,0) + $xp(1)]
184 set pv1(2) [expr $M2(1,0) + $xp(2)]
185 set pv1(3) [expr $M2(2,0) + $xp(3)]
186
187 set pv2(1) [expr -1.0*$M2(0,0) + $xp(1)]
```

Add vector of the periaxial bead to the axial bead, i.e. shift origin of $\vec{u}$ to **a**

Mirror the vector $-\vec{u}$

```
188 set pv2(2) [expr -1.0*$M2(1,0) + $xp(2)]
189 set pv2(3) [expr -1.0*$M2(2,0) + $xp(3)]
190
191 set pv3(1) [expr ($M2(1,0)*$ux(3)-$M2(2,0)* \
    $ux(2)) * 2.0]
192 set pv3(2) [expr ($M2(2,0)*$ux(1)-$M2(0,0)* \
    $ux(3)) * 2.0]
193 set pv3(3) [expr ($M2(0,0)*$ux(2)-$M2(1,0)* \
    $ux(1)) * 2.0]
194
195 set pv4(1) [expr -1.0*$pv3(1) + $xp(1)]
196 set pv4(2) [expr -1.0*$pv3(2) + $xp(2)]
197 set pv4(3) [expr -1.0*$pv3(3) + $xp(3)]
198
199 #the first periaxial bead
200 set xpv(1,$pid,1) $pv1(1)
201 set xpv(1,$pid,2) $pv1(2)
202 set xpv(1,$pid,3) $pv1(3)
203
204 #the second periaxial bead
205 set xpv(2,$pid,1) $pv2(1)
206 set xpv(2,$pid,2) $pv2(2)
207 set xpv(2,$pid,3) $pv2(3)
208
209 #the third periaxial bead
210 set xpv(3,$pid,1) [expr $pv3(1)+$xp(1)]
211 set xpv(3,$pid,2) [expr $pv3(2)+$xp(2)]
212 set xpv(3,$pid,3) [expr $pv3(3)+$xp(3)]
213
214 #the fourth periaxial bead
215 set xpv(4,$pid,1) $pv4(1)
216 set xpv(4,$pid,2) $pv4(2)
217 set xpv(4,$pid,3) $pv4(3)
218 }
219 #Generate periaxial particles
220  for {set j 1} {$j <= 4} {incr j} {
221   for {set pid 1} {$pid <= $NPART} {incr pid} {
222     part [expr ($j+1)*$NPART+$pid] pos\
    $xpv($j,$pid,1) $xpv($j,$pid,2) $xpv($j,$pid,3)\
    type 1 mass $massv gamma $gammav
223   }
224 }
```

The third vector of the solid cross is perpendicular to the first vector and vector of bond $\vec{v} = \vec{u} x \vec{w}$

The last vector is obtained by mirroring the third vector $-\vec{v}$

Create all beads with pre-calculated coordinates **v**$_1$, **v**$_2$, **v**$_3$, **v**$_4$, mass, gamma and set type to 1 for all virtual beads

Creating the Solid Cross

(13) The virtual particles are joined by strong bonds to stay in the position of a solid cross. Moreover, they are attached to the axial beads.

```
225 #Dihedral                                                  Parametrizing force
226 set mult 1                                                 interactions for fixing
227 set bend 10.0                                              a solid cross
228 set phase 0.
229 inter 3 dihedral $mult $bend $phase
230
231 # Bonded interactions params harmonic (axial)
232 inter 4 harmonic 1000. 0.50
233
234 # Periaxial                                                Set distance of
235 inter 5 harmonic 1000. $arm                                periaxial virtual bead
236                                                            from axis to $arm
237 # Bonded right angle
238 set stiffness 800.0
239 inter 6 angle $stiffness [expr 3.14159 / 2.]
240
241 # Bonded BENDING flat angle
242 set stiffness 800.0
243 inter 7 angle $stiffness 3.14159
244
245 #Virtuals :: Create cross                                  Bond the beads vi in
246  for {set pid 1} {$pid <= $NPART} {incr pid} {             Fig. 4 with a.
247   part [expr 2*$NPART+$pid] bond 5 [expr                   v1 with a
    $pid+\$NPART]
248   part [expr 3*$NPART+$pid] bond 5 [expr                   v2 with a
    $pid+\$NPART]
249   part [expr 4*$NPART+$pid] bond 5 [expr                   v3 with a
    $pid+\$NPART]
250   part [expr 5*$NPART+$pid] bond 5 [expr                   v4 with a
    $pid+\$NPART]
251 }
252 #Virtuals :: Attach cross to left and right bead           Attach the cross to
253  for {set pid 1} {$pid <= $NPART} {incr pid} {             flanking real beads
254   set p2 [expr $pid + 1]
255    if {$p2 > $NPART} then { set p2 1}
256   part [expr $NPART+$pid] bond 4 [expr $pid]               Bond a with ri
257   part [expr $NPART+$pid] bond 4 [expr $p2]                Bond a with ri+1
258 }
259
260 #Virtuals :: Flat angle
261  for {set pid 1} {$pid <= $NPART} {incr pid} {
262   part [expr 1*$NPART+$pid] bond 7 [expr                   Bond v1 with v2
    2*$NPART\+$pid] [expr 3*$NPART+$pid]
263   part [expr 1*$NPART+$pid] bond 7 [expr                   Bond v3 with v4
    4*$NPART\+$pid] [expr 5*$NPART+$pid]
```

264	`}`	
265	`#Virtuals :: Fix right angles` ` for {set pid 1} {$pid <= $NPART} {incr pid} {`	Fix angles in 90° between:
266	`  part [expr 1*$NPART+$pid] bond 6 [expr 0*$NPART\+$pid] [expr 2*$NPART+$pid]`	∠(rav_1)
267	`  part [expr 1*$NPART+$pid] bond 6 [expr 0*$NPART\+$pid] [expr 4*$NPART+$pid]`	∠(rav_3)
268	`  part [expr 1*$NPART+$pid] bond 6 [expr 4*$NPART\+$pid] [expr 2*$NPART+$pid]`	∠(v_1av_3)
269	`  part [expr 1*$NPART+$pid] bond 6 [expr 5*$NPART\+$pid] [expr 3*$NPART+$pid]`	∠(v_2av_4)
270	`}`	
271		
272	`#Virtuals :: Flat angle`	Fix 180° angle
273	`  for {set pid 1} {$pid <= $NPART} {incr pid} {`	
274	`   set p1 $pid`	
275	`   set p2 [expr $pid + 1]`	
276	`    if {$p2 > $NPART} then { set p2 1}`	∠(r_iar_{i+1})
277	`   part [expr 1*$NPART+$pid] bond 7 [expr\ 0*$NPART+$pid] [expr 0*$NPART+$p2]`	
278	`  }`	
279	`#Exclusions`	From the intermolecular interactions, we exclude the bonded monomers
280	` for {set pid 1} {$pid <= $NPART} {incr pid} {`	
281	`  set p1 [expr $pid - 1]`	
282	`  set p2 $pid`	
283	`  set p3 [expr $pid + 1]`	
284	`   if {$p1 < 1} then { set p1 $NPART }`	r_{i-1} and r_{i+1} from interacting with r_i
285	`   if {$p3 > $NPART} then { set p3 1 }`	
286	`  part $p2 exclude $p1 $p3`	
287	`}`	

3.3 The Gyrase

After generating coordinates of virtual beads and generating the particles for simulation, the axial swiveling can be restrained by applying dihedral lock along the chain. The dihedral lock is due to a dihedral potential defined by quadruplets of beads, where the axis of the dihedral angle is given by two consecutive axial virtual beads and the value of the angle is determined by the position of two periaxial beads attached to the respective axial beads (*see* Fig. 6).

The force constant of the dihedral angle determines the torsional stiffness. The parameter of the torsional stiffness in our model was set such that simulated molecules equilibrate reaching a Δtwist–to-Δwrithe ratio of 1:3 [23, 25].

By modifying the properties of the dihedral potential, we can simulate a range of phenomena during which the level of supercoiling in DNA changes.

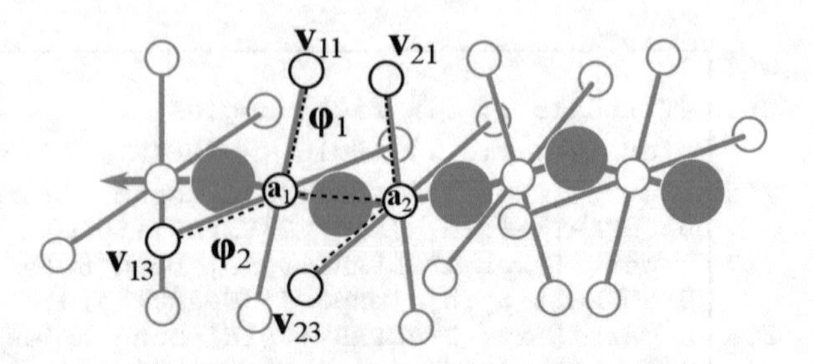

Fig. 6 Dihedral potential introduces torsional constraint into modeled DNA molecules. Between sequential crosses only, two dihedral angles $\phi_1(\mathbf{v}_{11}\mathbf{a}_1\mathbf{a}_2\mathbf{v}_{21})$ and $\phi_2(\mathbf{v}_{13}\mathbf{a}_1\mathbf{a}_2\mathbf{v}_{23})$ are used for the calculation of torsional deformation. The force constant $K1 = K2 = 10\varepsilon$ is set for both perpendicular dihedral angles in order to preserve symmetry during the axial deformation, and this value approximates real behavior of DNA molecules

For instance, DNA gyrase is a special type of topoisomerase which can continuously induce axial swiveling of DNA resulting in DNA supercoiling [26]. Action of gyrase in our model can be introduced by replacing the dihedral potential acting between two consecutive composite particles by oppositely oriented constant moments of force acting on periaxial beads belonging each to two consecutive composite particles. Another option is to maintain dihedral potential between two selected composite consecutive particles but vary the period in the dihedral angle definition with a constant change of angle over the time. We refer to these two options as a constant force and a constant speed motor, respectively. The constant force motor was built in the definition of the dihedral potential in `ESPResSo` source code (*see* Subheading 3.4).

We use here a harmonic dihedral potential, $U(r) = 1/2K(\phi - \phi_0)^2$, rather than frequently used *sine* form of the potential, which is periodic. The periodic potential may result in an undesirable relaxation of Pi turns around axis at large angles or at large torsional stress.

A special case of a dihedral potential with a force constant equal to zero simulates the situation at the place of a nick in the DNA, i.e., a place where one of DNA strands is interrupted. The virtual particles flanking the nick site rotate independently from each other as it is also the case of DNA base pairs flanking the site of a nick. This is a passive swivel, which unlike the active swivel permits the dissipations of the torsional energy and torsional relaxation of modeled DNA molecules.

(14) Next, we introduce dihedral constraints between consecutive composite beads. The axis of the dihedral angle is given by two consecutive axial virtual beads, and the value of the angle is determined by the position of two periaxial beads attached to the respective axial beads.

(15) To simulate the action of gyrase, place a special dihedral, e.g., at position 100, to include the molecular motor, into the model (see lines 299–302 in the simulation script).

(16) To simulate DNA nick, leave out two dihedral bonds at a given bead number, e.g., at position 250.

```
288 for {set pid 1} {$pid <= $NPART} {incr pid} {      Go along the chain
289     set pt1 [expr $pid+0]                           and generate dihedral
290     set pt2 [expr $pid+1]                           constraints
291
292  if {$pt1>$NPART} {set pt1 [expr $pt1-$NPART]}      In the circular DNA
293  if {$pt2>$NPART} {set pt2 [expr $pt2-$NPART]}      put last dihedrals
294                                                     connecting start and
                                                        end of molecule's
295 if {($pid!=100)&&($pid!=$nickedpart)} then {
296      part [expr $NPART+$pt1] bond 3 [expr\
    2*$NPART+$pt1] [expr $NPART+$pt2] [expr \
    2*$NPART+$pt2]
297     part [expr $NPART+$pt1] bond 3 [expr \
    4*$NPART+$pt1] [expr $NPART+$pt2] [expr \
    4*$NPART+$pt2]
298   } else {
299      if {($pid==100)} then {                         Place motor at
300       inter 78 dihedral 456 15.0 0.0                 position 100
301       part [expr $NPART+$pt1] bond 78 [expr\
302 2*$NPART+$pt1] [expr $NPART+$pt2] [expr \
    2*$NPART+$pt2]
303 puts "Bead $pid MOTOR"
304      } else {                                        'Do nothing' – don't
305        puts "Bead $pid NICK"                         create dihedral for
306     }                                                bead 250
307   }
308 }
```

3.4 Harmonic Dihedral Potential for ESPResSo

(17) The harmonic definition of dihedral potential, $U(r) = 1/2K (\phi - \phi_0)^2$, is not a part of standard ESPResSo package, hence at this point it's necessary to go into the source code. So find the file in the following location:

```
espresso-3.3.0/src/core/dihedral.hpp.
```

(18) Replace the content of the file by pasting the following code in c++.4

```
1  /*
2    Copyright (C) 2010,2011,2012,2013,2014 The ESPResSo project
3    Copyright (C) 2002,2003,2004,2005,2006,2007,2008,2009,2010
4      Max-Planck-Institute for Polymer Research, Theory Group
5    Copyright (C) 2016 Modified Dihedral Harmonic
6      Center of Integrative Genomics, Universite de Lausanne
7  */
8  #ifndef DIHEDRAL_H
9  #define DIHEDRAL_H
10 #include "utils.hpp"
11 #include "interaction_data.hpp"
12 #include "grid.hpp"
13 #define ANGLE_NOT_DEFINED -100
14 
15 int dihedral_set_params(int bond_type, int mult, double bend,
   double phase);
16 
17 inline void calc_dihedral_angle(Particle *p1, Particle *p2,
   Particle *p3, Particle *p4, double a[3], double b[3], double
   c[3], double aXb[3], double *l_aXb, double bXc[3], double
   *l_bXc, double *cosphi, double *phi)
18 {
19 int i;
20 get_mi_vector(a, p2->r.p, p1->r.p);
21 get_mi_vector(b, p3->r.p, p2->r.p);
22 get_mi_vector(c, p4->r.p, p3->r.p);
23 vector_product(a, b, aXb);
24 vector_product(b, c, bXc);
25 *l_aXb = sqrt(sqrlen(aXb));
26 *l_bXc = sqrt(sqrlen(bXc));
27 if ( *l_aXb <= TINY_LENGTH_VALUE || *l_bXc <=
   TINY_LENGTH_VALUE ) { *phi = -1.0; *cosphi = 0; return;}
28 for (i=0;i<3;i++) {
29  aXb[i] /= *l_aXb;
30  bXc[i] /= *l_bXc;
31 }
32 *cosphi = scalar(aXb, bXc);
33 if ( fabs(fabs(*cosphi)-1)  < TINY_SIN_VALUE  ) *cosphi =
   dround(*cosphi);
34 *phi = acos(*cosphi);
35 #ifdef OLD_DIHEDRAL
36 if( scalar(aXb, c) < 0.0 ) *phi =  - *phi;
37 #else
38 if( scalar(aXb, c) < 0.0 ) *phi = (2.0*PI) - *phi;
```

```
39  #endif
40  }
41  inline int calc_dihedral_force(Particle *p2, Particle *p1,
    Particle *p3, Particle *p4, Bonded_ia_parameters *iaparams,
    double force2[3], double force1[3], double force3[3])
42  {
43  int i;
44  double v12[3], v23[3], v34[3], v12Xv23[3], v23Xv34[3],
    v12Xv34[3], l_v12Xv23, l_v23Xv34, l_v12Xv34;
45  double v23Xf1[3], v23Xf4[3], v34Xf4[3], v12Xf1[3];
46  double phi, cosphi, sinmphi_sinphi;
47  double fac, f1[3], f4[3];
48  calc_dihedral_angle(p1, p2, p3, p4, v12, v23, v34, v12Xv23,
    &l_v12Xv23, v23Xv34, &l_v23Xv34, &cosphi, &phi);
49  if ( phi == -1.0 ) {
50  for(i=0;i<3;i++) { force1[i] = 0.0; force2[i] = 0.0; force3[i]
    = 0.0; }
51  return 0;
52  }
53  for(i=0;i<3;i++)  {
54  f1[i] = (v23Xv34[i] - cosphi*v12Xv23[i])/l_v12Xv23;;
55  f4[i] = (v12Xv23[i] - cosphi*v23Xv34[i])/l_v23Xv34;
56  }
57  vector_product(v23, f1, v23Xf1);
58  vector_product(v23, f4, v23Xf4);
59  vector_product(v34, f4, v34Xf4);
60  vector_product(v12, f1, v12Xf1);
61  #ifdef OLD_DIHEDRAL
62  fac = -iaparams->p.dihedral.bend;
63  #else
64  fac = -iaparams->p.dihedral.bend * iaparams->p.dihedral.mult;
65  #endif
66  if(fabs(sin(phi)) < TINY_SIN_VALUE) {
67  #ifdef OLD_DIHEDRAL
68  sinmphi_sinphi = cos( phi - iaparams-
    >p.dihedral.phase)/cos(phi);
69  #else
70  sinmphi_sinphi = iaparams->p.dihedral.mult*
71  cos(iaparams->p.dihedral.mult*phi - iaparams-
    >p.dihedral.phase)/cosphi;
72  #endif
73  } else {
74  #ifdef OLD_DIHEDRAL
75  if (iaparams->p.dihedral.mult==456) {
76  sinmphi_sinphi = -1.0/sin(phi);
```

```
77  } else {
78  sinmphi_sinphi = (phi- iaparams->p.dihedral.phase)/sin(phi);
79  }
80  #else
81  sinmphi_sinphi = sin(iaparams->p.dihedral.mult*phi - iaparams-
    >p.dihedral.phase)/sin(phi);
82  #endif
83  }
84  fac *= sinmphi_sinphi;
85  for(i=0;i<3;i++) {
86   force1[i] = fac*v23Xf1[i];
87   force2[i] = fac*(v34Xf4[i] - v12Xf1[i] - v23Xf1[i]);
88   force3[i] = fac*(v12Xf1[i] - v23Xf4[i] - v34Xf4[i]);
89   }
90    return 0;
91  }
92  inline int dihedral_energy(Particle *p1, Particle *p2,
    Particle *p3, Particle *p4,
93  {
94  double v12[3], v23[3], v34[3], v12Xv23[3], v23Xv34[3],
    l_v12Xv23, l_v23Xv34;
95  double phi, cosphi;
96  double fac;
97  calc_dihedral_angle(p1, p2, p3, p4, v12, v23, v34, v12Xv23,
    &l_v12Xv23, v23Xv34, &l_v23Xv34, &cosphi, &phi);
98  #ifdef OLD_DIHEDRAL
99  fac =pow(phi-iaparams->p.dihedral.phase,2)/2.0;
100 #else
101 fac = -cos(iaparams->p.dihedral.mult*phi -iaparams-
    >p.dihedral.phase);
102 fac += 1.0;
103 #endif
104 fac *=  iaparams->p.dihedral.bend;
105 *_energy = fac;
106 return 0;
107 }
108 #endif
```

(19) Go to the main ESPResSo directory and run precompiler.

```
./configure
```

(20) Compile the source code.

```
make -j 4
```

3.5 Output Files

(21) `ESPReSo` uses a standard output writing function VTF Trajectory Format, or .vtf files [27]. The files require initialization of a header which contains information on number of beads and their radius. In its simplest form it should look like "`atom0:299name DNA`

radius 1". The following script will create a more detailed header, which will allow to easily distinguish between virtual and real particles of DNA chain. The preformatted file named "trajectory.vtf" will be used to store the trajectory from main MD simulation.

```
309  set vtffile [open "trajectory.vtf" w]
310  puts $vtffile "atom 0:[expr $NPART*6-1] radius 1"
311  puts $vtffile "atom 0:[expr $NPART-1] radius 1\
      name DNA"
312  puts $vtffile \
     atom [expr $NPART]:[expr $NPART*2-1] radius 1\
     name V1
313  puts $vtffile "atom [expr $NPART*2]:[expr \
     $NPART*6-1] radius 1 name x2"
314  puts $vtffile "bond 0:[expr $NPART-1],0::\
     [expr $NPART-1]"
315  writevcf $vtffile
316  flush $vtffile
```

Create file and create simple heading

3.6 Thermostat

(22) In general, ESPResSo uses Langevin thermostat. The Langevin thermostat extends the classical Newton's equations of motion by the effect of solvent [28]. The solvent is represented by effective viscous drag and random kicking force which are both functions of an effective friction coefficient, gamma. The thermostat uses two parameters. The friction coefficient of the thermostat basically sets the viscosity of the media one wants to simulate. For aqueous solutions the standard setting of the friction coefficient is 1. The temperature of the bath has to be set to 2.5 for the room temperature, as indicated by the ESPResSo user guide.

```
317  set gammat 1.0
318  set temperature 2.5
319  thermostat langevin $gammat $temperature
320  puts "thermostat=[thermostat]"
321  integrate 0
```

Coupling with kick force

Initialize

3.7 Thermalization

(23) In the next step we have to thermalize constructed configurations before we can start collecting the data during a production run of the main simulation. The starting configurations of the DNA molecule might be quite far from a configuration that can be attained as a result of thermal fluctuation at ambient temperature. Despite the fact that the initial configurations were constructed with the intention to have all the beads close to their equilibrium bonding distances, there might have been artifacts from averaging and mathematical operations which may cause the overall energy of the system to be very large.

In addition, the initial structure originally has no kinetic energy, which has to be added by numerical thermostat. After random velocity vectors are generated, the potential and kinetic energy have to equilibrate while a desired temperature is maintained.

When thermalization process starts, the initial energy of the structure may be so large that the calculated moves of the beads can excess the permitted length of the bond—and some of the bonds would rip apart. In computer language, the calculated energy at the given distance would be so large that it exceeds the size of the register used to store floating point numbers, causing the so-called arithmetic overflow and crash of the program. In the simulation such process of the energy buildup can behave as a chain reaction, leading to an "explosion." For these reasons it is necessary to perform a special procedure called limited integration where the calculation of the potential energy is capped and set not to exceed some limiting value.

```
322 set save_warmingtrj "yes"                                  Save structures
323 puts "Warming up..."
324 set wcap 1.0
325 set i 1
326  while {$i < 100} {                                        Do 100 cycles
327   set i [expr $i+1]
328   set wcap [expr $wcap + 10]                               Remove the energy
329   inter forcecap $wcap                                     cap stepwisely
330   integrate 10000                                          Integrate 10000
331   puts -nonewline [format "Step: %4d" $i]\r                steps each cycle
332   flush stdout
333   if { $save_warmingtrj == yes } then {
334    writevcf $vtffile
335     flush $vtffile
336  }
337 }
```

3.8 Equilibration

DNA molecules are very flexible and they greatly change their overall shape under the effect of thermal fluctuations. Hence, in molecular dynamics simulations we are less interested in a given momentary configuration the molecule takes at the end of a simulation run, but we are rather interested in statistical averages of properties over the time <A> [29]. The <A> can stand for any experimental structural or thermodynamic observable, like gyration radius and other shape factors, energies, heat capacities, and diffusivities. The property average oscillates during the simulation until it reaches a steady or equilibrated value. Since we start simulations from rather idealized initial structures of DNA, one may need a long equilibration run. For equilibration the same code can be used as during the thermalization procedure, however without capping the potential function.

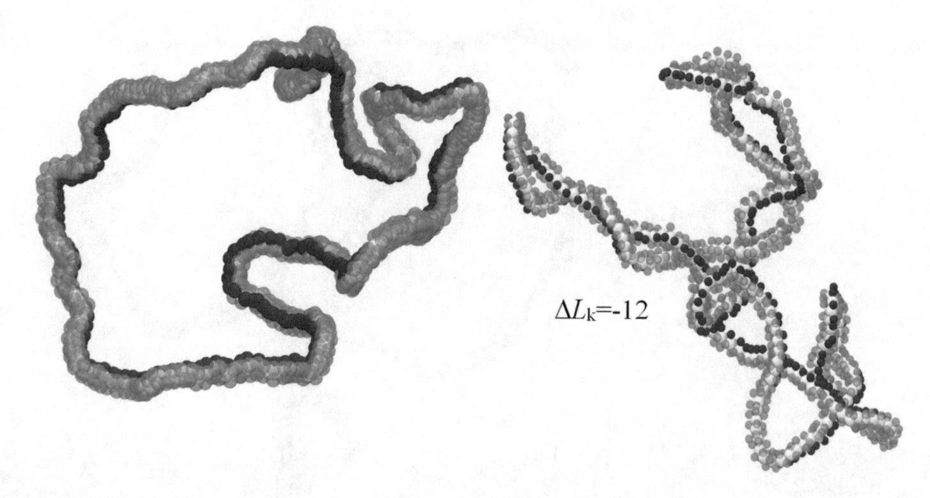

Fig. 7 Equilibrated structures of the circular DNA with ca 3000 bp and having $\Delta\text{Lk} = 0$ and $\Delta\text{Lk} = -12$. respectively

3.8.1 Running Simulation

(24) We end up with the script for the simulation of our beaded model of nicked and gapped DNA where gyrase acts as an active swivel. The following code runs 10,000 cycles of 10,000 integrations and saves a structure into the trajectory file every ten cycles.

```
338  set frame 0
339  for { } { $frame < 10000 } { incr frame } {          Do 10000 cycles
340    puts -nonewline [format "Step: %4d" $frame ]\r
341    flush stdout
342    integrate 10000                                   Integrate 10000 steps
343  if { $frame % 10 == 0 } then {
344    writevcf $vtffile                                 Update trajectory
345  }
346    flush $vtffile                                    Every 10 cycles
347  }
```

(25) To run the simulation, copy parts of the script into one file named `gyrase.tcl` in the main ESPResSo directory. In the terminal, enter the following command:

```
mpirun -n 4 ./Espresso gyrase.tcl
```

The command will run a parallel simulation job on four processors. This number of processors was found to be effective for the example we have shown here.

(26) You may see the result of the simulation in VMD (Figs. 7 and 8).

```
vmd trajectory.vtf
```

Fig. 8 The knotted DNA trefoil with an active and a passive swivel. The constant force motor mimicking the action of DNA gyrase was set to produce a torque $\tau = 15\ \varepsilon\sigma$. In nicked trefoil knot the action of DNA gyrase confines the knot and pushes it toward the place of the nick [25]

3.9 Analysis of Twist and Writhe

(27) Despite the optimized shape of the dihedral potential function, during the coarse-grained simulation a bead can "jump" over set energy barriers. Such jumps can cause loss of conservation of the linking number, which is undesirable in the simulations. Since the simulation is based on some randomness coming from the Langevin thermostat, such jumps can be reversed when stepping back in the simulation and repeating last dozen time units of simulation with a different setting of random number generator (*see* Subheading 4.2). Although such passages are very rare and unlikely, it is recommended to check the conservation of linking number during the simulation, especially in the case of long simulation runs.

Monitoring twist and writhe, and calculating the linking number can be also important when simulating the action of DNA gyrase. In this case the linking number is expected to change continuously.

For this purposes you may download a routine for calculating twist and writhe from the public library on the webpage of the git hub project [30]. You may download the necessary files from terminal by typing the following:

```
git clone https://github.com/fbenedett/polymer-libraries
```

(28) Copy the following code into a file "analyse.cpp" and compile.

```
g++ analyse.cpp Tstatistics.cpp -llapack -lfftw3 -fopenmp -o analyse
```

```
1  # include "polymer_lib.h"
2  int main() {
3  ifstream infile;
4  string namef,s;
5  ofstream outfile;
6
7  int n_loops=1, nrealatoms, coords=3;
8  long num_atoms_sim, num_frames;
9  s="output.dat";
10 frames_atoms_size(s, num_frames, num_atoms_sim);
11 double *temp_data=new double[num_atoms_sim*num_frames*coords];
12 load_matrix(temp_data, s);
13 int num_atoms=3*num_atoms_sim/6;
14 double *data=new double[num_atoms*num_frames*coords];
15 for(int i=0; i<num_frames; ++i)
16  for(int j=0; j<3*num_atoms;++j)
17   data[i*num_atoms*coords + j]=temp_data[i*num_atoms_sim*coords
   +j];
18 delete[] temp_data;
19 nrealatoms=num_atoms/3;
   double epsilons_v[] = {2.0, 3.0, 4.0, 6.0, 8.0, 10.0, 12.0,
   15.0, 20.0};
20 vector<double> epsilon (epsilons_v, epsilons_v + 9 );
21 for(long l=0;l<epsilon.size();++l)
22   write_CM(0, nrealatoms, num_frames, num_atoms, epsilon[l],
   data);

23 infile.open("current_f.txt",ios::in);
   long current_f=0;
24 if(infile)
25   {infile>>current_f;
26   infile.close(); infile.clear();
27   outfile.open("current_f.txt");
28   current_f+=num_frames;
29   outfile<<current_f;
30   }
31 else
32   {infile.close(); infile.clear();
33    outfile.open("current_f.txt");
34    outfile<<num_frames;}
35 outfile.close(); outfile.clear();
36
37 infile.open("rg_vals.txt", ios::in);
38 if(!infile)
```

```
39     {
40     infile.close(); infile.clear();
41     outfile.open("rg_vals.txt", ios::out);
42     if(outfile)
43      {
44       outfile<<"Current_frame Rg_value"<<endl;
45       outfile.close(); outfile.clear();
46      }
47     }
48   else
49     {infile.close(); infile.clear();}
50   outfile.open("rg_vals.txt", ios::app);
51   outfile<<current_f<<" "<<radius_gyration(num_frames-
     1,num_atoms,data,0,nrealatoms)<<endl;
52   outfile.close(); outfile.clear();
53
54   vector < vector<double> > finalvect;
55   vector <double> a_tst;
56   vector< string > names;
57   double awr, atw;
58   vector <double> c_x(nrealatoms), c_y(nrealatoms),
     c_z(nrealatoms), p_x(nrealatoms), p_y(nrealatoms),
     p_z(nrealatoms),
     t_x(nrealatoms), t_y(nrealatoms), t_z(nrealatoms);
59
60   for(long i=0; i<nrealatoms;++i){
61   c_x[i]=getx(num_frames-1, num_atoms, i, data);
62   c_y[i]=gety(num_frames-1, num_atoms, i, data);
63   c_z[i]=getz(num_frames-1, num_atoms, i, data);
64     p_x[i]=getx(num_frames-1, num_atoms, i+nrealatoms, data);
65   p_y[i]=gety(num_frames-1, num_atoms, i+nrealatoms, data);
66   p_z[i]=getz(num_frames-1, num_atoms, i+nrealatoms, data);
67     t_x[i]=getx(num_frames-1, num_atoms, i+2*nrealatoms, data);
68     t_y[i]=gety(num_frames-1, num_atoms, i+2*nrealatoms, data);
     t_z[i]=getz(num_frames-1, num_atoms, i+2*nrealatoms, data);
     }
69   awr=get_writhe(0, nrealatoms, c_x, c_y, c_z);
70   atw=get_twist(0, nrealatoms+1, c_x, c_y, c_z, p_x, p_y, p_z,
     t_x, t_y, t_z);
71   infile.open("writhe_twist_chunk.txt", ios::in);
72   if(!infile)
73     {
74     infile.close(); infile.clear();
75     outfile.open("writhe_twist_chunk.txt", ios::out);
76     if(outfile)
```

```
77       {
78        outfile<<"Current_frame writhe twist old_twist"<<endl;
79        outfile.close(); outfile.clear();
80       }
81     }
82   else
83     {infile.close(); infile.clear();}
84   outfile.open("writhe_twist_chunk.txt", ios::app);
85   outfile<<current_f<<" "<<awr<<" "<<atw<<endl;
86   outfile.close(); outfile.clear();
87   for(int nl=0;nl<n_loops;++nl){
88     s="rg_ring_"+num_to_str(nl);
89     names.push_back(s);
90       for(long i=0;i<num_frames;++i)
91   a_tst.push_back( radius_gyration(i,num_atoms,data,
     nl*nrealatoms*3 , nl*nrealatoms*3 + nrealatoms )  );
92   finalvect.push_back(a_tst); a_tst.clear();
93   }
94   observable_tstat(finalvect, names, current_f);
     finalvect.clear();
95   return 0;
96   }
```

(29) Convert .vtf trajectory into a xyz format (*see* Subheading 4.3).

```
vmd output.vtf -dispdev text -eofexit <dmd_to_coord.tcl> test.log
```

(30) Run the analysis.

```
./analyse
```

4 Notes

4.1 Constant Speed Motor

For a constant speed motor use the following simulation script replacing the script shown in Subheading 3.8.1 between code lines 338 and 347.

```
338 set frame 0
339 for { } { $frame < 100000 } { incr frame } {            Do 100,000 cycles
340  puts -nonewline [format "Step: %4d" $frame ]\r
341  flush stdout
342  integrate 1000                                          Integrate 1,000 steps
343  set phase [expr 2.0* 3.14 / 360.0 * $frame]
344  inter 78 dihedral 0 20.0 $phase                         Change the phase of
345 if { $frame % 100 == 0 } then {                          dihedral by 1 degree
346  writevcf $vtffile                                       Save every 100
347 }                                                        cycles
348  flush $vtffile
349 }
```

4.2 Configuring Random Number Generator

For random events ESPReSo implements pseudorandom number generator (PRNG) [12]. It is a deterministic random number generator which generates sequences of random values which depend on initial values of the seed. If a given simulation is repeated and the seed is not reset, the second simulation will keep reproducing exactly the same trajectory as the first one. The value of seed can be set manually, but to prevent bias it is convenient to implement certain automation. The seed can be set using for example the date, temperature of CPU, or number of messages in mailbox. In the following script the CPU time in seconds is taken as the seed and set for each CPU used in parallelized calculation.

```
1 set n_cpu 8                                               Adjust according the
2 set ran_seed [clock seconds]                              number of CPU's
3 set ran_seed [expr $ran_seed - 1407918000]                used
4 expr srand($ran_seed)
5 set cmd "t_random seed"
6 for {set i 0} {$i < $n_cpu} { incr i } { lappend          Set seed for each
  cmd [expr $ran_seed+$i] }                                 CPU as number of
7 eval $cmd                                                 Seconds+ID of each
8 puts "$ran_seed"                                          CPU
```

4.3 Converting Trajectories

In order to run and use the C++ routine for monitoring twist and writhe successfully, the input data must be in a prescribed format. One can make sure and preformat his/her data from any trajectory readable in VMD by using a simple `.tcl` code and the `tcl/tk` interface of the VMD.

```
1  set file [open output.dat w]                          Save into output.dat
2  set rings [atomselect 0 "all"]
3  set nfram [molinfo 0 get numframes]
4  set natoms [$rings num]
5  puts $file "Num_of_atoms $natoms"                     Write heading of the
6  puts $file "Num_of_frames $nfram"                     file
7  for {set j 0} {$j < $nfram } { incr j } {
8    $rings frame $j
9    $rings update
10   set coords [$rings get {x y z} ]
11   puts $file "$coords"                                Write coordinates
12 }
13 close $file
```

Acknowledgments

We acknowledge financial support granted by the Swiss National Science Foundation (31003A_138367 and 31003A_166684 to A.S.) and by the Leverhulme Trust (RP2013-K-017 to A.S.). We thank Dimoklis Gkountaroulis for his comments and proofreading.

References

1. Irobalieva RN, Fogg JM, Catanese DJ Jr, Sutthibutpong T, Chen M, Barker AK, Ludtke SJ, Harris SA, Schmid MF, Chiu W, Zechiedrich L (2015) Structural diversity of supercoiled DNA. Nat Commun 6:8440
2. Schvartzman JB, Stasiak A (2004) A topological view of the replicon. EMBO Rep 5:256–261
3. Wolfe SL (1993) Molecular and cellular biology. Wadsworth Publishing Company, Belmont. ISBN 0-534-12408-9
4. Spengler SJ, Stasiak A, Cozzarelli NR (1985) The stereostructure of knots and catenanes produced by phage-lambda integrative recombination—implications for mechanism and DNA-structure. Cell 42:325–334
5. Shimokawa K, Ishihara K, Grainge I, Sherratt DJ, Vazquez M (2013) FtsK-dependent XerCD-dif recombination unlinks replication catenanes in a stepwise manner. Proc Natl Acad Sci U S A 110:20906–20911
6. Wasserman SA, Cozzarelli NR (1991) Supercoiled DNA-directed knotting by T4 topoisomerase. J Biol Chem 266:20567–20573
7. The KnotPlot site. http://www.knotplot.com
8. Adams CC (1994) The knot book—elementary introduction to the mathematical theory of knots. WH Freeman and Company, New York. ISBN 0-7167-2393-X
9. The Knot Server page (2003). http://www.colab.sfu.ca/KnotPlot/KnotServer/
10. Limbach HJ, Arnold A, Mann BA, Holm C (2006) ESPResSo—an extensible simulation package for research on soft matter systems. Comput Phys Commun 174:704–727
11. ESPResSo extensible simulation package for research on soft matter (2016). http://espressomd.org
12. The list of resources for User documentation of ESPResSo (2016). http://espressomd.org/wordpress/documentation/
13. Rubinstein M, Colby RH (2003) Polymer physics. Oxford University Press, Oxford
14. Weeks JD, Chandler D, Andersen HC (1971) Role of repulsive forces in determining the equilibrium structure of simple liquids. J Chem Phys 54:5237–5247
15. Di Stefano M, Tubiana L, Di Ventra M, Micheletti C (2014) Driving knots on DNA with

AC/DC electric fields: topological friction and memory effects. Soft Matter 10:6491–6498

16. Cifra P, Benková Z, Bleha T (2008) Effect of confinement on properties of stiff biological macromolecules. Faraday Discuss 139: 377–392
17. Bednar J, Furrer P, Katritch V, Stasiak A, Dubochet J, Stasiak A (1995) Determination of DNA persistence length by cryo-electron microscopy. Separation of the static and dynamic contributions to the apparent persistence length of DNA. J Mol Biol 254:579–594
18. Racko D, Cifra P (2013) Segregation of semi-flexible macromolecules in nanochannel. J Chem Phys 138:184904
19. Tubiana L, Rosa A, Fragiacomo F, Micheletti C (2013) Spontaneous knotting and unknotting of flexible linear polymers: equilibrium and kinetic aspects. Macromolecules 46 (9):3669–3678
20. Arnold A, Suckjoon J (2007) Time scale of entropic segregation of flexible polymers in confinement. Phys Rev E 76:031901
21. Marenduzzo D, Orlandini E, Stasiak A, Sumners DW, Tubiana L, Micheletti C (2009) DNA–DNA interactions in bacteriophage capsids are responsible for the observed DNA knotting. Proc Natl Acad Sci U S A 106 (52):22269–22274
22. Benjamin Renner C, Doyle PS (2014) Untying knotted DNA with elongational flows. ACS Macro Lett 3:963–967
23. Benedetti F, Japaridze A, Dorier J, Racko D, Kwapich R, Burnier Y, Dietler G, Stasiak A (2015) Effects of physiological self-crowding of DNA on shape and biological properties of DNA molecules with various levels of supercoiling. Nucleic Acids Res 43:2390–2399
24. Brackley CA, Morozov AN, Marenduzzo D (2014) Models for twistable elastic polymers in Brownian dynamics, and their implementation for LAMMPS. J Phys Chem 140:135103
25. Racko D, Benedetti F, Dorier J, Burnier Y, Stasiak A (2015) Generation of supercoils in nicked and gapped DNA drives DNA unknotting and postreplicative decatenation. Nucleic Acids Res 43(15):7229–7236
26. Gore J, Bryant Z, Stone MD, Nöllmann M, Cozzarelli NR, Bustamante C (2006) Mechanochemical analysis of DNA gyrase using rotor bead tracking. Nature 439:100–104
27. The VTF format. https://github.com/olenz/vtfplugin/wiki/VTF-format
28. Schlick T (2002) Molecular modeling and simulation. Springer, Berlin. ISBN 0-387-95404-X
29. Frenkel D, Smit B (2002) Understanding molecular simulations, 2nd edn. Academic, London
30. Benedetti F (2015) Polymer public library at github. https://github.com/fbenedett/polymer-libraries. Accessed May 2016

Erratum to: Sequential Super-Resolution Imaging of Bacterial Regulatory Proteins, the Nucleoid and the Cell Membrane in Single, Fixed *E. coli* Cells

Christoph Spahn, Mathilda Glaesmann, Yunfeng Gao, Yong Hwee Foo, Marko Lampe, Linda J. Kenney, and Mike Heilemann

Erratum to:
Chapter 20 in: Olivier Espeli (ed.), *The Bacterial Nucleoid: Methods and Protocols*, Methods in Molecular Biology, vol. 1624, DOI 10.1007/978-1-4939-7098-8_20

The original version of Chapter 20 was inadvertently published without carrying out author corrections. The corrections that were missed out have now been incorporated. The corrections are listed below:

On Page 269, "...Regulatory Proteins: The Nucleoid..." was changed to "...Regulatory Proteins, the Nucleoid..."

On Page 275, line 233: "...workflow (*see* **Note 4**)" was changed to "...workflow (*see* **Note 3**)"

On Page 274, line 179: "...using rapidSTORM v3.31 [31]." was changed to "...using rapi*d*STORM v3.31 [31]."

On Page 283, line 446: "...in rapidSTORM." was changed to "...in rapi*d*STORM."

The updated original online version for this chapter can be found at https://dx.doi.org/10.1007/978-1-4939-7098-8_20

Olivier Espeli (ed.), *The Bacterial Nucleoid: Methods and Protocols*, Methods in Molecular Biology, vol. 1624, DOI 10.1007/978-1-4939-7098-8_25, © Springer Science+Business Media LLC 2017

INDEX

Olivier Espéli (ed.), *The Bacterial Nucleoid: Methods and Protocols*, Methods in Molecular Biology, vol. 1624, DOI 10.1007/978-1-4939-7098-8, © Springer Science+Business Media LLC 2017

Printed in the United States
By Bookmasters